MARINE BIOLOGY

COPEPODS

DIVERSITY, HABITAT AND BEHAVIOR

MARINE BIOLOGY

Additional books in this series can be found on Nova's website
under the Series tab.

Additional e-books in this series can be found on Nova's website
under the e-book tab.

MARINE BIOLOGY

COPEPODS

DIVERSITY, HABITAT AND BEHAVIOR

LAURENT SEURONT
EDITOR

nova publishers
New York

For permission to use material from this book please contact us:
Telephone 631-231-7269; Fax 631-231-8175
Web Site: http://www.novapublishers.com

Library of Congress Cataloging-in-Publication Data

Copepods : diversity, habitat and behavior / editor, Laurent Seuront
(School of Biological Sciences, Flinders University, Australia).
pages cm. -- (Marine biology)
Includes index.
ISBN 978-1-63117-846-7 (hardcover)
1. Copepoda. I. Seuront, Laurent.
QL444.C7C676 2014
595.3'4--dc23

2014013782

Published by Nova Science Publishers, Inc. † New York

Contents

PREFACE

COPEPODS: TINY BUGS IN A BROAD CONTEXT

Laurent Seuront[1,2*]

[1]Centre National de la Recherche Scientifique, Laboratoire d'Océanologie et de Géosciences, UMR LOG CNRS 8187, Station Marine de Wimereux, Wimereux, France

[2]School of Biological Sciences, Flinders University, Adelaide, Australia

Copepods are among the most abundant — if not actually the most abundant — metazoans (i.e., multi-celled organisms) on Earth and can literally be found everywhere where there is (even not so much) water. For instance, copepods have been found in the minute volumes of water collected from bromeliad leaves [1], leaf litter [2], *Sphagnum* moss [3], and even ant nests [4]. They have colonized virtually every habitat from subterranean environments such as caves [5] and groundwater [6] and deep oceanic trenches 10,000 meters down [7] to high-altitude lakes 5,000 meters up in the Himalayas [8]. Their thermal and haline habitats respectively range from subzero polar waters and ice-water interface [9] to hot springs [10] and active hydrothermal vents [11], and from freshwater (including streams, rivers and lakes) [12] to hypersaline ponds and lakes [13].

Copepods are a very diverse group of crustaceans, exhibiting a wide range of forms and sizes (Figure 1); the length of adults is typically in the range 0.5 to 15 mm, but the largest copepods are parasitic, reaching as much as 30 cm in length; see e.g., [14]. Copepods comprise 210 family-level groups, over 1,600 genera [15], and *ca.* 14,000 species including 2,814 freshwater ones [16]. The total number of known species is fast growing, but probably still underestimated [17,18]. Copepods may be free-living, either in the open water or in various types of sediments ranging from sand to fine mud and ooze, and symbiotic. They are also often found as both internal and external parasites of most phyla of animals in water,

* E-mail address: laurent.seuront@univ-lille1.fr.

ranging from sponges to reptiles, fish and mammals; nearly one-third of known copepod species are parasitic, nearly evenly hosted by invertebrates and fishes [18].

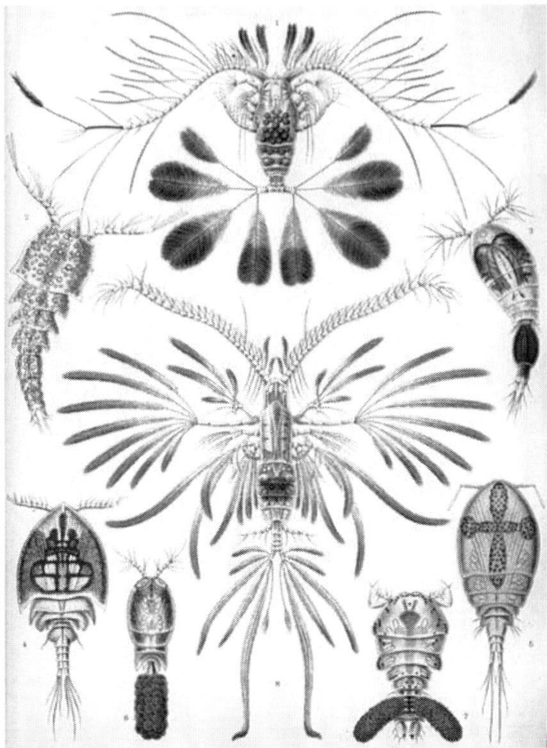

Figure 1. Illustration of the variety of shapes and sizes found in copepods. Top, from left to right: *Clytemnestra scutellata* (female), *Calanus pavo* (male) and *Oncaea venusta* (male). Bottom, from left to right: *Cryptopontius thorelli* (female), *Corycaeus venustus* (female), *Augaptilus filigerus* (male), *Sapphirina darwinii* (female) and *Acontiophorus scutatus* (female). From Ernst Haeckel (1904) *Kunstformen der Natur*, 56th plate.

In the ocean, minute copepods are by far the most dominant form of zooplankton that also include fish larvae, and large jellyfish and salps. To answer the infamous question of Sir Alister Hardy "*Who will dare to estimate the probable numbers of copepods in the world at any one time?*" [19], GA Boxshall [20] provided the following calculation; estimating the open water of the world ocean as *ca.* 1.35×10^9 km^3 and assuming an average density of 1 copepod per liter of seawater (a gross underestimation of their actual density in many marine waters) would give a world population of 1.35×10^{21} copepods. This humongous number of tiny crustaceans represents only a fraction of copepod diversity in the ocean as only *ca.* 2500 species are free-living marine planktonic species. From an ecological standpoint, planktonic copepods are, however, an essential component of pelagic ecosystems. They are the classical herbivorous link between the primary production of phytoplankton and the larvae and juveniles of fishes — hence ultimately whales and fisheries — in most pelagic ecosystems (Figure 2a). In oligotrophic waters, copepods also play an essential role in transferring (i) the organic carbon released by phytoplankton (i.e., up to 50% of the carbon fixed through photosynthesis) and subsequently assimilated by heterotrophic bacteria, and (ii) the inorganic carbon fixed by prokaryotic and eukaryotic picoplankton (i.e., up to 70 % of the carbon fixation in oligotrophic systems) towards higher trophic levels through copepod grazing on microzooplankton (i.e., heterotrophic flagellates and ciliated protozoans; Figure 2b).

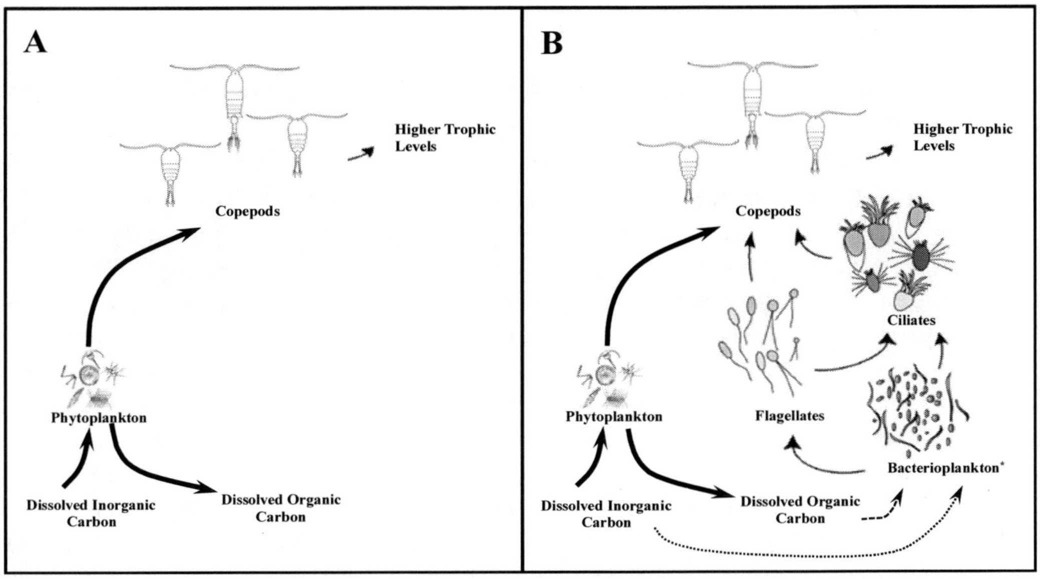

Figure 2. Schematic illustration of planktonic food chains: the classical herbivorous food chain (A) and the microbial loop (B). *: the term bacterioplankton is here used for the sake of simplicity, and includes (i) heterotrophic bacteria that recycle the dissolved organic carbon released by phytoplankton cells (dashed arrow) and both the prokaryotic (*Synechoccocus* and *Prochlorococcus*) and eukaryotic picoplankton that incorporate dissolved inorganic carbon into higher trophic levels through the grazing activity of heterotrophic flagellates and ciliated protozoans, essentially in oligotrophic systems.

Copepods have been known for more than two millennia. The first description of a copepod goes back to the time of Aristotle, who noticed in the Chapter 19 of his *History of Animals* (*ca.* 350 B.C.) that "*tunny* [i.e., tuna] and *swordfish are infested as a dog by a gadfly, by a worm under the fins, which looks like a scorpion, has the size of a spider.*" Pliny the Elder provided a similar description in his *Historia Naturalis*, where he described an animal that "*is small, looks like a scorpion, has the size of a spider.*" These early descriptions of parasitic copepods were followed by numerous reviews and monographs, including the first description of a free-living marine copepod in 1770 and the first published book dealing solely with copepods [21]; see [22] for a biographical and a bibliographical review. Beyond these historical considerations and their ubiquitous nature, copepods are still at the core of a very active research area. The Web of Science (accessed February 20, 2014) hence returns a grand total of 8,971 and 17,880 papers including the word *copepod* respectively in their title and topic, 2,924 and 8,106 over the last decade, and 314 and 813 for the year 2013 only. In addition, the World Association of Copepodologists claims *ca.* 57,000 published works on the '*History of Copepodology*' page of their website (www.monoculus.org).

Despite the plethora of information available on them, copepods are still, and probably more than ever, at the center of the scientific debate related to e.g., their taxonomy [15], biology and ecology [23], behavior [23,24], distribution, diversity and biogeography [25,26], with a particular focus on the role of climate variability and global change [27], and their role in biogeochemical cycles [28]. In this context, this book was conceived for a wide audience — typically from undergraduate students to experienced researchers without a necessary strong background in either plankton ecology or zooplankton ecology — with the intent to bring some insights into relevant aspects of the "*Diversity, Habitat and Behavior*" of copepods. As such this book is built around a series of comprehensive reviews written by some of the leading experts in the areas of copepod diversity and habitat (Chapters 1, 2 and 4) and behavior (Chapters 6, 8 and 10). Some chapters are devoted to research topics that may have been less widely researched, but that are believed to be of critical relevance for the future of the field, such as the community composition of copepods living under stressful conditions (here Harpacticoid living in either temporarily or permanently hypoxic environments; Chapter 3), the role of solar radiation in marine copepod ecology (Chapter 5), the improvement in our understanding of copepod behavior using new technologies such as high-speed imaging (Chapter 7), the widely understated relevance of non-Newtonian fluid properties on a range of processes occurring on copepod surface (Chapter 9) and the ecological role and relevance of copepod carcasses (Chapter 11). Even if most of the eleven chapters presented hereafter are derived from research conducted in the marine environment, the concepts and ideas they contain can be easily extrapolated and applied to freshwater copepods, in particular the chapters dealing with various aspects of copepod behavioral ecology (Chapters 6-10). Note that, conversely, the chapter dealing with solar radiation (Chapter 5) builds up on previous research conducted in freshwater systems to discuss the relevance of this research topic that is still in its infancy in the marine environment.

Specifically, the first chapter presents a review of both natural climatic variability and anthropogenically-driven global warming on pelagic species and ecosystems, with a focus on phenological and biogeographical shifts, modifications in key ecosystem attributes, and long-term community and ecosystem shifts. A specific attention is also given to ocean acidification, and Grégory Beaugrand shows that although calcifying organisms are strongly affected by the effects of ocean acidification, temperature remains the main driver of the distribution and diversity of North Atlantic calcifying plankton.

In the second chapter, McKinnon & Duggan first discuss the taxonomic mayhem related to tropical copepod diversity, stress that "*taxonomy is a dying art*" and consider this as a serious threat for the future of routine identification of tropical copepods, despite the fast growing development and use of molecular techniques. They subsequently review the habitat-related nature of the community composition of tropical copepods from their work in tropical estuaries, coral reefs, and coastal and shelf waters of Australia and Indonesia, and compare them to other tropical systems. In particular, they stress that the trophic determinants of community composition may be as important as swarming and emergence behaviors in shaping tropical copepod communities. They finally stress that the observed *tropicalization* of temperate copepod communities as a result of global warming may end up in an increase in the distribution range of these communities

The third chapter investigates the abundance and community composition of harpacticoid copepods in various conditions of hypoxia encountered either (i) seasonally in shallow coastal sediments and shallow benthic mats driven by active methane seeps, and (ii) permanently in deep waters. Kolesnikova et al. show a clear distinction between shallow and deep-water species, identify species tolerant to a range of hypoxic conditions, and found a new species

(*Darcythompsonia fairlensis*) that is suggested as an indicator of hypoxic conditions in the Black Sea.

In Chapter 4, Swadling & Gibson provide a review of the knowledge on Antarctic lacustrine copepods, and discuss the origin of the eight species that have been identified so far. Calanoid copepods have been observed in both marine-derived and freshwater lakes, while cyclopoids are presently identified only from freshwater lakes and harpacticoids from saline lakes. They suggests that some freshwater species such as *Gladioferens antarcticus* and *Boeckella poppei* may be the remnants of an ancient fauna that was either *in situ* prior to the Last Glacial Maximum — hence stress the possibility that "*vicariance rather than dispersal may prove to be the most parsimonious and reasonable explanation*" of their origin — or colonized from lakes along the Antarctic Peninsula by copepods from South America during the Holocene. Populations of marine-derived copepods are found living in saline lakes, particularly in oases along the margins of East Antarctica. These populations have been present for thousands of years, following isolation of the lakes from the ocean *via* isostatic rebound approximately 5000 years BP. It is finally stressed that much work is needed to assess the potential impact environmental changes, e.g., climate change, may have on those copepod populations.

In Chapter 5, Gonçalvez & Hylander stress that most of our knowledge of the effects of solar radiation on copepods comes from freshwater zooplankton (copepods, but mainly cladocerans), whereas marine studies are comparatively scarce. In this context, these authors summarize the current knowledge in marine copepods and provide an overview of knowledge gaps and areas that may warrant further research. They show that the indirect effects of ultraviolet radiation are likely to be the most important — and less predictable due to interactions with other factors across trophic levels — than direct effects. They also stress that no information is available yet on their effects on copepods in particular and marine life in general when considered in combination with other stressors such as global warming and ocean acidification.

Chapter 6 provides a review of the behavioral ecology of copepods, with a focus on swimming, feeding and mating behaviors. First, swimming behavior is described *via* the different swimming mechanisms and patterns that are shown to be species-dependent, but also sex- and stage-related. Feeding behavior is then described in terms of feeding modes and related foraging strategies, and mating behavior is described through its main components, i.e., mate search, mate recognition and copulation. In a final section, Sabia et al. briefly discuss the different techniques now available to study copepod behavior in the field, in the laboratory, and theoretically through computer simulations and theoretical models.

The behavior of copepods is further explored in Chapter 7, where Yugi Tanaka describes how high-speed imaging technology now allows us to *see the invisible* — indeed "*Hearing a hundred times is not as good as seeing once*" — and more specifically details (i) the motion of both the mouthparts and swimming legs of a tethered *Calanus pacificus*, and (ii) the free-swimming behavior of *Paraeuchaeta elongata*, with a thorough description of their aerial jumps, discussed as a predator-avoidance strategy.

In Chapter 8, David M. Fields applies current knowledge about the mechanosensory structures of marine copepods as a case study for understanding sensory ecology from a structure-function vantage point. The role of mechanoreception is critical in planktonic copepods because fluid signals often act as the final proximate cue for copepod behavioral responses, and the basic properties of the copepod mechanosensory system are relatively easy to identify, and may be broadly general across a diverse range of species. Specifically, after a brief review of the properties of fluid disturbance, the author shows how sensor design shapes sensor response properties (i.e., physical, physiological and neurological) and thus the behavioral and/or ecological function of particular sensor types. The author also discusses the mathematical modeling of the interaction between flows and mechanoreception, provides details on spatial detection of fluid signals by copepods, and finally addresses the role of fluid viscosity on copepod mechanoreception.

Chapter 9 uniquely reviews a neglected aspect of fluid properties in copepod research in particular and plankton research in general, i.e., the non-Newtonian properties of the fluid at scales of pelagic copepods (< *ca.* 1 mm). Non-Newtonian properties (i.e., excess viscosity and elasticity) are mainly driven by the exopolymeric substances (EPS) secreted by algae and bacteria, hence likely to be most marked in blooms and chlorophyll maxima. In this context, Ian R. Jenkinson reviews seawater rheology and the subsequent possible effects of EPS producing stick layers of finite thickness, particularly on copepods' prey. A range of possible slip layers are subsequently considered and their effect in increasing *vs.* decreasing water velocity and shear rate close to a surface are discussed in the context of copepod feeding. The differences between the relatively mucus-free calanoid copepods and mucus-rich harpacticoids are explored, and suggested as a future area of research. Finally, the author makes a suggestion for interdisciplinary capacity building activities for rheology and nano/microfluidics to be understood better by teams investigating copepod dynamics and all aspects of plankton ecology and biogeochemistry.

In Chapter 10, Woodson et al. provide a thorough review of how copepods behaviorally respond to the information available in their environment as a range of cues — typically gradients in light, temperature, salinity, fluid density, fluid motion and highly soluble chemicals, and their potential interactions — to bridge the gap between dispersed prey and fast-moving predators. They subsequently explore the implications of these behaviors on copepod distributions, trophic interactions, and overall marine ecosystem production. These aspects are illustrated on the basis of a case study, how copepod ecology relates to harmful algal blooms. Future directions of research are suggested, and include (i) developing an understanding of how copepods may employ horizontal gradients in turbidity for position maintenance in regions where horizontal gradients are of appropriate scale (estuaries, coastal ocean fronts), (ii) elucidating the proximate cues for diel vertical migration as it relates to fine-scale foraging behavior, (iii) assessing the role of turbulence in structuring pelagic communities, and (iv) quantifying how the natural variability of predator-prey covariance affects the role of copepods in complex food webs.

Because everything has an end, especially life, the last chapter reviews the occurrence, fate and ecological importance of copepod carcasses, a critically neglected component of copepod life cycle and contribution to biogeochemical cycles. Conventional field sampling indeed typically does not distinguish between live and dead copepods in samples, hence predation is often assumed to be the dominant cause of copepod mortality to the extent that non-predatory mortality is ignored. In this chapter, on the basis of a meta-analysis of the literature showing that globally, non-predatory factors account for up to one-third of total mortality among marine epipelagic copepods, Tang & Elliott describe newly developed techniques allowing to easily quantify carcasses in marine and freshwater samples, discuss the implications of numeration errors related to ignoring non-predatory mortality in e.g., the production of realistic population growth predictions and vertical fluxes estimates.

Specifically, the authors stress that further research into the occurrence and fate of copepod carcasses is needed to improve our understanding of not only copepod mortality and population dynamics, but also the link between copepods, microbial food web and biogeochemical fluxes.

I finally want to acknowledge all the authors who contributed to this book for their trust in the project and their invaluable contributions. Without them this work would not exist. I would also like to thank all the reviewers that contributed to improve the quality of the chapters for their time and effort. Without them this work would not exist either. Last but not least, on a more personal level, I would like to thanks my two wonders, Dorine and Leena, for being such a source of happiness, energy and inspiration that is renewed every single day. Without you this work would definitely not exist.

REFERENCES

[1] Suárez-Morales E, Mendoza F & Mercado-Salas N (2010) A new *Allocyclops* (Crustacea, Copepoda, Cyclopoida) from bromeliads and records of freshwater copepods from Mexico. *Zoosystema*, *32*, 393-407.

[2] Fiers F & Jocque M (2013) Leaf litter copepods from a cloud forest mountain top in Honduras (Copepoda: Cyclopidae, Canthocamptidae). *Zootaxa*, *3630*, 270-290.

[3] Reid JW (1986) Some usually overlooked cryptic copepod habitats. *In*: Schriever G, Schminke HK & Shih CT (eds.), *Proceedings of the Second International Conference on Copepoda*. Syllogeus No. 58, National Museums of Canada, National Museum of Natural Sciences, Ottawa, 594-598.

[4] Jocque M, Fiers F, Romero M & Martens K (2013) Curstacea in phytotelmata: a global overview. *Journal of Curstacean Biology*, *33*, 451-460.

[5] Suárez-Morales E, Mercado-Salas NF & Barlow R (2013) A new stygobiotic species of *Diacyclops* Kiefer, 1927 (Copepoda, Cyclopoidae) from caves in New Mexico, United States. *Crustaceana*, *86*, 1155-1177.

[6] Branceli A (2011) Copepoda from a deep-groundwater porous aquifer in contact with karst: description of a new species, *Paramorariopsis brigitae* n. sp. (Copepoda, Harpacticoida). *In*: Defaye D, Suarez-Morales E & von Vaupel Klein JC (eds.), *Studies on Freshwater Copepoda*. Koninklijke Brill, Leiden, 85-104.

[7] Laakmann S, Auel H & Kochzius M (2013) Evolution in the deep sea: Biological traits, ecology and phylogenetics of pelagic copepods. *Molecular Phylogenetics and Evolution*, *65*, 535-546.

[8] Shah JA & Pandit AK (2013) Relation between physic-chemical limnology and crustacean community in Wular lake of Kashmir Himalaya. *Pakistan Journal of Biological Sciences*, *16*, 976-983.

[9] Hop H, Mundy CJ, Gosselin M et al. (2011) Zooplankton boom and ice amphipod bust below sea ice in the Amundsen Gulf, Arctic Canada. *Polar Biology*, *34*, 1947-1958.

[10] Dash A, Palita SK & Patra HK (2012) Diversity of zooplankton in hot water spring of Atri, Odisha, India. *International Journal of Environmental Science and Technology*, *1*, 111-119.

[11] Ivanenko VN, Corgosinho PHC, Ferrari F et al. (2011) Microhabitat distribution of *Smacigastes micheli* (Copepoda: Harpacticoida: Tegastidae) from deep-sea hydrothermal vents at the Mid-Atlantic Ridge, 37° N (Lucky Strike), with a morphological description of its nauplius. *Marine Ecology*, *33*, 246-256.

[12] Defaye D, Suarez-Morales E & von Vaupel Klein JC (2011, eds.) *Studies on Freshwater Copepoda*. Koninklijke Brill, Leiden.

[13] Alonso M (2010) Branchiopoda and Copepoda (Crustacea) in Mongolian saline lakes. *Mongolian Journal of Biological Sciences*, *8*, 9-16.

[14] Williams EH Jr & Bunkley-Williams L (1996) Parasites of offshore big game fishes of Puerto Rico and the western Atlantic. Puerto Rico Department of Natural and Environmental Resources, San Juan, PR, and the University of Puerto Rico, Mayaguez, PR.

[15] Boxshall GA & Halsey SH (2004) *An Introduction to Copepod Diversity*. The Ray Society, London.

[16] Boxshall GA & Defaye D (2008) Global diversity of copepods (Crustacea: Copepoda) in freshwater. *Hydrobiologia*, *595*, 195-207.

[17] Bowman TE & Abele LG (1982) Classification of the recent crustacea. *In*: Bliss DE (editor-in-chiefà, *The biology of Crustaea*. Vol. 1. Abele LG (ed.), *Systematics, the Fossil Record, and Biogeography*. Academic Press, New York, 1-27.

[18] Humes AG (1994) How many copepods? *Hydrobiologia*, *292/293*, 1-7.

[19] Hardy A (1956) *The Open Sea: Its Natural History. I. The World of Plankton*. Collins, London.

[20] Boxshall GA (1998) Preface. *Philosophical Transactions of the Royal Society of London B: Biological Sciences*, *353*, 669-670.

[21] Claus C (1863) *Die Frei Lebenden Copepoden: Mit Besonderer Berücksichtigung der Fauna Deutschlands, der Nordsee und des Mittelmeeres*. Verlag von Wilhelm Engelmann, Leipzig.

[22] Damkaer DM (2002) *The Copepodologist's Cabinet: a Biographical and Bibliographical History*. American Philosophical Society, Philadelphia.

[23] Kiørboe T (2008) *A Mechanistic Approach to Plankton Ecology*. Princeton University Press, Princeton.

[24] Lenz P Hartline DK, Purcell J & Macmillian D (1997) *Zooplankton: Sensory Ecology and Physiology*. CRC Press, Boca Raton.

[25] Stenseth NChr, Ottersen G, Hurrell JW & Belgrano A (2005) *Marine Ecosystems and Climate Variation: The North Atlantic: A Comparative Perspective*. Oxford University Press, Oxford.

[26] McQuatters-Gollop A, Burkill P & Beaugrand G (2010) *Atlas of Calcifying Plankton: Results from the North Atlantic Continuous Plankton Recorder Survey*. Sir Alister Hardy Foundation for Ocean Science (SAHFOS), Plymouth.

[27] Beaugrand G (2014) *Marine Biodiversity, Climatic Variability and Global Change*. Routledge, Earthscan Oceans Series.

[28] Miller CB & Wheeler PA (2012) *Biological Oceanography*. Wiley-Blackwell, Chichester.

In: Copepods: Diversity, Habitat and Behavior
Editor: Laurent Seuront
ISBN: 978-1-63117-846-7

Chapter 1

PELAGIC DIVERSITY AND CLIMATE CHANGE

Grégory Beaugrand*
Centre National de la Recherche Scientifique,
Laboratoire d'Océanologie et de Géosciences, UMR LOG CNRS 8187,
Station Marine, Université des Sciences et Technologies de Lille 1, Wimereux, France

ABSTRACT

Since life diversified, climate change has exerted a strong influence on biological and ecological systems of the oceans and it is likely that species and ecosystems have developed mechanisms to absorb these fluctuations. However, the human pressure has strongly increased over the last decades and both direct (e.g., fishing, pollution, eutrophication) and indirect anthropogenic (e.g., global climate change) factors are now influencing the oceanic biodiversity. In this chapter, I specifically review the effects of both natural climatic variability and anthropogenic global warming on marine species and ecosystems.

INTRODUCTION

The state of our planet has continuously evolved for 4.6 billion years and fluctuations have always taken place over time scales, ranging from intra-seasonal to decadal and millennial. In the last decades, anthropogenic activities have exerted huge pressures on living systems and have altered their biodiversity, structure and functioning [1]. In addition, anthropogenic global warming has started to influence both species and ecosystems. All these different influences are difficult to tease apart and we have only started to recognize strong potential interactive effects between these stressors [2,3].

Anthropogenic global warming is having a discernible influence on many natural systems of the planet. Among the responses of species and ecosystems to anthropogenic climate change observed in the marine environment, phenological, biogeographical and long-term

* E-mail address: gregory.beaugrand@univ-lille1.fr.

community shifts (including abrupt community shifts) have been frequently reported. Despite their profound consequences for ecosystem functioning, services and the global carbon cycle, the understanding of the mechanisms underlying these changes remains rudimentary. In this chapter, I will review the different effects of both natural climatic variability and global warming on pelagic ecosystems. Changes commonly identified range from phenological [4] and biogeographical [5] to long-term community/ecosystem shifts [6-8]. Modifications in key ecosystem attributes caused by climate have also been reported recently [9]. Such changes may have severe consequences on ecosystem functioning and both regulating and provisioning services [10]. I will also briefly tackle the effects of ocean acidification but will show that calcifying organisms remain also strongly affected by the effects of temperature [11].

BIODIVERSITY AND HYDRO-CLIMATIC VARIABILITY

Many studies have provided evidence for a strong influence of natural climatic variability on marine ecosystems [4,12-14]. Among the most affected taxonomic groups, plankton seem to amplify even subtle environmental variations [15]. One commonly proposed explanations is that these organisms are ectotherms and that metabolic rates, growth, reproduction, activity and species interactions are all influenced by temperature [16,17]. Their short life cycles also make this trophic group highly reactive to climate change. Fish also exhibit many changes that are sometimes synchronous with other groups and can be explained by climate variability or climate change. In this section, I show how natural climatic variability may affect ecosystems and their biodiversity using three examples: (i) the Bohuslän herring periods and the North Atlantic Oscillation, (ii) the Russell's cycle and the Atlantic Multidecadal Oscillation, and (iii) the effects of the El Niño Southern Oscillation (ENSO) on tuna. I also detail the climatic mechanisms by which these large-scale climatic phenomena may influence ecosystems and their species.

Bohuslän Periods

The investigation of historical records provides strong evidence for an effect of climate variability on pelagic ecosystems [18]. Since Viking time, large fluctuations in the abundance of herring have been observed in a small region called the Bohuslän and located close to the Swedish west coast. Periods of elevated herring abundance (hereafter called Bohuslän period) have alternated with periods where the fish were absent, overwintering in other regions. Since 1100 AD, nine Bohuslän episodes have been identified because of their effects on the regional economy [18]. Up to 50,000 people came in the region when herring was abundant. These periods of large herring abundance coincided with cold winters, whereas the opposite situation corresponded to warmer winters. It has been hypothesized that the severity of the winter and therefore Bohuslän periods was under the control of the North Atlantic Oscillation (NAO) [18]. The NAO is the most prominent source of atmospheric variability in the Northern Hemisphere, identified in all months of the years. This oscillation represents a basin-scale alternation of atmospheric masses over the North Atlantic between the

Subtropical and Subarctic regions [19]. The conventional index of NAO activity is based on the pressure difference between the Azores (Lisbon or Gibraltar) high-pressure and the Icelandic (Stykkisholmur or Reykjavik) low-pressure center [20]. The NAO is more evident during the boreal winter because the atmosphere is more dynamically active. A high positive value indicates a reinforced contrast between low pressures around Iceland and high pressures over the Azores, whereas a significant negative value corresponds to a weakened pressure gradient (Figure 1). Figure 1 shows anomalies of both zonal and meridional winds, precipitation and sea surface temperatures (SST) between a positive and a negative phase of the NAO.

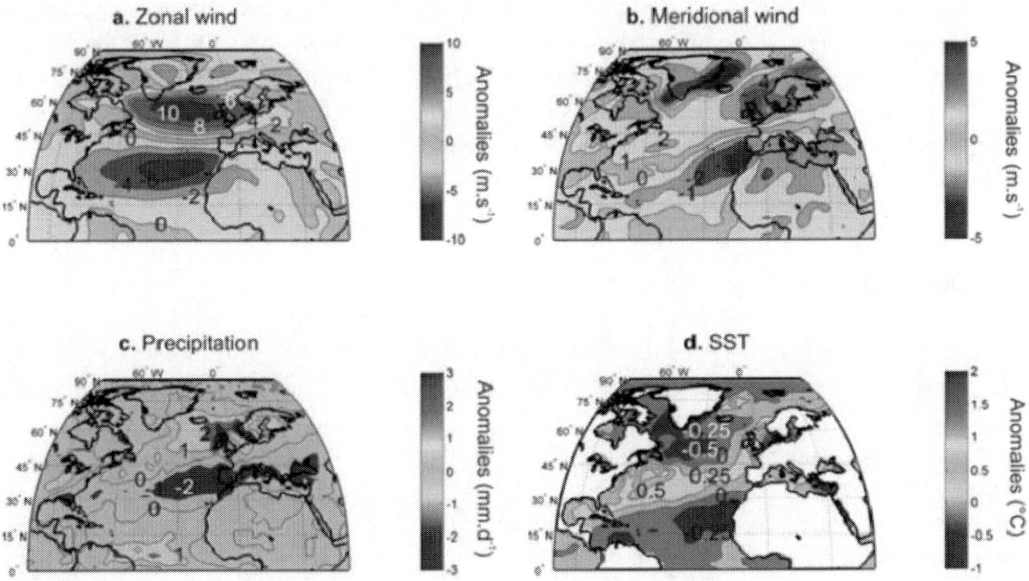

Figure 1. Differences in some winter (December to March) climatic variables (precipitation, a; sea surface temperature, b; zonal wind, c; meridional wind, d) anomalies between a positive and a negative phase of the North Atlantic Oscillation (NAO) over the period 1979-2012. Thresholds used to determine a positive and a negative phase are 1 and -1, respectively. Winter precipitation (GPCP Version 2.2 Combined Precipitation Dataset) data are from the Polar Satellite Data Centre. Winter SST (ERSST V3) data are from the National Climatic Data Center. Both zonal and meridional wind data (NCEP/DOE AMIP-II Reanalysis-2 Monthly Averages) are from the Earth System Research Laboratory. Winter NAO data are from the Climate Prediction Center.

When the NAO is in its positive phase, both meridional (west to east) and zonal (south to north) circulation (i.e., the westerlies) clearly reinforce between 45°N and 65°N all over in the North Atlantic, whereas decreasing zonal circulation occurs over the Mediterranean Sea and in the subtropical gyre (Figure 1a,b). Easterly trade winds reinforce over the tropical Atlantic during a positive phase of the NAO because of the increase in the Azores High and its associated stronger clockwise flow. The meridional component of the wind is correlated positively to the NAO over the path of the Gulf Stream but mainly north of the Oceanic Polar Front between 45°N and 65°N and negatively in the Labrador Sea. This change in the atmospheric circulation brings more warm moist maritime air during winter, but therefore more precipitation over the northern part of Europe (Figure 1c). It influences the position of the North Atlantic storm tracks, shifted to the north when the NAO is positive [19,21]. During a positive phase of the NAO, winters are therefore anomalously warm and wet over the northern part of Europe, colder and drier than usually in the Mediterranean Sea, parts of the Middle East and colder and drier in the eastern part of North America and the western part of Greenland (Figure 1d). Opposite effects occur during a negative phase of the NAO.

A large range of physical and biological responses to the NAO have been documented [22]. In the North Sea, Fromentin & Planque [23] provided evidence that year-to-year fluctuations in the state of the NAO and the abundance of the key-structural subarctic species *Calanus finmarchicus* around the British Isles were correlated negatively [23]. However, this correlation broke down at the time of the North Sea abrupt ecosystem shift at the end of the 1980s [24]. This example shows that the correlation between marine ecosystem or species and a climatic index may also be affected by other factors. It has been shown that the correlation between the state of the NAO and the abundance of *C. finmarchicus* is modulated by the thermal regime of the North Sea. When the thermal regime is high, the NAO has less effect on *C. finmarchicus* because the sensitivity of the species to the regional climatic effects of the NAO diminishes [24].

The Russell's Cycle

As early as the beginning of the 1930s, Russell and colleagues [25] already observed a shift in the English Channel from a community dominated by the cold-water chaetognath *Sagitta elegans* to a community dominated by the warm-water species *Sagitta setosa.* A decline in large zooplankton also paralleled a collapse in the number of young herring (*Clupea harengus*) replaced after 1935 by pilchard (*Sardina pilchardus*). These biological shifts were apparently associated to an alteration of the chemical properties of seawater (e.g., reduction in winter nutrient concentration). The change reversed in the middle of the 1960s. The chaetognath *Sagitta elegans* returned and was accompanied by other cold-water species such as the medusa *Aglantha* and the siphonophore *Nanomia* in summer and by the euphausid crustaceans *Meganyctiphanes* and *Thyssanoessa* in winter [26]. Copepods of the genus *Calanus*, the pteropod *Spiratella retroversa* and larvae of the starfish *Luidia sarsi* increased. Phytoplankton also shifted after 1967 with pronounced increased in the diatoms *Biddulphia sinensis* and *Coscinodiscus concinnus* [25]. The alteration in the plankton composition propagated towards higher trophic levels. When *Sagitta elegans* was abundant, high levels of herring egg concentration were observed. In contrast, when *Sagitta setosa* dominated pilchard egg concentration was elevated. Changes in both plankton and fish composition paralleled a shift in sea temperature and nutrient concentrations (e.g., phosphorus) [25]. Edwards and colleagues [27] have related the Russell's cycle to the state of the Atlantic Multidecadal Oscillation (AMO). The AMO is a large-scale oceanic phenomenon, source of a natural variability in the range of 0.4°C in the oceanic regions of the North Atlantic [28]. The AMO is also detected in paleoclimatic reconstructions with a cycle ranging from 60 to 100 years [29].

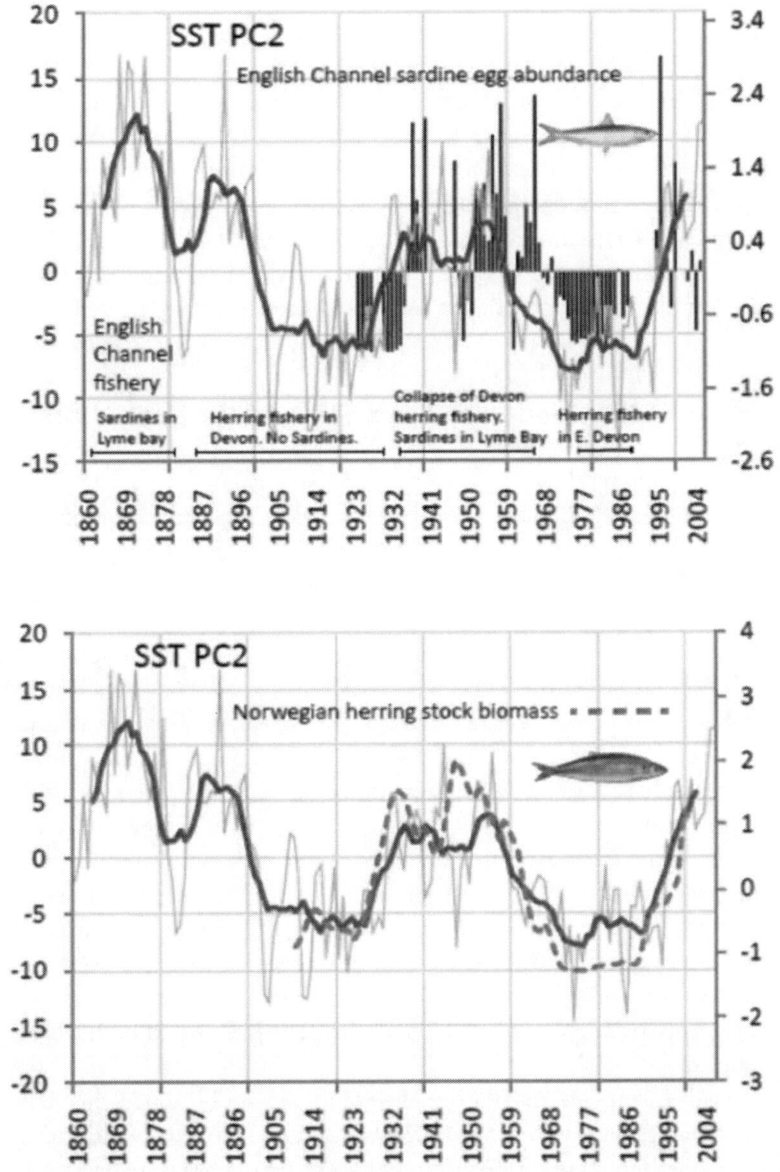

Figure 2. Long-term changes in fish abundance in the North Atlantic. (a) Detrended changes in annual sea surface temperature (SST) that reflect long-term changes in the Atlantic Multidecadal Oscillation (AMO) and sardine egg abundance (bars) from the English Channel from 1924- 2004. Sardine egg abundance values standardized to zero mean and unit variance (bars used as time-series is not continuous). Fluctuations in the fisheries of the English Channel are superimposed. The period after 1995 is dominated by sardines in Western English Channel. (b) Detrended changes in annual sea surface temperature that reflect long-term changes in the AMO and long-term trend in Norwegian herring stock biomass (dashed line) between 1908-2000. Norwegian stock biomass values were standardized to zero mean and unit variance. The correlation between detrended annual SST reflecting the AMO signal and both species were significantly positive (Sardine: $r = 0.49$; Herring: $r = 0.91$). Modified from Edwards et al. [27].

Multi-decadal changes in both herring and sardine exhibited very similar trends to the AMO signal (Figure 2). The English Channel is close to a biogeographic boundary between the cold waters of the Atlantic Polar Biome and the warm waters of the Atlantic Westerly Winds Biome [30]. In the English Channel, warming periods observed during the positive phases of the AMO were associated to the dominance of sardine whereas cold periods observed during the negative phases of the AMO were characterized by the dominance of herring (Figure 2a). It is interesting to note that, in the Norwegian and Barents Seas, the herring stocks are close to their northern distributional range and warming was associated to a high level of herring abundance (Figure 2b). The AMO has also been involved in some plankton changes in the Northeast Atlantic and its adjacent seas [5,31].

The Influence of El Niño Southern Oscillation on the Spatial Distribution of Tuna

The El Niño Southern Oscillation (ENSO) is the most prominent source of hydro-climatic variability in the global climate system [32]. An El Niño event corresponds to a warming of the East Pacific along the northern part of the Peruvian coast and the southern part of Ecuador.

This phenomenon has been known for many centuries. Fishermen noticed the arrival of a southward warm current along the Peruvian coast, associated to a reduction of trade winds. As the event appeared every year around the Christmas season, this warm current was named El Niño, as a reference to the Chris child. Because the intensity of the warm current varied from year to year, the term El Niño was later used to qualify the year when this phenomenon was particularly pronounced. When the pressure difference between the western and the eastern part of the Pacific decreases, the intensity of the trade winds reduces to such an extent that the reserve of warm water of the western Pacific starts to flow eastwards (Figure 3a,b).

This moves further the low-pressure center towards the central Pacific, which emphasizes the weakening in the pressure difference between the east and the west and thereby the southern oscillation. The propagation of the warm waters accelerates and the warm waters reach the eastern part of the Equatorial Pacific a few months after the beginning of the phenomenon (Figure 3c).

This deepens the depth of the thermocline and the trade winds that continue to flow eastwards along the Peruvian coast are not strong enough to reach the thermocline and to trigger upwelling. The movement of low-pressure center towards the central Pacific reduces precipitation over the northern part of Australia and Indonesian where severe droughts occur (Figure 3d).

The interaction between the ocean and the atmosphere is permanent and the movement of warm waters triggers mechanisms of convection along its way. The Hadley cell intensifies transporting more energy towards the poles. The progression of the water eastwards implicates an increase in sea level that is detected by the networks of tide gauges set up along the Equatorial Pacific (Figure 3d).

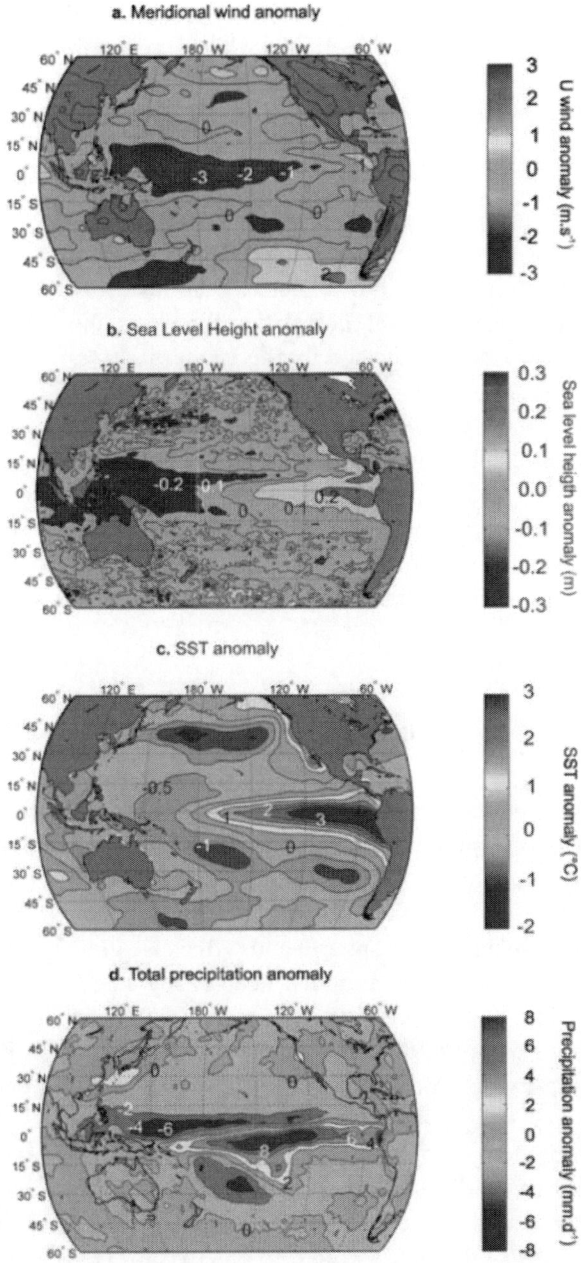

Figure 3. Hydrologic and meteorological changes associated to the El Niño of 1997-1998. (a) Monthly anomalies in the meridional component of the wind in 1997. (b) Monthly anomalies in sea level height anomalies between July 1997 and March 1998. (c) Monthly anomalies in sea surface temperature (SST) anomalies between July 1997 and March 1998. (d) Total precipitation monthly anomalies between January and June 1998. U-wind data (1979-2012) are from the Earth System Research Laboratory (NCEP/DOE AMIP-II Reanalysis-2). Sea level height data (1958-2007) are from CARTON-GIESE Simple Ocean Data Assimilation (v2p0p2-4). SST data (1979-2012) are from the ERA-INTERIM dataset. Total precipitation data are from the GPCP Version 2.2 Combined Precipitation Dataset.

The phenomenon may have a very large impact on marine biodiversity. For example, it is thought that the super El Niño event of 1997-1998 killed 16% of the coral reef of the planet [33]. The warm pool of the western part of the Equatorial Pacific provides a suitable habitat for the Skipjack Tuna (*Katsuwonus pelamis*). A Convergence Zone is detected at the eastern boundary of the warm pool. This zone results from the advection of cold and saline waters from the eastern part of the central Equatorial Pacific that encounters warmer and less saline waters in the west due to sporadic bursts of westerly wind [34]. This Convergence Front is indicated by a salinity front where nutrients concentrate and are at the origin of higher plankton abundance, which propagates to higher trophic level, including tuna. Lehodey and colleagues [34] observed a significant correlation between the ENSO index, the mean location of the Convergence Zone indicated by the isotherm 29°C and the mean longitudinal position of the Skipjack Tuna. During the El Niño event of 1991-1992, a zonal movement of more than 40° of longitude eastwards of the warm pool was associated to the zonal displacement of the relative abundance of the Skipjack Tuna with the same magnitude. These zonal displacements were consistent with observed movements of tagged Skipjack Tuna. During the El Niño of 1991-1992, the eastward movements of tagged Skipjack Tuna were prominent [34]. These results had important implications for the fishing industry because it enabled the mean position of the relative abundance of the species to be forecasted several months in advance.

Global Warming

In contrast to hydro-climatic variability, which is natural, there is now unequivocal evidence that the world is warming rapidly, and very high confidence that the warming is due to human forcing from greenhouse gas emissions, land use changes and factors linked to economic development and population growth [35,36]. Global warming will imbalance natural systems (e.g., cryosphere, hydrosphere) and will cause global climate change. Because water takes up more than 4000 times as much heat as the same volume of air, the oceans covering more than 71% of the surface of the Earth are the major heat reservoir of our planet and play a crucial role in damping down the impact of global warming [37]. Warming was more pronounced during two periods: the period 1920-1940 and the period after the end of the 1970s [38].

Figure 4 shows global long-term changes in sea surface temperatures for the period 1960-2008. Increasing values of the first principal component, reflecting rising temperature, was not continuous, but augmented in steps *circa* 1977, 1988 and 1998 after the super El Niño event (Figure 4a). Mapping of the first normalized eigenvector (EV) shows areas where long-term changes were correlated with the first principal component (PC; Figure 4a). High positive values are concentrated in the tropics and especially the Indian Ocean with extratropical extensions to the north and south in all oceans. The main regions with negative EVs are located in the North and South Pacific. The second principal component exhibited three periods (Figure 4b): (i) lower positive anomalies at the beginning, (ii) an increase that follows an abrupt change in 1976 [7,39], and (iii) another sharp change after the super El Niño of 1997/98 [40] to negative anomalies, reflecting a cooler eastern Pacific. El Niño years clearly influenced this second principal component. The second normalized eigenvectors show positive correlations to the second PC over much of the central equatorial Pacific, in many cases reflecting El Niño years, and a V shaped area of negative correlations in the western Pacific (Figure 4b).

Warming has been more important in the Northern ($\Delta T = 0.62°C$ between 1900-1920 and 1990-2007) than in the Southern Hemisphere ($\Delta T = 0.59°C$ between 1900-1920 and 1990-2007). Levitus and colleagues [41] assessed that the ocean heat content augmented by 14.5×10^{22} J, while Beltrami and colleagues [42] calculated that the continental lithosphere only gained 0.91×10^{22} J during the period 1950-2000. The ocean is by far the functional unit that has increased most. Levitus and colleagues [43] reported an increase of 0.66×10^{22} J in the global atmosphere during the period 1955-1996, a rise of 0.81×10^{22} J in continental glaciers for the same period, an augmentation of 0.32×10^{22} J in Antarctic sea ice between the 1950s and the 1970s and an increase of 1.1×10^{22} J in mountain glaciers for the period 1961-1997.

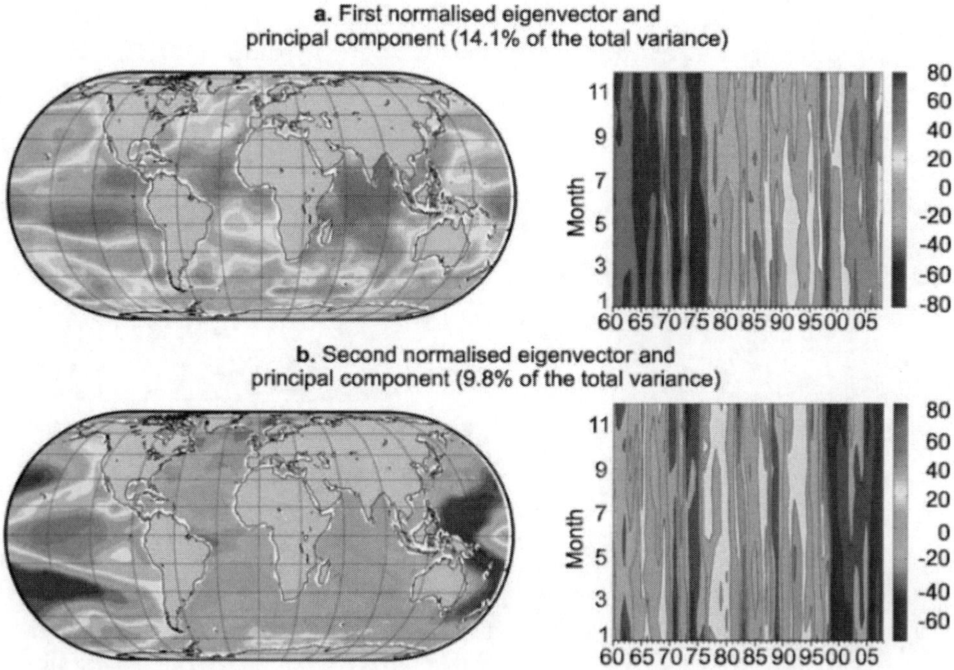

Figure 4. Principal Component Analysis (PCA) of global sea surface temperature (1960-2008). (a) First normalized eigenvector (left) and associated principal component (right). (b) Second normalized eigenvector (left) and associated principal component (right). Normalized eigenvectors reveal the correlation of sea surface temperature with the corresponding principal component for each geographical cell. Modified from Reid & Beaugrand [37].

Consequences of Global Warming on Ecosystems and Their Species

Several types of consequences of global warming have been documented: (i) phenological shifts, (ii) biogeographical shifts, (iii) long-term community/ecosystem shifts, including abrupt shifts (also called regime shifts) and trophic cascades, and (iv) changes in key ecosystem attributes such as diversity and species or community size. I also tackle hereafter the potential effect of ocean acidification, resulting from the increase in CO_2, but show that changes in temperatures may also affect strongly calcifying organisms.

Phenological Shifts

Phenology is the study of periodic biological phenomena, the seasonal peak of occurrence or abundance of species. Many works in both the terrestrial and the marine realms have shown that climate variability can influence the phenology of a species [4,44-46]. Processes behind phenological shifts are complex but the influence of temperature has been frequently invoked in both the terrestrial and the marine realms [47]. For example, phenological shifts in many planktonic species (e.g., dinoflagellates and copepods) have been reported in the North Sea and attributed to regional warming [4]. Temperature increases the rapidity of gonad development and trigger reproduction of a large number of species [4,48]. It is therefore not surprising that this parameter is often correlated with species phenology. Figure 5 shows spatial patterns in the peak of maximum abundance of *Calanus helgolandicus* around the British Isles calculated using the centre of gravity index [49]. Associated with temperature rise, the seasonal maximum of the species has advanced by nearly one month between the 1960s and the beginning of the 21st century (2000-2005).

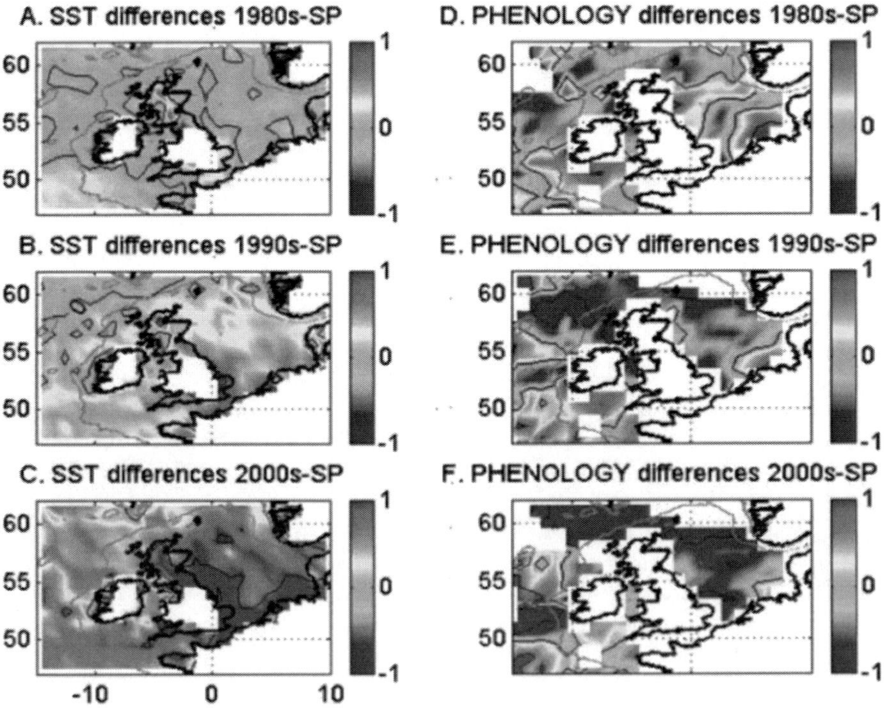

Figure 5. Changes in sea surface temperature (left panels) and in the phenology of *Calanus helgolandicus* (right panels; numbers are months) in the North Sea. SP: period 1960-1979; 1980s: period 1980-1989; 1990s: period 1990-1999; 2000s: period 2000-2005. For temperature, the unit of difference is in °C. For phenology, the unit of difference is the month. Modified from Beaugrand [92].

Mackas and colleagues [50] also propounded that temperature may be an important driver, although the authors stressed that other factors such as day-length or mixed-layer depth may also play a major role [4]. Edwards and Richardson [4] observed no phenological shift for diatoms in spring in the North Sea. In higher latitudes, photoperiod becomes important and may control the phenology of many species.

Species-dependent phenological responses may also deeply alter species interactions and involve trophic mismatch or abrupt ecosystem shifts [51,52]. Beaugrand and colleagues [52] showed a strong correlation between an index of larval cod survival and cod recruitment in the North Sea. This plankton index explained 48% of the variability in cod recruitment at age 1. The negative correlation between plankton changes and cod recruitment was explained by a reduction in the abundance of euphausiids (source of vitamin A for young cod), a reduction in the mean size of prey and a reduction of *Calanus finmarchicus* in the North Sea. The calanoid has been virtually replaced by its congeneric species *C. helgolandicus*: 80% of *C. finmarchicus versus* 20% of *C. helgolandicus* in 1960 and 20% *versus* 80% in 2005. Because the cold-water species has a seasonal maximum in spring and the warmer-water species in late summer, the change in the dominance of *Calanus* in the North Sea has altered *Calanus* phenology from spring to late summer. This shift may have caused a trophic mismatch [52]. It has been suggested that *Pseudocalanus* spp. may be consumed by cod larvae instead of *Calanus* [53]. However, a recent macroecological study has shown that areas where *Pseudocalanus* spp. are abundant are also areas where cod occurs in low abundance [54].

Biogeographical Shifts

Biogeographical movements are generally regarded as latitudinal species shifts in response to regional warming [55-58]. In the ocean, biogeographical shifts have taken place, consistent with latitudinal changes expected under climate warming and hydro-climatic variability. Major biogeographical shifts in zooplankton biodiversity have been observed in the North-east Atlantic in response to regional warming (about 1°C from 1960 to 2009 in the North Sea and between 0.5 and 1°C in the Northeast Atlantic) [5,56]. Using data from the Continuous Plankton Recorder (CPR) survey, the biodiversity of all copepods were first divided into nine species assemblages by means of multivariate analyses [59]. Beaugrand and colleagues [56] found an northwards increase in warm-water species assemblages that paralleled a reduction in the number of cold-water species (Figure 6).

All assemblages exhibited comprehensible long-term shifts that mimicked a movement of marine ecosystems towards a warmer dynamic regime. The rapidity with which these biogeographical shifts took place was unexpected in comparison to what was usually detected in the terrestrial realm [57]. Warm-temperate pseudo-oceanic species experienced a movement of about 10° of latitudes (52–62°N; 10°W), which represented a mean poleward movement of 23.16 km yr^{-1} for the period 1958-2005 [5]. The magnitude of the species shifts were however comparable to the northward movements of some isotherms (e.g., the 10°C isotherm moved northwards by an approximate rate of 21.75 km yr^{-1}) in the North Sea. As a result of these biogeographical shifts, calanoid copepod biodiversity increased in the Northeast Atlantic and the North Sea [10,60]. Such an increase in diversity has also been identified for other taxonomic groups such as dinoflagellates [10] and fish [61].

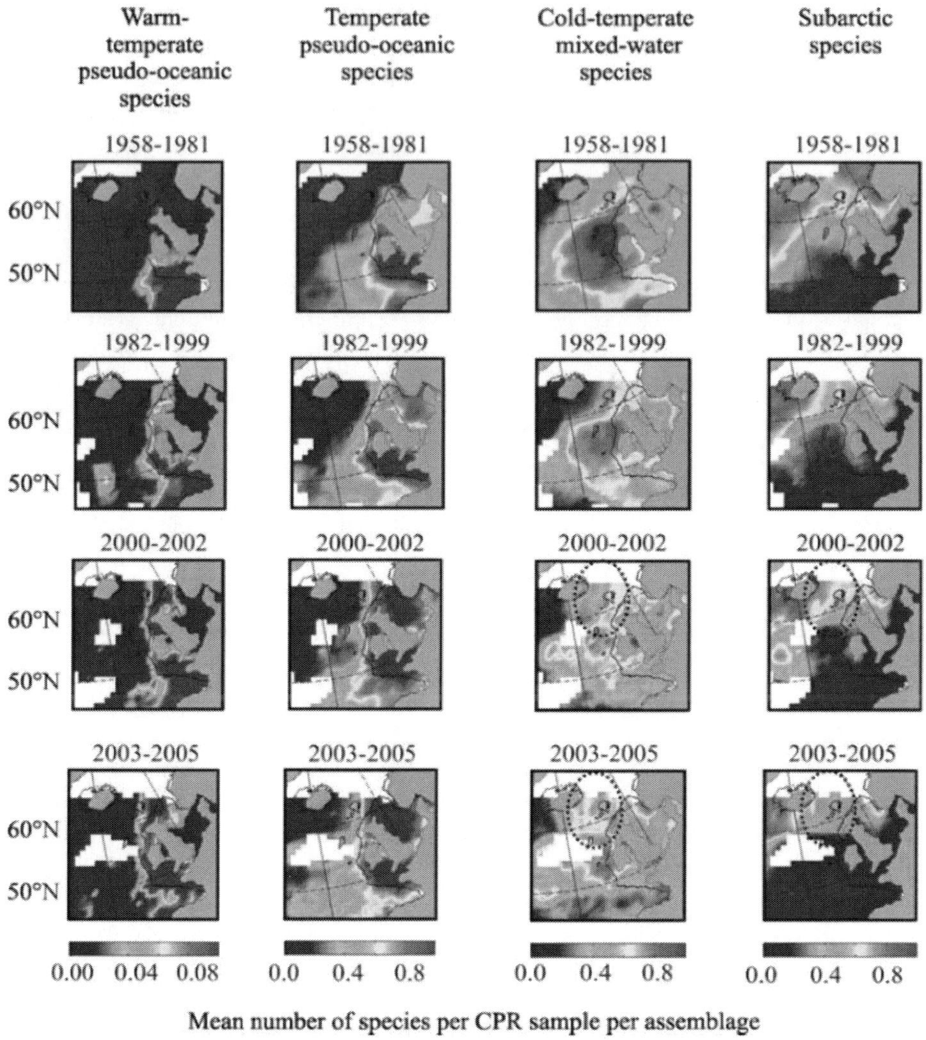

Figure 6. Long-term changes in the mean number of species per assemblage for the periods 1958-1981, 1982-1999, 2000-2002 and 2003-2005. Modified from Beaugrand [5].

Long-Term Abrupt Community/Ecosystem Shifts

At the community level, substantial changes have become frequently identified [6,7,62,63]. Abrupt community shifts (ACS) and abrupt ecosystem shifts (AES), commonly referred to as regime shifts, are characterized by a sudden, temporally persistent and substantial change in the state of an ecosystem and involve major biological modifications, often with implications for exploited resources. These phenomena remain generally poorly understood, and most of these abrupt changes have been reported years or decades after they had actually occurred (e.g., the Pacific Ocean and the North Sea). AESs have been observed to occur around specific time periods. For example during the late 1970s, AESs were observed in the Pacific and the Atlantic Oceans.

In the late 1980s, AESs have been documented in the North Sea [6], in the Mediteranean Sea [64], in the Baltic Sea [65] and in the western part of the North Atlantic [8]. In the late

1990s, shifts have been observed in the Northeast Pacific and the northeast Atlantic [62,66]. Reid and Beaugrand [37] noticed that in many cases the reported ecological regime shifts coincided with major temporal changes seen in marine temperature anomalies, and suggested that temperature may be an important synchronizing agent.

Recently, a new shift at the end of the 1990s has started to be documented in the north-east Atlantic and in its adjacent seas [11,62]. Luczak and colleagues [62] detected an AES *circa* 1996 in the northeast Atlantic (Figure 7). An increase in the critically endangered Balearic shearwater (*Puffinus mauretanicus*) paralleled an increase in annual SST, a diminution in total calanoid copepod reflecting a decrease in the size of calanoids and an augmentation of the modeled probability of occurrence of both anchovy and sardine. Their results provide evidence for both a direct and an indirect effect of climate change on the spatial distribution of post-breeding Balearic shearwater through trophic amplification or trophic cascade. Trophic cascade is the movement of the climatic signal from bottom to the top of the food web or inversely [67]. Trophic amplification (TA) is the amplification through the food web of the climatic signal, shown in practice by a bottom-up or a top-down rise in the correlation between trophic level [51]. TA is the result of a direct relationship between temperature and each trophic level and the indirect effect of temperature through the food web [51]. TA may represent a positive feedback that drives an ecosystem towards a new dynamic regime.

In lakes and in some marine ecosystems [68-70], some processes leading to abrupt shift, have been identified. For example, the shift from highly diverse corals to algal overgrowth in the Caribbean Sea was probably caused by (i) nutrient loading stimulated algal growth and (ii) overfishing that decreased the number of herbivorous fish controlling algal proliferation [69]. Studies in other systems (kelp forest ecosystems) found similar conclusions [71]. However in large marine ecosystems, long-term community shifts remain difficult to understand [72]. While temperature is often correlated to the ecosystem state in pelagic ecosystems such as those of the Pacific Ocean [7], north-east Atlantic [62,63] and the North Sea [6,73], processes by which the climate signal may cause substantial ecosystem changes remain unclear.

Changes in Key Functional Ecosystem Attributes

Biodiversity are composed of interconnected species within food webs that exchange matter, energy and information and form complex systems. Such complex systems are said to be self-regulating stable systems. Biodiversity is expected to reorganize as climate warms [10,61,74]. Beaugrand and colleagues [10] showed a pronounced and rapid increase in diversity of phytoplankton and zooplankton in the extratropical regions of the North Atlantic that paralleled a decrease in the mean size of copepods. These results were consistent with the latitudinal ecogeographical pattern of diversity of copepods already observed in the world's oceans [75]. The change in plankton diversity is probably already propagating to other trophic compartments (e.g., fish), mostly in extratropical regions where both the mean and seasonality of temperature limit diversity [61].

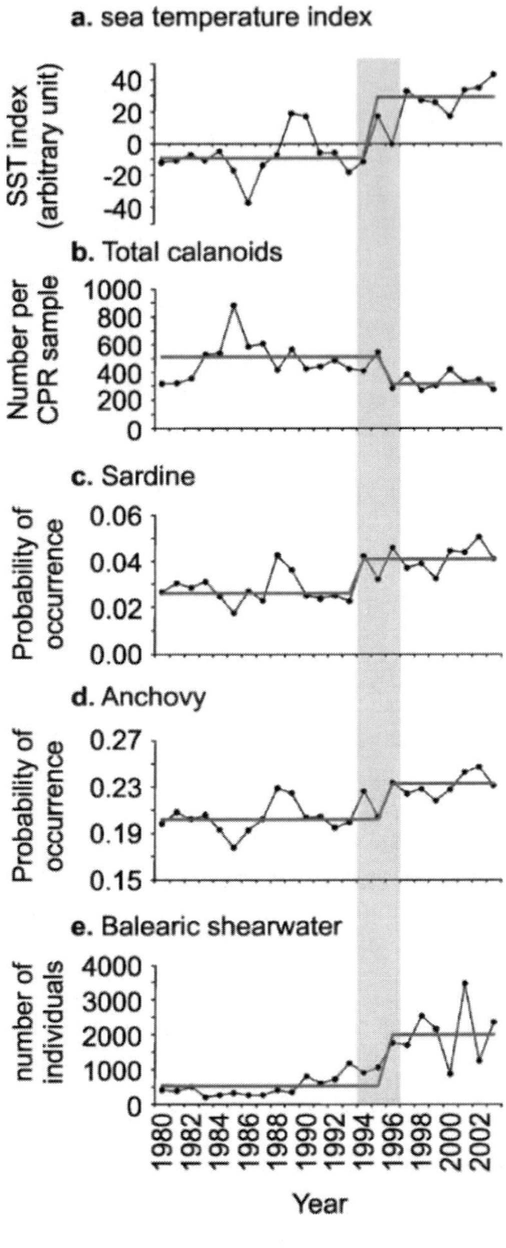

Figure 7. Long-term changes in (a) sea surface temperature, (b) total calanoids, (c) anchovies, (d) sardine and (e) Balearic shearwater. The timing of the shift, identified by the Rodionov's test, is superimposed. The red bar indicates the period 1994-1996. Modified from Luczak et al. [62].

Individual-level traits, such as body size, are important with respect to microevolution, population dynamics, and for ecosystem functioning and both provisioning and regulating services. Body size is a super-parameter, capturing a range of associated physiological and ecological traits. Changes in the mean size of any organisms can have profound consequences for the functioning of ecosystems [76]. Community body size also largely determines the types and strengths of the flows of energy and materials in ecosystems, affecting ecological networks and the way ecosystems are structured and function [77,78]. Biomass, production, predator/prey interactions, and carbon export are among processes that respond to the size structure of an ecosystem [77-79]. The mean size of individual species and communities is expected to diminish with climate warming [10], a phenomenon sometimes called dwarfism [80]. When diversity increases, size diminishes and significant negative relationships have been revealed at both spatial and temporal scales in the North Atlantic (Figure 8).

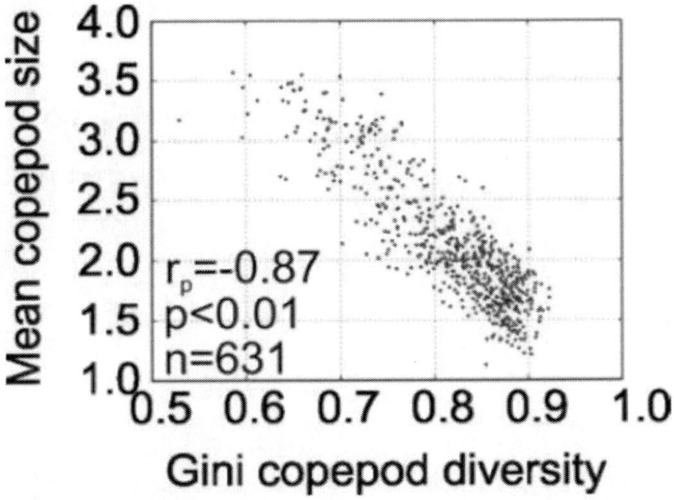

Figure 8. Relationships between latitudinal (36°N and 64°N) and year-to-year (1958-2007) changes in the diversity (Gini coefficient) and the mean size of calanoid copepods in the Northeast Atlantic. Linear correlation (rp), probability of H0 (p) and degrees of freedom (n) are indicated. Modified from Beaugrand et al. [10].

From the size of organisms, it is possible to assess the minimum turnover time of biogenic carbon (inversely proportional to the size) and the mean residence time of sinking particles from organisms in surface waters (higher for large species) [10,79]. The two processes alter the functioning of pelagic ecosystems and the biological carbon pump, respectively. Because of the inverse relationships between diversity and size of calanoids (Figure 8), highly significant correlations between copepod diversity and both the minimum turnover time of carbon in copepods (negative) and the mean residence time of sinking copepod particles (positive) have been revealed at both spatial and decadal scales (Figure 9). Figure 9 shows that the northward increase in diversity observed in the North Atlantic (Figs. 9a and 9d) has paralleled a reduction in the minimum turnover time of biogenic carbon (Figs. 9b and 9e). The increase in diversity also paralleled an increase in the mean residence time above 50 m of sinking copepod fecal pellets (Figs. 9c and 9f). Although it remains debated whether fecal pellets of copepods contribute substantially to the sinking of biogenic materials in the oceans [81], food-web controlled carbon export could be smaller in warmer oceans.

The extratropical oceanic regions of the North Atlantic play a key role in carbon export [82], and studies have suggested that the biological carbon pump may be less efficient because of changes in phytoplankton composition and reduced upward mixing of nutrients due to an increase in oceanic stratification [82,83]. In addition, the biological carbon pump may also be altered because oceanic warming increases the residence time of organic carbon in surface waters [9]. A reduction in the mean size of phytoplankton in relation to more stable and warmer environments has also been revealed in the Atlantic [84]. Similar findings have been found for North Sea fish [61]. It therefore appears that increasing temperatures lead to smaller-sized community assemblages across multiple pelagic trophic levels. Although the most conspicuous factors that seem to explain the observed changes are the increase in temperature and the associated reduction in the seasonal variability, these modifications are probably driven by both climate change and the complex interactions between ocean variability and climate [10].

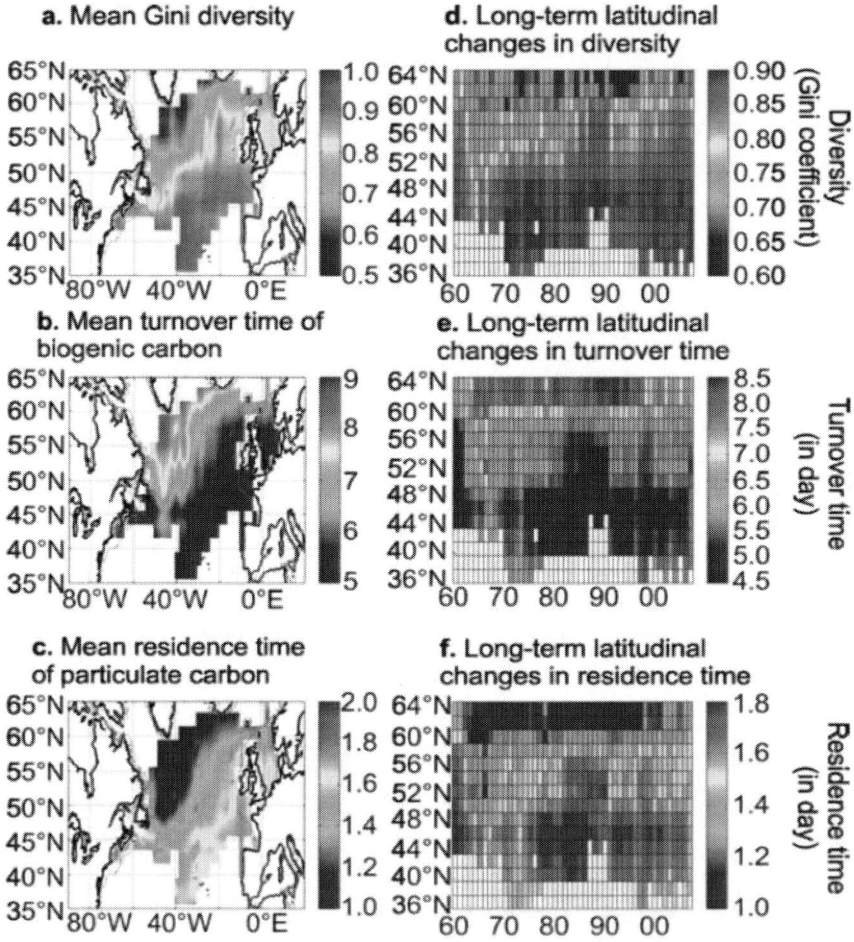

Figure 9. Relationships between the spatial distribution and long-term latitudinal changes in the diversity, the minimum turnover time of carbon incorporated in copepods and the mean residence time above 50 m of sinking copepod particles in the extratropical North Atlantic. Diversity was measured by first-order jackknife performed on the Gini coefficient. Mean spatial distributions (1960-2007) of (a) copepod diversity, (b) minimum turnover time of carbon incorporated in copepods (in days), and (c) mean residence time above 50 m of sinking copepod particles (in days). Long-term latitudinal changes in (d) copepod diversity, (e) minimum turnover time of carbon incorporated in copepods, and (f) mean residence time above 50 m of sinking copepod particles. Modified from Beaugrand et al. [10].

Acidification

The oceans contain around 38,000 Gt of carbon, which is 50 times more than the atmosphere and 20 times more than the carbon stored by the terrestrial biosphere and the soils. It is not only a crucial reservoir but also the most long-term carbon sink.

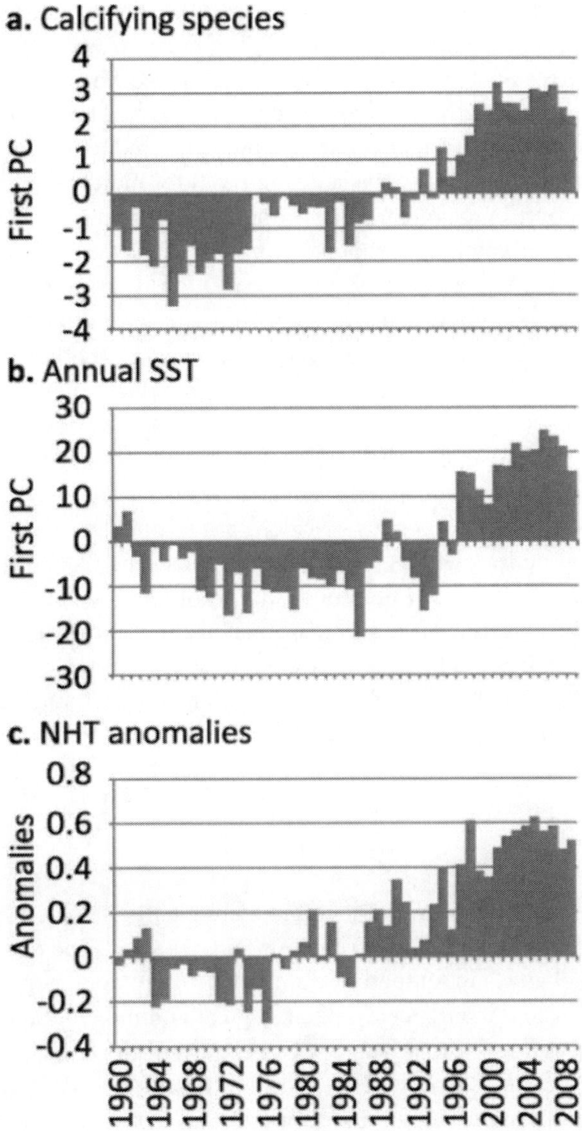

Figure 10. Decadal changes (1960-2009) in North Atlantic calcifying plankton in relation to large-scale hydro-climatological forcing. (a) Long-term changes in an index reflecting the abundance or frequency of calcifying plankton. (b) Long-term changes in annual North Atlantic sea temperature. (c) Long-term changes in Northern Hemisphere Temperature (NHT) anomalies. The index of the abundance or the frequency of calcifying plankton was calculated using a standardized Principal Component Analysis (PCA) on the first principal components reflecting the long-term changes in foraminifers, coccolithophores, Limacina spp., Clione limacina, echinoderm larvae and bivalves inferred from the Continuous Plankton Recorder (CPR) survey. Modified from Beaugrand et al. [11].

However since 1750, the atmospheric concentration in CO_2 has risen from 280 ppm to *ca.* 390 ppm due to the combustion of fossil fuels, cement production, agriculture and deforestation. Both biogeochemical models and observations suggest that about 140 Gt C on the 500 Gt C emitted in the atmosphere by human activities since pre-industrial time have been absorbed by the ocean. Increase in CO_2 have been observed mainly in the first 1000 m of the oceans but in some regions the penetration of CO_2 has been deeper [85]. This has been the case in the North Atlantic where a rise in CO_2 has been observed down to 3000 m due to deep convection. As a result, the aragonite saturation horizon has also shoaled to 50-200 m below the surface in all oceanic basins [86]. The global pH has been reduced from 8.2 to 8.1 since industrialization began. In logarithmic pH unit, this reduction of 0.1 unit might seem negligible, but in absolute terms, that translates into a 30% increase in surface-ocean acidity. Plankton are a major contributor to carbon cycling and ocean-atmosphere exchange and can generate important flux of calcium carbonate from the surface to the bottom of the oceans [87]. Most (> 80%) of the biogenic carbonate precipitation is accomplished by plankton [88]. Among calcifying plankton, blooms of the coccolithophores *Emiliania huxleyi* cover large oceanic area, about 1.4×10^6 km² per year [89]. Hofmann and colleagues [90] reviewed the effect of ocean acidification on five main types of ecosystems: (i) coral reef ecosystems, (ii) open-ocean ecosystems, (iii) coastal ecosystems, (4) deep-sea ecosystems, and (v) high-latitudes ecosystems. Coral reef ecosystems may cease to grow and undertake a net dissolution worldwide when atmospheric concentration of CO_2 reaches 560 ppm [91]. However, many aspects of the responses of calcifying species to ocean acidification remain elusive. Beaugrand and colleagues [11] provided evidence that calcifying organisms such as foraminifers, coccolithophores, both pteropod and non-pteropod molluscs and echinoderms exhibited an abrupt shift *circa* 1996 at a time of a substantial increase in temperature (Figure 10). Although acidification is potentially a serious threat to marine calcifying organisms, their results showed that the primary driver of North Atlantic calcifying plankton was oceanic temperature and not acidification.

CONCLUSION

Climate change is unambiguous and its effects are clearly detected in all functional units of the Earth System. Temperature has increased in many parts of the oceans and responses of pelagic ecosystems to changes in temperature start to be more documented. Among responses exhibited by ecosystems and their species, phenological and biogeographical shifts have been reported. At the ecosystem level, many abrupt community shifts (ACS) and abrupt ecosystem shifts (AES) have been identified around the world at the end of the 1970s, 1980s and 1990s, at time of pronounced changes in global temperatures. Key functional attributes such as biodiversity and size have also been altered by ocean warming and their study has revealed potential strong effects of global warming on ecosystem functioning and both regulating (biological carbon pump) and provisioning (fisheries) services ecosystems provide. Although ocean acidification is a serious threat for calcifying organisms, they are also strongly affected by changes in the thermal regime of the oceanic regions. Climate variability remains elevated and should always be considered when potential relationships between both ecosystems and species and global warming are investigated.

REFERENCES

[1] Halpern BS, Walbridge S, Selkoe KA et al. (2008) A global map of human impact on marine ecosystems. *Science*, 319, 948-952.

[2] Kirby RR, Beaugrand G & Lindley JA (2009) Synergistic effects of climate and fishing in a marine ecosystem. *Ecosystems*, 12, 548-561.

[3] Luczak C, Beaugrand G, Lindley JA et al. (2012) North Sea ecosystem changes from swimming crabs to seagulls. *Biology Letters*, 8, 821-824.

[4] Edwards M & Richardson AJ (2004) Impact of climate change on marine pelagic phenology and trophic mismatch. *Nature*, 430, 881-884.

[5] Beaugrand G, Luczak C & Edwards M (2009) Rapid biogeographical plankton shifts in the North Atlantic Ocean. *Global Change Biology*, 15, 1790-1803.

[6] Reid PC, Borges M & Svenden E (2001) A regime shift in the North Sea circa 1988 linked to changes in the North Sea horse mackerel fishery. *Fisheries Research*, 50, 163-171.

[7] Hare SR & Mantua NJ (2000) Empirical evidence for North Pacific regime shifts in 1977 and 1989. *Progress in Oceanography*, 47, 103-145.

[8] Beaugrand G, Edwards M, Brander K et al. (2008) Causes and projections of abrupt climate-driven ecosystem shifts in the North Atlantic. *Ecology Letters*, 11, 1157-1168.

[9] Li B-L, Gorshkov VG & Makarieva AM (2004) Energy partitioning between different-sized organisms and ecosystem stability. *Ecology*, 85, 1811-1813.

[10] Beaugrand G, Edwards M & Legendre L (2010) Marine biodiversity, ecosystem functioning and the carbon cycles. *Proceedings of the National Academy of Sciences of the United States of America*, 107, 10120-10124.

[11] Beaugrand G, McQuatters-Gollop A, Edwards M & Goberville E (2013) Long-term responses of North Atlantic calcifying plankton to climate change. *Nature Climate Change*, 3, 263-267.

[12] Roemmich D & McGowan J (1995) Climatic warming and the decline of zooplankton in the California Current. *Science*, 267, 1324-1326.

[13] Richardson AJ & Schoeman DS (2004) Climate impact on plankton ecosystems in the northeast Atlantic. *Science*, 305, 1609-1612.

[14] Mackas DL, Batten S & Trudel M (2007) Effects on zooplankton of a warmer ocean: recent evidence from the northeast Pacific. *Progress in Oceanography*, 75, 223-252.

[15] Taylor AH, Allen JI & Clark PA (2002) Extraction of a weak climatic signal by an ecosystem. *Nature*, 416, 629-632.

[16] Brown JH, Gillooly JF, Allen AP et al. (2004) Toward a metabolic theory of ecology. *Ecology*, 85, 1771-1789.

[17] Atkinson D (1994) Temperature and organism size - a biological law for ectotherms? *Advances in Ecological Research*, 3, 1-58.

[18] Alheit J & Hagen E (1997) Long-term forcing of European herring and sardine populations. *Fisheries Oceanography*, 6, 130-139.

[19] Hurrell JW, Yochanan K, Visbeck M (2001) The North Atlantic Oscillation. *Science*, 291, 603-605.

[20] Dickson RR & Turrell WR (2000) The NAO: the dominant atmospheric process affecting oceanic variability in home, middle and distant waters of european atlantic

]salmon. *In*: Mills D (ed.), *The Ocean Life of Atlantic salmon Environmental and Biological Factors Influencing Survival*. Bodmin: Fishing News Books, 92-115.

[21] Dickson RR, Kelly PM, Colebrook JM et al. (1988) North winds and production in the eastern North Atlantic. *Journal of Plankton Research*, 10, 151-159.

[22] Ottersen G, Planque B, Belgrano A et al. (2001) Ecological effects of the North Atlantic Oscillation. *Oecologia*, 128, 1-14.

[23] Fromentin JM & Planque B (1996) *Calanus* and environment in the eastern North Atlantic. II. Influence of the North Atlantic Oscillation on *C. finmarchicus* and *C. helgolandicus*. *Marine Ecology Progress Series*, 134, 111-118.

[24] Beaugrand G (2012) Unanticipated biological changes and global warming. *Marine Ecology Progress Series*, 445, 293-301.

[25] Russell FS, Southward AJ, Boalch GT & Butler EI (1971) Changes in biological conditions in the English Channel off Plymouth during the last half century. *Nature*, 234, 468-470.

[26] Southward AJ (1974) Changes in the plankton community of the western English Channel. *Nature*, 249, 180-181.

[27] Edwards M, Beaugrand G, Helaouët P & Coombs SH (submitted) Marine ecosystem response to the Atlantic Multidecadal Oscillation. *PLos ONE*, 8, e57212.

[28] Enfield DB, Mestas-Nunez AM, Trimble PJ (2001) The Atlantic Multidecadal Oscillation and its relationship to rainfall and river flows in the continental U.S. *Geophysical Research Letters*, 28, 2077-2080.

[29] Gray ST, Graumlich LJ, Betancourt JL & Pederson GT (2004) A tree-ring based reconstruction of the Atlantic Multidecadal Oscillation since 1567 A.D. *Geophysical Research Letters*, 31, L12205.

[30] Longhurst A (2007) *Ecological Geography of the Sea*. Elsevier, Amsterdam.

[31] Goberville E, Beaugrand G & Edwards M (2013) Synchronous response of marine plankton ecosystems to climate in the Northeast Atlantic and the North Sea. *Journal of Marine Systems*, 129, 189-202.

[32] An SI (2009) A review of interdecadal changes in the nonlinearity of the El Niño-Southern Oscillation. *Theoretical and Applied Climatology*, 97, 29-40.

[33] Wilkinson C (2004) *Status of the Coral Reefs of the World 2004*. Australian Institute of Marine Sciences, Townsville.

[34] Lehodey P, Bertignac M, Hampton J et al. (1997) El Niño Southern Oscillation and tuna in the western Pacific. *Nature*, 389, 715-718.

[35] Bindoff N, Willebrand J, Artale V et al. (2007) Observations: Oceanic climate change and sea level. *In*: Solomon S, Qin D, Manning M et al. (eds.), *Climate Change 2007: The Physical Science Basis Contribution of Working Group I to the Fourth Assessment Report of the Intergovernmental Panel on Climate Change*. Cambridge University Press, Cambridge, 385-432.

[36] Hansen J, Ruedy R & Sato M (2010) Global surface temperature change. *Reviews of Geophysics*, 48, RG4004.

[37] Reid PC & Beaugrand G (2012) Global synchrony of an accelerating rise in sea surface temperature. *Journal of the Marine Biological Association of the United Kingdom*, 92, 1435-1450.

[38] Intergovernmental Panel on Climate Change WGI (2007) *Climate Cchange 2007: The Physical Science Basis*. Cambridge University Press, Cambridge.

[39] Guilderson TP & Daniel PS (1998) Abrupt shift in subsurface temperatures in the tropical Pacific associated with changes in El Nino. *Science*, 281, 240-243.

[40] Takayabu YN, Iguchi T, Kachi M et al. (1999) Abrupt termination of the 1997-98 El Niño in response to a Madden-Julian oscillation. *Nature*, 402, 279-282.

[41] Levitus S, Antonov J & Boyer T (2005) Warming of the world ocean, 1955-2003. *Geophysical Research Letters*, 32, LO2604, doi: 2610.1029/2004GLO21592.

[42] Beltrami H, Smerdon JE, Pollack HN & Huang S (2002) Continental heat gain in the global climate system. *Geophysical Research Letters*, 29, 10.1029/2001GL014310.

[43] Levitus S, Antonov JI, Wang J et al. (2001) Anthropogenic warming of Earth's climate system. *Science*, 292, 267-270.

[44] Crick HQP, Dudley C, Glue DE & Thomson DL (1997) UK birds are laying eggs earlier. *Nature*, 388, 526.

[45] Myneni RB, Keeling CD, Tucker CJ et al. (1997) Increased plant growth in the northern high latitudes from 1981 to 1991. *Nature*, 386, 698-702.

[46] Reid PC, Edwards M, Hunt HG & Warner AJ (1998) Phytoplankton change in the North Atlantic. *Nature*, 391, 546.

[47] Root TL & Hughes L (2005) Present and future phenological changes in wild plants and animals. *In*: Lovejoy TE & Hannah L (eds.), *Climate Change and Biodiversity*. Yale University Press, New Haven, 61-74.

[48] Kirby RR, Beaugrand G, Lindley JA et al. (2007) Climate effects and benthic-pelagic coupling in the North Sea. *Marine Ecology Progress Series*, 330, 31-38.

[49] Colebrook JM & Robinson GA (1965) Continuous Plankton Records: seasonal cycles of phytoplankton and copepods in the north-eastern Atlantic and the North Sea. *Bulletin of Marine Ecology*, 6, 123-139.

[50] Mackas DL, Greve W, Edwards M et al. (2012) Changing zooplankton seasonality in a changing ocean: Comparing time series of zooplankton phenology. *Progress in Oceanography*, 97-100, 31-62.

[51] Kirby RR & Beaugrand G (2009) Trophic amplification of climate warming. *Proceedings of the Royal Society London B: Biological Sciences*, 276, 3053-3062.

[52] Beaugrand G, Brander KM, Lindley JA et al. (2003) Plankton effect on cod recruitment in the North Sea. *Nature*, 426, 661-664.

[53] Heath MR & Lough RG (2007) A synthesis of large-scale patterns in planktonic prey of larval and juvenile cod (*Gadus morhua*). *Fisheries Oceanography*, 16, 169-185.

[54] Beaugrand G & Kirby RR (2010) Climate, plankton and cod. *Global Change Biology*, 16, 1268-1280.

[55] Thomas CD & Lennon JJ (1999) Birds extend their ranges northwards. *Nature*, 399, 213.

[56] Beaugrand G, Reid PC, Ibañez F et al.(2002) Reorganisation of North Atlantic marine copepod biodiversity and climate. *Science*, 296, 1692-1694.

[57] Parmesan C & Yohe G (2003) A globally coherent fingerprint of climate change impacts across natural systems. *Nature*, 421, 37-42.

[58] Perry AI, Low PJ, Ellis JR & Reynolds JD (2005) Climate change and distribution shifts in marine fishes. *Science*, 308, 1912-1915.

[59] Beaugrand G, Ibañez F, Lindley JA & Reid PC (2002) Diversity of calanoid copepods in the North Atlantic and adjacent seas: species associations and biogeography. *Marine Ecology Progress Series*, 232, 179-195.

[60] Beaugrand G & Ibañez F (2002) Spatial dependence of pelagic diversity in the North Atlantic Ocean. *Marine Ecology Progress Series*, 232, 197-211.

[61] Hiddink JG & Hofstede ter R (2008) Climate induced increases in species richness of marine fishes. *Global Change Biology*, 14, 453-460.

[62] Luczak C, Beaugrand G, Jaffré M & Lenoir S (2011) Climate change impact on Balearic shearwater through a trophic cascade. *Biology Letters*, 7, 702-705.

[63] Beaugrand G, McQuatters-Gollop A, Edwards M, Goberville E (2012) Long-term changes in calcifying plankton in the North Atlantic and climate. *Nature Climate Change*, 3, 263-267.

[64] Conversi A, Umani SF, Peluso T et al. (2010) The Mediterranean Sea regime shift at the end of the 1980s, and intriguing parallelisms with other European basins. *PLoS ONE*, 5, 1-15.

[65] Mollmann C, Diekmann R, Müller-Karulis B et al. (2009) Reorganization of a large marine ecosystem due to atmospheric and anthropogenic pressure: a discontinuous regime shift in the Central Baltic Sea. *Global Change Biology*, 15, 1377-1393.

[66] Cloern JE, Hieb KA, Jacobson T et al. (2010) Biological communities in San Francisco Bay track large-scale climate forcing over the North Pacific. *Geophysical Research Letters*, 37, L21602.

[67] Cury P, Shannon L & Shin Y-J (2003) The functioning of marine ecosystems: a fisheries perspective. *In*: Sinclair M & Valdimarsson G (eds.), *Responsible Fisheries in the Marine Ecosystem*. FAO and CAB international, Rome, 103-123.

[68] Scheffer M & van Nes EH (2004) Mechanisms for marine regime shifts: can we use lakes as microcosms for oceans? *Progress in Oceanography*, 60, 303-319.

[69] Nystrom M, Folke C & Moberg F (2000) Coral reef disturbance and resilience in a human-dominated environment. *Trends in Ecology and Evolution*, 15, 413-417.

[70] Scheffer M (2009) *Critical Transitions in Nature and Society*. Princeton University Press, Princeton.

[71] Casini M, Hjelma J, Molinero J-C et al. (2009) Trophic cascades promote threshold-like shifts in pelagic marine ecosystems. *Proceedings of the National Academy of Sciences of the United States of America*, 106, 197-202.

[72] de Young B, Barange M, Beaugrand G et al. (2008) Regime shifts in marine ecosystems: detection, prediction and management. *Trends in Ecology and Evolution*, 23, 402-409.

[73] Weijerman M, Lindeboom H & Zuur AF (2005) Regime shifts in marine ecosystems of the North Sea and Wadden Sea. *Marine Ecology Progress Series*, 298, 21-39.

[74] Sheridan JA & Bickford D (2011) Shrinking body size as an ecological response to climate change. *Nature Climate Change*, 1, 401-406.

[75] Rombouts I, Beaugrand G, Gasparini S et al. (2009) Global latitudinal variations in marine copepod diversity and environmental factors. *Proceedings of the Royal Society London B: Biological Sciences*, 276, 3053-3062.

[76] Peters RH (1983) *The Ecological Implications of Body Size*. Cambridge University Press, Cambridge.

[77] Daufresne M, Lengfellner K & Sommer U (2009) Global warming benefits the small in aquatic ecosystems. *Proceedings of the National Academy of Sciences of the United States of America*, 106, 12788-12793.

[78] Woodward G, Ebenman B, Emmerson M et al. (2005) Body size in ecological networks. *Trends in Ecology and Evolution*, 20, 402-409.

[79] Legendre L & Michaud J (1998) Flux of biogenic carbon in oceans: size-dependent regulation by pelagic food webs. *Marine Ecology Progress Series*, 164, 1-11.

[80] Smith JJ, Hasiotis ST, Kraus MJ & Woody DT (2009) Transient dwarfism of soil fauna during the Paleocene–Eocene Thermal Maximum. *Proceedings of the National Academy of Sciences of the United States of America*, 106, 17655-17660.

[81] Turner JT (2002) Zooplankton fecal pellets, marine snow and sinking phytoplankton blooms. *Aquatic Microbial Ecology*, 27, 57-102.

[82] Sarmiento JL, Slater R, Barber R et al. (2004) Response of ocean ecosystems to climate warming. *Global Biogeochemical Cycles*, 18, 1-23.

[83] Bopp L, Aumont O, Cadule P et al. (2005) Response of diatoms distribution to global warming and potential implications: A global model study. *Geophysical Research Letters*, 32, L19606, doi:19610.11029/12005GL023653.

[84] Li WKW (2002) Macroecological patterns of phytoplankton in the northwestern North Atlantic Ocean. *Nature*, 419, 154-157.

[85] Sabine CL, Feely RA, Gruber N et al. (2004) The oceanic sink for anthropogenic CO2. *Science*, 305, 367-371.

[86] Orr JC, Fabry VJ, Aumont O et al. (2005) Anthropogenic ocean acidification over the twenty-first century and its impact on calcifying organisms. *Nature,* 437, 681-686.

[87] Riebesell U, Zondervan I, Rost B et al. (2000) Reduced calcification of marine plankton in response to increased atmospheric CO_2. *Nature*, 407, 364-367.

[88] Milliman JD (1993) Production and accumulation of calcium carbonate in the ocean: budget of a nonsteady state. *Global Biogeochemical Cycles*, 7, 927-957.

[89] Brown CW (1995) Global distribution of coccolithophore blooms. *Oceanography*, 8, 59-60.

[90] Hofmann GE, Barry JP, Edmunds PJ et al. (2010) The effect of ocean acidification on calcifying organisms in marine ecosystems: an organism to ecosystem perspective. *The Annual Review of Ecology, Evolution, and Systematics*, 41, 127-147.

[91] Silverman J, Lazar B, Cao L et al. (2009) Coral reefs may start dissolving when atmospheric CO_2 doubles. *Geophysical Research Letters*, 36, L05606.

[92] Beaugrand G (2009) Decadal changes in climate and ecosystems in the North Atlantic Ocean and adjacent seas. *Deep-Sea Research II*, 56, 656-673.

In: Copepods: Diversity, Habitat and Behavior
Editor: Laurent Seuront
ISBN: 978-1-63117-846-7

Chapter 2

COMMUNITY ECOLOGY OF PELAGIC COPEPODS IN TROPICAL COASTAL WATERS

A. D. McKinnon and S. Duggan*
Australian Institute of Marine Science, Townsville M.C., Queensland, Australia

ABSTRACT

Tropical copepod communities are more diverse than those of temperate ecosystems, in large part due to the occurrence of multiple species of small copepods belonging to the families Oithonidae and Paracalanidae. We contrast the communities found in tropical estuaries, coasts and the epipelagic of shelf seas based on our own studies in Australian and Indonesian waters and compare these to other tropical systems. Trophic factors related to the predominance of microbial food webs in tropical seas are important in shaping tropical copepod communities, as are swarming and emergence behaviors. "Tropicalization" of plankton resulting from global warming may result in these communities becoming more widespread.

INTRODUCTION

Studies of the community ecology of copepods began in Europe and North America using sampling gear developed to better understand the distribution of keystone species such as *Calanus finmarchicus*, a key determinant of the recruitment success of herring and other commercially important fishes. Consequently, our knowledge of the ecology of pelagic copepods is largely based upon temperate ecosystems, with the calanoid genera *Calanus* and *Acartia* widely regarded as model copepods. Despite the tropical ocean comprising approximately 50% of the area of all open water and 30% of the total area of continental shelf [1], it is less well known than temperate and even polar regions. The copepod communities occurring in the tropics are more diverse than those in temperate regions and differ in species composition. Tropical copepods are small in comparison to the more widely studied copepods

* E-mail address: d.mckinnon@aims.gov.au

of temperate ecosystems, consistent with Bergmann's rule that states that body size is smaller in warmer climates both at the interspecific and intraspecific level [2]. Consequently, the use of sampling equipment developed for temperate plankton has meant that the smaller fraction of tropical plankton has been undersampled. Thus, the full diversity of tropical copepod communities has only become apparent since the use of fine mesh nets (<200 μm) has become more widely adopted.

In this chapter we review the community ecology of tropical planktonic copepods. Our purpose is two-fold. Firstly, we seek to describe the copepod communities of characteristic coastal marine ecosystems in the tropics, ranging from those occurring in turbid and turbulent ultra-trophic mangrove-dominated coastal systems to those of clear oligotrophic shelf waters. In the first instance this is based on our own work in tropical Australia (Figure 1) for which methods and taxonomic resolution are comparable, but we draw on comparable studies from elsewhere in the tropics where appropriate. Secondly, we briefly consider the behavioral and trophic factors that influence community structure. We confine ourselves to copepod communities in the epipelagic (<200 m depth) between the tropics of Cancer and Capricorn. Copepod populations in the field are dominated by juvenile stages: nauplii outnumber copepodites that in turn outnumber adults. However, since it is only possible to identify adult copepods to species, this review is based only on the adults.

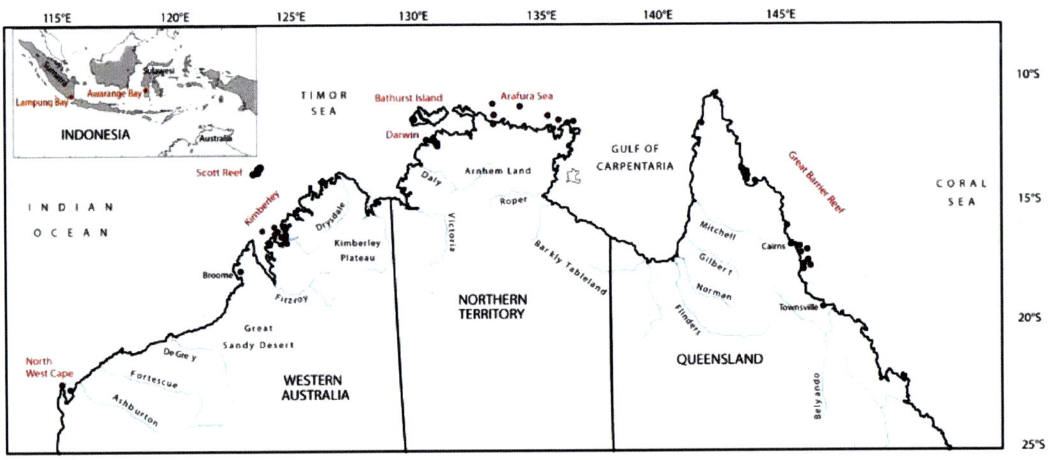

Figure 1. Location of Australian and Indonesian studies referred to in the present work.

Taxonomic Mayhem

The importance of taxonomy cannot be understated when attempting to understand community composition. Unfortunately, tropical ecosystems tend to be understudied in comparison to those in temperate regions, and this is made worse by the small size of the common pelagic copepods: most are <800 μm in total length. Three families dominate tropical pelagic marine copepod communities: the cyclopoid families Oithonidae and Oncaeidae, and the calanoid family Paracalanidae. Indo-Pacific forms of Oithonidae are comparatively well described [3], though many of the smaller species (especially those in the *Paroithona* group) remain undescribed, and there seems little doubt that widely occurring taxa such as *Oithona simplex*, amongst others, are plural. Epipelagic forms of Oncaeidae are also quite well known, e.g., [4]. This family is, however, characterized by tiny forms that become more diverse in the mesopelagic and for which the taxonomy is much less mature. The family Paracalanidae comprises superficially similar copepods belonging to a number of genera (*Paracalanus*, *Acrocalanus*, *Parvocalanus*, *Bestiolina*, *Delibus*, *Calocalanus*) in which there are many closely related species. Almost without exception, taxonomic descriptions are old and do not reach the standard or conventions of modern copepod taxonomy [5,6]. Consequently, there are many form variants and regional morphotypes that are difficult if not impossible to resolve using the available taxonomy, as a result of which recent taxonomic effort has been redirected toward the use of molecular tools [7]. For instance, *Paracalanus parvus* (Claus, 1863) was widely regarded as being a cosmopolitan species. With increasing frequency, new species have been erected to accommodate local variants of this "superspecies", usually on the basis of minor differences in character states and without consideration of the level of variation within the genus as a whole. For the Australian material of this keystone species, we have adopted the resurrection of the name *P. indicus* for material from New Zealand, which our specimens most closely match based on the comparison of collections from a range of locations [8]. We recognize, however, for this species and for others in the family, that there is an enormous range of regional variation that in all likelihood represent complexes of sibling species, as has been confirmed in molecular studies underway (Cornils, personal communication).

Taxonomic problems are not confined to these families. We believe other common genera also contain cryptic species. There are new records of undescribed or poorly described species with each new study in a previously unstudied area of the tropics. Unfortunately, taxonomy is a dying art, and though molecular tools are improving our knowledge of diversity there appears no easy solution to resolving the routine identification of the basic units of ecosystems.

Habitat Differences in Copepod Community Composition

The community composition of tropical marine copepods varies depending on habitat. Inshore, shallow-water communities are characterized by high abundances but low diversity. For instance, in the mangrove-dominated macrotidal system of Bathurst Island, northern Australia, abundances of adult copepods are high (>12,000 adults m^{-3}) but are comprised of only 21 species, with three clear dominants (Table 1). In contrast, at Scott Reef, the location furthest offshore, abundance was low (730 adult copepods m^{-3}), but diversity was high (119 species). Though less oceanic in character, waters adjacent to Australia's North West Cape (NWC) had similar abundances to Scott Reef but had more species (136), probably as a result of the combined effects of greater sampling effort and smaller mesh size (73 μm net used compared to 100 μm net at Scott Reef).

Table 1. The proportion of copepods in total zooplankton, and the abundance (adults m^{-3}) of the 10 most abundant copepod species in 8 tropical environments. Where species reached this criterion in only one environment, they are listed at the bottom of the column, with their abundances in brackets

Species Name	Bathurst Island	Indonesia	Darwin Harbour	Arafura Sea	Kimberley	Great Barrier Reef	North West Cape	Scott Reef
Proportion adult copepods (%)	9	10	11	23	35	12	9	18
Proportion adult copepods + copepodites (%)	98	93	92	83	87	75	72	76
Total copepod abundance	12,235	12,719	5,112	3,692	1,620	1,655	1,092	731
Oithona simplex	1,391	1,697	278	259	138	292	73	118
Euterpina acutifrons	708	889	353	269	200	226	48	
Microsetella spp.	560	351	434	441		65	70	93
Oithona nana		1,723	43	1,021	186	359	126	121
Oithona attenuata	359		368	118	180	89	42	
Parvocalanus crassirostris	3,283	1,743	1,175			274	31	
Bestiolina similis	186		157			17		
Oithona aruensis	5,132	176	1,456					
Oithona rigida		4,706		74			76	
Paracalanus indicus	83					47	84	
Oithona plumifera				109				25
Oncaea spp.						108		108
Parvocalanus dubia		822					32	
Parvocalanus sp. 2				598	462			

Table 1. Continued

Species Name	Bathurst Island	Indonesia	Darwin Harbour	Arafura Sea	Kimberley	Great Barrier Reef	North West Cape	Scott Reef
	Acartia pacifica (189)	*Bestiolina* sp. (188)	*Acartia sinjiensis* (53)	*Oncaea clevei* (177)	*Acrocalanus gibber* (47)	*Ditricho-corycaeus dahli* (39)	*Oncaea waldemari* (32)	*Calocalanus* sp. 1 (19)
	Onychocorycaeus catus (102)	*Spinoncaea* sp. (127)	*Oithona nishidai* (663)	*Paracalanus aculeatus f:minori* (101)	*Bestiolina* sp. 1 (52)			*Calocalanus* sp. 2 (13)
					Canthocalanus pauper (35) *Parvocalanus* sp. 1 (109)			*Delibus* sp.(37) *Oithona* sp. 4 (23) *Oithona* sp. T (15)

Copepod communities become more diverse with distance offshore. There are three patterns of rank abundance curves in Figure 2, indicating differences in community structure in mangrove, coastal and shelf environments. Mangrove estuaries in Indonesia and Darwin Harbour follow a similar pattern to those at Bathurst Island but are more speciose because of greater exchange with coastal waters. Coastal plankton communities of the Great Barrier Reef (GBR) and Kimberley are similar, but include a greater proportion (1/3) of rare species (i.e., those with abundances <1 m^{-3}), possibly as a result of mixing of offshore water. In contrast, shelf environments (NWC and Scott Reef) have more attenuated rank abundance curves and a much larger proportion (approximately half) of rare species. The Arafura Sea is a special case of coastal community, similar in pattern to the shelf communities but truncated in species number. By contrast, temperate coastal plankton, using the example of data from the L4 Station near Plymouth in the English Channel, has similar abundance of the most common species but is less diverse, though some genera have not been speciated [9].

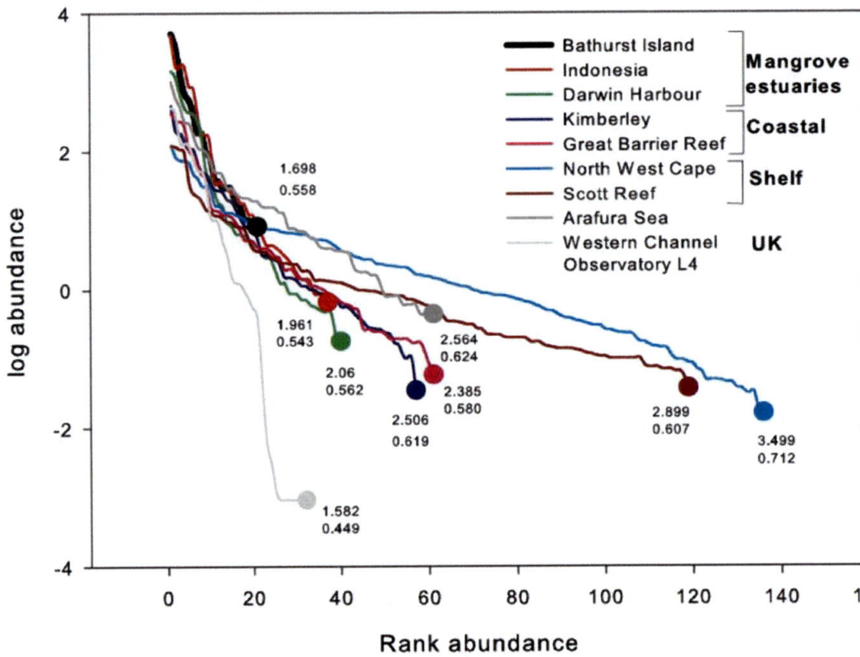

Figure 2. Rank abundance curves ("Whittaker plots") for adult copepods collected in the top 60 m of the water column at 8 locations in tropical Australia and Indonesia. Samples were collected with 73 μm plankton nets, except for the Arafura Sea, Kimberley and Scott Reef studies where 100 μm nets were used. For comparison with temperate communities, the corresponding pattern for 200 μm coastal plankton in the English Channel is included [9]. The top number adjacent to the terminal symbol of each line is the Shannon-Weiner diversity index (H'), and the bottom number is Pielou's evenness (J).

What species characterize these environments? Table 1 lists the average abundance of adults for the 10 most abundant species in each environment, across a depth gradient from inshore to offshore environments. *Oithona simplex* occurred in all environments, and *O. nana* and the harpacticoids *Euterpina acutifrons* and *Microsetella* spp. were present in all but one. Of the 33 species that accounted for the 10 most abundant species in all habitats, 9 belonged to the cyclopoid family Oithonidae, and 13 to the calanoid family Paracalanidae. In inshore environments, the dominance of *Parvocalanus crassirostris* is remarkable, with abundances up to 50,000 individuals m^{-3} [10]. *O. aruensis* is restricted to inshore environments, and *O. rigida* can reach very high abundances in seagrass dominated systems such as those in the Indonesian Bays. In the English Channel *Oithona*, *Oncaea* and *Paracalanus* are also the most abundant genera, but the importance of *Pseudocalanus* and *Calanus* is distinct from tropical communities. Overall, the abundance of adult copepods at L4 is approximately 1,400 individuals m^{-3}, which is similar to the coastal environments in tropical Australia (Table 1). In other respects, the abundant genera (*Ditrichocorycaeus*, *Acartia*, *Temora*, *Clausocalanus* and *Centropages*) are in common with tropical communities, but differ in usually being represented by only a single species.

Estuaries

In Australia, many tropical estuaries are inverse estuaries in the dry season (i.e., more saline in the inner regions of the waterway) and are dominated by tidal exchange [11]. The biodiversity of estuaries can be conceptualized as a Remane diagram in which the number of species changes along the salinity gradient from freshwater to marine, and in which the truly estuarine component is restricted to only a few species (Figure 3). This concept has recently been revised [12] to more realistically depict estuarine fauna rather than that of brackish seas.

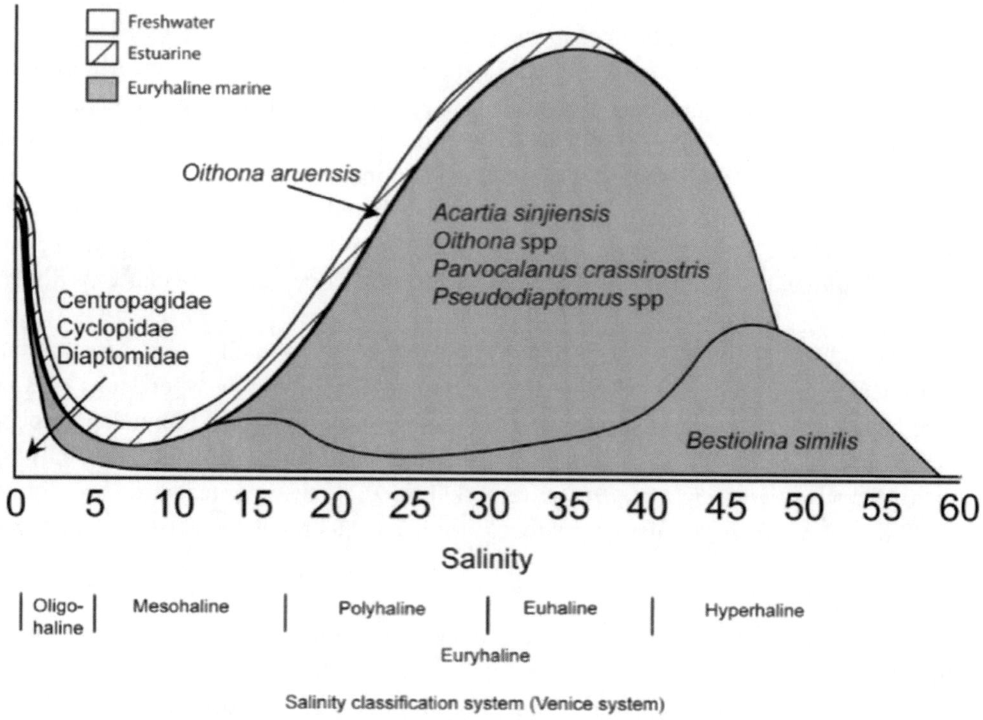

Figure 3. A modified Remane-style diagram [12] for copepod communities in the tropical estuaries of Australia, indicating the dominant species within each community.

There is little information on copepods in the oligohaline region of Australian estuaries (salinity <5), but Australian freshwater zooplankton is dominated by cyclopoid copepods of the family Cyclopidae and calanoid copepods of the family Centropagidae [13]. To our knowledge, cyclopoid communities of northern Australian freshwaters have not been studied. In northern Australia, freshwater calanoids are represented by the centropagids *Boeckella triarticulata, Calamoecia lucasi* and *C. canberra* [13]. *C. ultima* and *C. trifida* also occur, as well as the diaptomids *Eodiaptomus lumholtzi* and *Tropodiaptomus australis* [14]. In periods of flood, these species may be flushed into the upper reaches of estuaries, but there are simply too few studies to confirm this.

The mesohaline region (salinity 5–18) is dominated by the truly estuarine component of the fauna, typically *Pseudodiaptomus annandalei* and *P. griggae*, and small species of *Oithona* spp. In North Queensland and Darwin Harbour, the dominant oithonid was *O. nishidai* [10,15], but in Cape York rivers north of the McIvor River the dominant oithonid was *O. robertsoni* (as *Oithona* sp. 1 and *Oithona* sp. 2 respectively, [10]). Consequently, there appears to be a faunal divide in the mesohaline sections of Cape York estuaries, reflected not only in differences in dominant oithonids, but also in the occurrence of *Gladioferens pectinatus* south of the McIvor River [10], a species common in estuaries of SE Australia. Amazingly, *O. nishidai* extended its distribution to the oligohaline region of Darwin Harbour where it reached an abundance maximum of 23,000 individuals m^{-3} [15]. At salinities of 8–25 in Darwin Harbour, the truly estuarine *O. aruensis* was the most common oithonid, reaching densities higher than 6,000 individuals m^{-3} [15].

Euryhaline marine species are important in tropical estuaries, especially during the dry season. In polyhaline and euhaline waters (salinity 18–40) the oithonid species *Oithona simplex, O. nana,* and *O. attenuata* co-occur with *O. aruensis*, and the family Paracalanidae (mostly represented by *Parvocalanus crassirostris* and *Bestiolina similis*) clearly dominates the calanoid community. In euhaline waters *P. crassirostris* is generally more abundant than *B. similis* [10]. However, *B. similis* appears to be more euryhaline than *P. crassirostris* since it is found in salinities lower than 3 [15] and is dominant in hyperhaline waters [16]. In addition, the calanoid *Acartia sinjiensis* is widely distributed in tropical Australian estuaries, occurring in salinities as low as 8 in Darwin Harbour [15], and though not generally numerically dominant is often one of the most common copepods, especially as juveniles [10].

The copepod composition of Australian tropical estuaries is broadly similar in structure to that of tropical regions in other parts of the world, based on studies where nets of similar mesh size have been used. Small species of *Oithona* are usually the most abundant copepods, though the species composition depends on the location and may change seasonally. The ubiquity of *Parvocalanus crassirostris* is remarkable, and other paracalanids are common. Though *Acartia sinjiensis* is characteristic of northern Australian estuaries, elsewhere the genus is represented by other species, such as *A. spinicauda* in Malaysia [17] and *A. lilljeborgi* in Brazil [18].

In Indian estuaries, many of the calanoid copepods are marine in origin, reflecting high rates of exchange with coastal waters especially during the dry season [19]. In total, 67 species occur, which may be an underestimate because many studies used coarse mesh nets (> 300 μm). Ten species of Diaptomidae occur in oligohaline waters [19], but the most significant difference between Indian and Australian estuaries is the diversity of Acartiidae and Pseudodiaptomidae. The family Acartiidae has 14 representatives in Indian estuaries. In the Cochin Backwater of Kerala, India, 9 species of Acartiidae co-occur, with their co-existence facilitated by catastrophic change associated with seasonal flooding preventing the competitive exclusion of "fugitive species" that find refuge in marginal environments [20]. By contrast, the Acartiidae of Australian tropical estuaries is generally represented by one species (*A. sinjiensis*), though others can occur (*A. pacifica*, *A. erthyraea* and *A. fossae*) [15]. The genus *Pseudodiaptomus* is extremely speciose (79 species) [21] and has species that range from fully marine waters to freshwater; 12 species occur in Indian estuaries alone [19] and 5 in northern Australia [15]. Because of its strongly demersal behavior, *Pseudodiaptomus* is generally under-represented in plankton samples collected during the day. Few studies sample widely at night, but some indication of the presence of the genus can be gauged by the occurrence of the nauplii, which are distributed throughout the water column and are distinctive in body form. Similarly, *Temora turbinata* is a common constituent of estuarine zooplankton, and is abundant and sometimes dominant in mangrove estuaries in Brazil [18,22]. This species was also common in plankton samples collected in mangrove estuaries of SE Queensland [23], but in Australia the occurrence of *T. turbinata* is ephemeral.

Coral Reefs

Early coral reef researchers noticed that resident zooplankton communities within coral reefs are different from those in surrounding waters, as a result of the removal of the larger organisms such as larvaceans and large copepods [24,25]. This is because each cubic meter of water that passes over a reef face is visually inspected by *ca.* 500 planktivorous fishes [26], and results in the removal of most of the larger planktonic organisms. Consequently, there is an almost complete absence of large copepods within the waters immediately downstream from coral reefs. However, some copepod species have developed behaviors to avoid the intense predation on coral reefs, including swarming, emergence, bottom avoidance and surface aggregation. Swarming and emergence are not necessarily limited to coral reef environments and we discuss those phenomena in the Behavior section below.

Numerous studies have described patterns in zooplankton abundances near coral reefs such as near-bottom depletion [27], bottom avoidance [28] and surface accumulation [29]. However, these studies have not identified copepods further than to the level of Order, and the distribution of specific taxa has not yet been described. The relative importance of copepods in waters immediately adjacent to coral reefs varies abruptly depending on time of day. Copepods comprised 72–83% of the zooplankton at a fringing reef at Moorea, Tahiti, during the day, but only 26–49% during the night [29]. In the coral reef lagoons of the Laccadives, copepod communities were dissimilar to those of surrounding seas, and abundances were greater at night because of the emergence of copepods from the sediments [30]. Dominant copepods differed between lagoons, with *Dioithona oculata*, *Acartia dweepi* and a species of *Calanopia* most abundant in net tows. In emergence traps, the calanoid *Pseudocyclops* spp. was important [30]. On Caribbean reefs, *Oithona colcarva* is the dominant copepod [31], but is seldom reported where nets with mesh >200 μm are used.

Most coral reefs of the GBR are located within the GBR "lagoon", which is partially isolated from the open waters of the Coral Sea. *Parvocalanus crassirostris* was the most abundant copepod in the plankton at Davies Reef, a mid-shelf platform reef [32]. The reef associated plankton at this location closely resembled that of open water plankton elsewhere within the lagoon of the GBR [33]. However, the occurrence of swarming species such as *Dioithona oculata*, *Acartia australis* and *Centropages orsinii* caused the community structure within the lagoon to be distinct from that in surrounding waters during summer. The latter two species occur widely in inter-reefal waters, but their increased abundance within Davies Reef lagoon may be the result of some environmental cue that caused them to form swarms [32]. Since swarms form during the day and disperse at night, light cues have been frequently invoked to explain their occurrence [34, 35]. At Davies Reef, emergent taxa such as *Pseudodiaptomus* spp., *Pseudocyclops* sp. and *Stephos* sp. are also present [32].

The presence of small copepods such as *Parvocalanus crassirostris* and species of *Acartia* and Oithonidae are characteristic of the coral reef associated zooplankton in the Indian Ocean, Indo-Pacific and Caribbean Sea, with demersal species of genera such as *Pseudodiaptomus*, *Stephos* and *Pseudocyclops* further distinguishing these communities from those of surrounding waters.

Coastal and Shelf Waters

Copepods account for the majority of the mesozooplankton abundance in open seas of the tropics, though 42% of the zooplankton in the Gulf of Guinea were cladocerans and in the Indian Ocean a similar proportion were ostracods [36]. Overall though, copepods comprise 70% of the abundance and 30% of the biomass of zooplankton in the open tropical ocean [36]. The prevalence of pelagic tunicates such as salps and larvaceans account for a larger proportion of non-copepod zooplankton community biomass in the tropics than in the temperate ocean, probably since these mucous-net feeders are better able to feed on the dominant picoplankton. Moreover, this conclusion [36] was based on plankton samples from the top 250 m of the ocean as sampled by nets with meshes of 200–300 μm. The conclusions therefore ignore the fact that tropical copepods are small, and that in many cases the dominant copepod taxa in tropical waters pass through a 200 μm mesh. The effects of plankton net mesh size on various aspects of copepod community composition were tested in the South China Sea [37], and the abundance estimated with a 100 μm mesh net was more than two-fold higher than that estimated with a 200 μm mesh net. Surprisingly, there were no substantial differences in the abundances of most adult copepods, despite an earlier study finding that up to 92% of oithonids passed through 333 μm mesh nets [38]. There is a widespread lack of appreciation in the literature of the role of small copepods in marine pelagic ecosystems [39,40], particularly cyclopoid copepods such as *Oithona.* The family Oncaeidae is particularly important in tropical waters and is thought to be closely associated with larvaceans. Oncaeid copepods are mostly tiny, with most species having body lengths larger than 400 μm, and are most abundant in the mesopelagic. To fully account for all copepods living in the tropical ocean, fine mesh nets are required, and the use of nets of 50 μm mesh has been recommended [41].

Table 2. Composition of tropical coastal copepod communities

	Bay of Bengal [42]	South China Sea [37]	Jamaica [44]	Straits of Malacca [45]	Great Barrier Reef [46]	North West Cape [33]	Average
Mesh (μm)	200	100	64	140	73	73	
No. of copepod species	17	78	69	117	60	122	
Abundance (ind. m^{-3})	1,151	382	2,450	2,238 – 2,927	1,847	1,092	
% Abundance by Order							
Calanoida	73	67	42	33	28	35	46
Cyclopoida	19	29	53	41	53	51	41
Harpacticoida	8	5	2	14	18	10	10
% Abundance by Family							
CALANOIDA							
Acartiidae	2	5	*	2	1	1	2
Calanidae	2	6	2		1	1	2
Candaciidae		2	1		*	*	1
Centropagidae	4	1	*		1	2	1
Clausocalanidae	2	17	15			*	9
Eucalanidae	1	2	*	2	1	1	1
Lucicutiidae		4	*		*	1	1
Mecynoceridae		0	7		*	1	2
Metridinidae		2	1				2
Paracalanidae	49	25	14	28	21	19	26
Pontellidae	5	1	*		0	1	2
Pseudodiaptomidae	*			*		2	1
Temoridae	8	1	2		2	3	3
Tisbidae							

Table 2. Continued

	Bay of Bengal [42]	South China Sea [37]	Jamaica [44]	Straits of Malacca [45]	Great Barrier Reef [46]	North West Cape [33]	Average
CYCLOPOIDA							
Oithonidae	8	3	26	27	43	34	23
Corycaeidae	6	18	11	2	4	4	8
Oncaeidae	5	7	16	12	6	12	10
HARPACTICOIDA							
Ectinosomatidae	*	4	2	1	4	6	3
Euterpinidae	2	*		11	14	4	6
Miraciidae	6	1	*	2		*	2

Families are only listed where they constituted 2% of the community in at least one habitat, and where species were present but in less than 1% of total copepod abundance they are indicated by *.

To summarize the composition of coastal copepods in various regions of the tropics, we have made a comparison of 6 regions (Table 2). Such comparisons are always prone to bias because of methodological differences, taxonomic biases and differences in oceanographic regimes between environments. All 6 of the studies we have chosen used fine mesh nets (< 200 μm) except in the case of the Bay of Bengal. Copepod abundance was low in the South China Sea because this study spanned coastal and open sea communities, but has been retained because the multivariate analyses in that study demonstrated that the community composition in these environments was similar [37].

Overall, in these 6 examples calanoids comprised 46% of copepod communities, and cyclopoids 41% (Table 2). Calanoids dominated the copepod community in the Bay of Bengal and South China Sea, whereas cyclopoids were equally or more abundant elsewhere. Harpacticoids were most abundant in the GBR, primarily as a result of the prevalence of *Euterpina acutifrons* (Euterpinidae). The Paracalanidae and Oithonidae were the most important families, comprising 26% and 23% of the copepod community on average. The Paracalanidae were most important in the Bay of Bengal [42], where they comprised almost half the total copepod abundance, and even so are probably under-represented in that study because genera such as *Parvocalanus* and *Delibus* are not efficiently sampled by the 200 μm mesh net used. The comparatively high abundances of Clausocalanidae and Corycaeidae in the South China Sea and Caribbean Sea (Jamaica) are probably a reflection of the more open water environments in these studies. In the Makassar Strait of Indonesia, copepods comprised 41% of the fraction of the coastal zooplankton larger than 200 μm [43]. Here, calanoids constituted 66% of the coastal copepod community, and 89 species were recorded. The Paracalanidae accounted for more than 60% of the coastal calanoid community and included over 14 species. The Temoridae were also important, constituting 17% of the calanoid fauna.

The east and west tropical coasts of Australia differ markedly in nature, but their copepod communities are similar. Unusually, both have poleward flowing boundary currents, in the west the Leeuwin Current, and in the east the East Australian Current. The west coast has an extensive shallow shelf and is macrotidal (up to 11 m in tidal range), whereas the east is characterized by the GBR and is mesotidal (up to 5 m tidal range). Shelf waters on the west are strongly physically forced, well mixed and turbid, whereas the waters of the GBR are comparatively clear. Amazingly, given these environmental differences, mean adult copepod abundances on each side of the continent are virtually identical (GBR 1,660 individuals m^{-3}, Kimberley 1,620 individuals m^{-3}). The highest mean copepod abundances from our campaigns occurred in the Arafura Sea (3,690 individuals m^{-3}; McKinnon unpublished), and the lowest at Scott Reef (731 individuals m^{-3}). There appear to be some regional differences in communities however, with the GBR having a richer corycaeid community. Overall, this family constituted more than 20% of the total copepod count in surface waters of the GBR lagoon [47]. From our studies, *Onychocorycaeus agilis*, *O. pacificus*, *Corycaeus speciosus* and *Farranula carinata* occurred in the GBR, but not in Western Australia or the Arafura Sea.

The GBR is 2,300 km in length and up to 330 km in width and extends over 15° of latitude (9 – 24°S). Approximately 90% of the shelf area comprises lagoonal habitat [48], which means exchange with the open ocean is limited by the matrix of coral reefs. The GBR thus provides a good model for pelagic copepod communities of a coastal shelf system. Over 200 species of pelagic copepods have been recorded within this ecosystem [46], and their ecology has been the subject of a number of studies [46,47,49-52]. The study of the Great

Barrier Reef Expedition of the 1920's is focused on the area around Low Isles and includes samples taken adjacent to the outer margin of the GBR, where copepod communities are the most diverse [49]. More recent work concentrated on inshore waters and used 73 μm mesh nets that allow a better account of the smaller fraction of the community [46]. However, the most complete areal coverage of copepod communities is contained in an unpublished study that aimed to parameterize the relationship between zooplankton biomass/abundance and chlorophyll on the GBR, for which an underway pumping system was developed where water was pumped from 2 m below the surface through a 200 μm mesh net [47]. Overall, copepod abundances were patchy (Figure 4), and were positively correlated with chlorophyll concentration. This study was unable to demonstrate any latitudinal change or seasonal change in copepod abundances. However, in the Central zone of the GBR off Townsville, there was a pronounced cross-shelf gradient in abundance: the geometric mean abundance inshore was 543 copepods m^{-3}, on the inner shelf 273 copepods m^{-3}, outer shelf 130 copepods m^{-3}, and in the Coral Sea 73 copepods m^{-3}. We believe these cross-shelf patterns are typical of tropical copepod communities.

All studies to date have identified similar patterns of community structure in waters of the GBR lagoon based on multivariate analyses of species abundances. In each case a distinctive inshore community has been identified, juxtaposed against an offshore (outer GBR or Coral Sea) community, and sometimes (depending on the study) one or two transitional communities in between. Most pertinently, during the wet season, the inshore community extends further out into the GBR lagoon [52,53], abruptly changing the mid-shelf community composition. Similarly, the abundance of zooplankton doubled 16 days after floods associated with the passage of a tropical cyclone [47].

The percentage of total zooplankton attributable to trophic role changes with position across the GBR shelf (Figure 5). On average, copepods constituted 80% of the zooplankton in that study, but the pronounced difference in the proportion of herbivores inshore was attributed to small herbivorous copepods (mostly Paracalanidae) and larvaceans [47]. The higher proportion of omnivores mid-shelf was attributable to increased abundance of *Acartia australis* and *Centropages orsinii*, both of which are associated with coral reefs (see Behavior section). The higher proportion of carnivores in the Coral Sea resulted from an increased abundance of Corycaeidae.

Copepod communities of tropical coastal waters vary in their structure depending upon location, but the overwhelming importance of two families, the Paracalanidae and the Oithonidae is characteristic of all. Each of these families comprise mostly small copepods that are not efficiently sampled by nets of 200 μm or more, which has led to an under-appreciation of the nature of tropical zooplankton, which in turn has compromised our understanding of ecosystem function [54]. In general, tropical coastal plankton communities show little seasonal pattern compared to those in temperate latitudes but can change markedly depending upon events such as flooding and cyclonic activity.

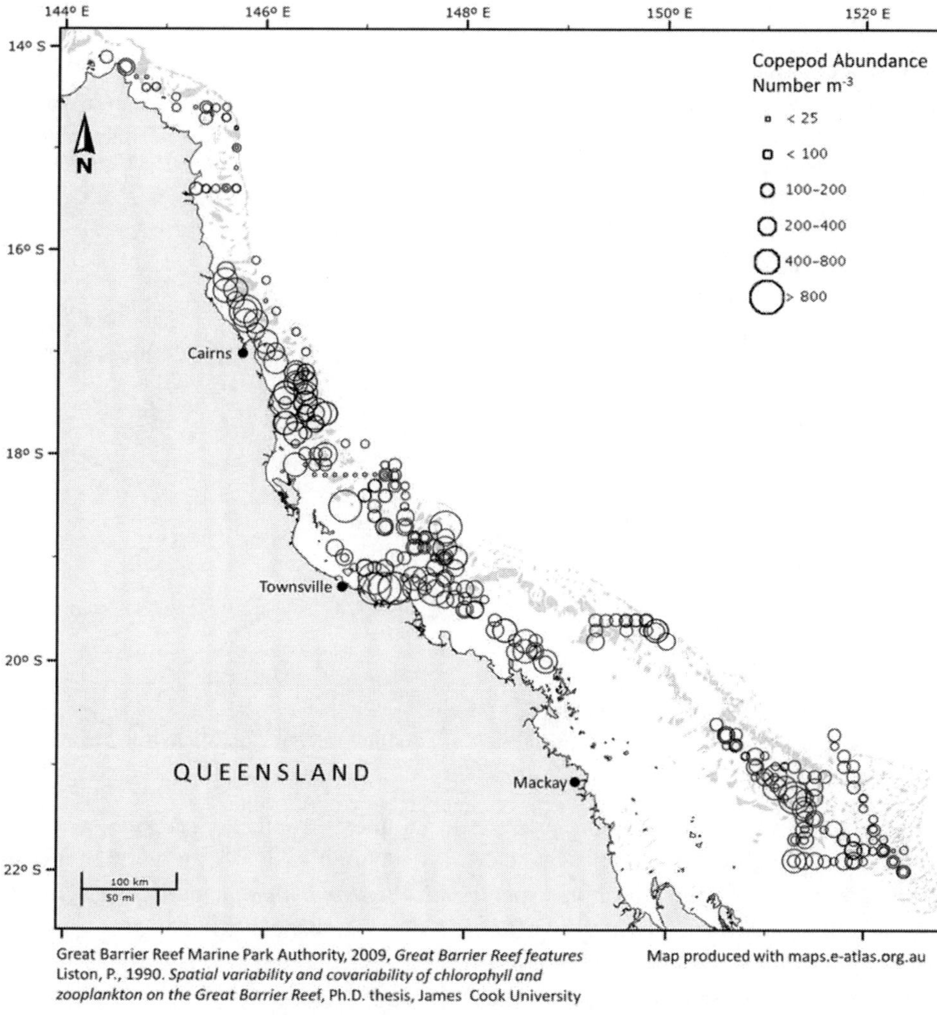

Figure 4. The relative abundance of copepods in the waters of the GBR lagoon [47].

BEHAVIOR

Copepod behavior can greatly influence the community composition of tropical planktonic copepod communities. The formation of swarms (dense aggregations of individuals not organized by orientation) is a feature of some tropical copepod species [34]. *Dioithona oculata* occurs both in coral reef and mangrove environments, where it forms swarms with densities up to 91,000 individuals l^{-1}, apparently in response to light cues [55]. In the mangroves of Belize, swarms of *D. oculata* form in light shafts between mangrove prop roots, whereas on the GBR *D. oculata* forms discrete ball-shaped swarms behind coral heads [32,56]. To maintain position in swarms, *D. oculata* swims at speeds up to 2.0 cm s^{-1}, corresponding to 25 body lengths s^{-1} [55]. A number of tropical species of the calanoid genus *Acartia* have been observed forming swarms. These include *A. (Acanthacartia) spinata*, *A.*

(Acanthacartia) tonsa, *A. (Acanthacartia) fossae (as A. hamata)*, *A. (Odontacartia) bispinosa* and *A. (Odontacartia) australis* [34]. On the GBR, *A. australis* forms a blanket in shallow well-lit waters over coral sand on the bottom of the reef lagoon [32,56]. In addition, *Centropages orsinii* forms near-bottom swarms [32]. There is circumstantial evidence to suggest that *Dioithona rigida* also forms swarms, based on highly overdispersed plankton counts, especially in seagrass habitats such as beds of *Enhalus* in Indonesia [57].

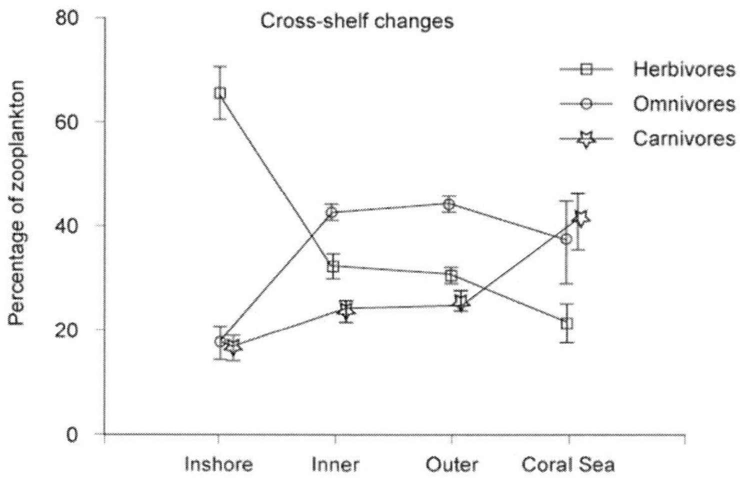

Figure 5. Cross shelf differences in the proportion of zooplankton attributable to three trophic types [47].

Patterns of vertical migration are subtle in most tropical copepods, and there has generally been insufficient sampling at scales fine enough to detect the vertical movement of any but the most obvious migrators, species of *Pleuromamma* in particular. On the GBR, there are only significant day-night differences in abundances of the larger copepod taxa (*Canthocalanus pauper*, *Undinula vulgaris*, *Temora turbinata*) [46]. It is likely that the small copepods dominating shallow water tropical communities do not have sufficient swimming power to overcome bottom-generated mixing and turbulence. Our knowledge of the fine scale vertical distribution of small copepods is limited by the small number of studies where sampling has been undertaken at sufficient resolution to resolve the comparatively small amplitudes of migratory movements. When a Longhurst Hardy Plankton Recorder was used to map distribution of copepods in oceanic waters of the eastern tropical Pacific it was demonstrated that a large number of tropical pelagic copepods have distinct depth preferences and that there are detectable day-night differences in the depth of maximum abundance [58]. In oceanic waters, small amplitude vertical migrations have been demonstrated for a number of taxa, and can also extend to an amplitude of approximately 80 m in the case of *Triconia conifera* [59]. At Scott Reef, depth stratified sampling indicated that some small copepod taxa (5 species of *Oncaea* and 2 of *Triconia*) were undertaking diel migrations [60].

The most pronounced vertical movement of copepods in shallow tropical waters is the emergence of copepods from sediments into the water column. At Heron Reef in the southern GBR, four emergent species of *Stephos* and 3 emergent species of *Pseudodiaptomus* occur [61]. Because these animals are seldom caught using conventional plankton sampling techniques, only one species of each of these genera was previously known to science (*S.*

tropicus and *P. colefaxi* respectively). The behavior of emergent species is complex, depending on substrate preferences, the stage of the moon and seasonal differences [61,62]. The genus *Pseudodiaptomus* demonstrates emergent behavior in a wide range of habitats [63]. In addition, the harpacticoid *Metis holothuriae* and the calanoid *Pseudocyclops* spp. also showed emergent behavior at Heron Reef [62]. In the tropical southwestern Atlantic, *Calanopia americana* shows pronounced patterns of emergence [64]. Overall, emergence appears strongly under the control of light, as emergence and reentry patterns are cued by nightfall and appear stronger on non-moonlit nights [65].

TROPHIC DETERMINANTS OF COMMUNITY COMPOSITION

Most primary production (> 80%) in tropical marine waters originates from cells < 2 μm in size, i.e., the picophytoplankton [66]. Cells of this size range are too small to be consumed by copepods. Though copepods, especially many of the small calanoids, are traditionally regarded as herbivores, this is unlikely to be the case in tropical environments because phytoplankton cells sufficiently large to be grazed on by copepods only occur in ephemeral blooms such as those arising from upwelling or storm activity. Consequently, the most important grazers of picoplankton production are the protistan community (heterotrophic nanoflagellates, ciliates and hetero/mixotrophic dinoflagellates), in combination grazing 75% of primary production in the tropics [67]. It is these organisms that constitute most of the diet of pelagic copepods. However, there are a couple of other linkages that are very important for the trophic ecology of tropical pelagic copepods. Firstly, because of the dominance of microbial food webs in tropical seas, mucous net feeders, especially pelagic tunicates, are very important. Larvaceans are at times as equally abundant as copepods and have fast growth rates. They discard multiple houses each day, which are rich in organic matter and form the basis of aggregates that are colonized by a wide range of microbiota, from bacteria to ciliates. Many species of copepods are known to utilize these resources, but particularly the family Oncaeidae [68], which is most abundant in the tropical mesopelagic zone. Secondly, and related to the first, is the predominance of marine snow in the water column, which combines small particles into aggregates of sufficient size to be fed upon by mesozooplankton. The importance of this detritus in the diets of tropical pelagic copepods has yet to be fully understood but is likely to be significant.

The Oithonidae are usually regarded as raptorial carnivores, with a preference for ciliate prey [69], but can also be herbivorous [70], carnivorous [71], detritivorous [72], and coprophagous [73]. Generally, *Oithona* appears to be broadly omnivorous [54]. Similarly, paracalanid copepods have been regarded as herbivorous [74] but can also be carnivorous [75], and we therefore conclude that the broad dietary preferences shown for *Oithona* are also likely to be the case for other common small copepods such as *Parvocalanus crassirostris*. The dominance of these taxa in turbid and detritus-rich mangrove environments supports their close linkage to microbial foodwebs associated with the breakdown of organic material originating from the forests. Pelagic harpacticoids of the genera *Macrosetella*, *Miracia* and *Oculosetella* graze on the cyanobacterium *Trichodesmium* in the Caribbean [76], and we believe this is also the case for *Microsetella* in waters of the GBR. *Trichodesmium* forms huge surface slicks in the waters of the GBR [77], and zooplankton samples in these conditions are dominated by *Microsetella* (McKinnon, pers. obs.). There are some small copepods, however, such as the Corycaeidae, which appear to be purely carnivorous [78, 79].

The nature of tropical copepod communities, then, is determined by the predominance of microbial and detrital food webs, in contrast to the stronger influence of the "classical" food web in temperate regions and the greater importance of herbivory within these.

THE "TROPICALIZATION" OF TEMPERATE COPEPOD COMMUNITIES

Marine plankton are useful indicators of climate-driven changes in marine ecosystems, and calanoid copepods in particular have proven useful in demonstrating changes in distribution primarily attributable to effects of warming [80]. In temperate seas such as the Northeast Atlantic, warm-water species such as *Calanus helgolandicus* have extended their range at the expense of cooler water congeners such as *C. finmarchicus* [81]. In the Northeast Pacific, subtropical species (*Acartia danae*, *Candacia bipinnata*, *Clausocalanus* spp., *Corycaeus* spp., *Mesocalanus tenuicornis*, *Pleuromamma abdominalis* and *Undeuchaeta major*) are becoming more prevalent [82]. It stands to reason, therefore, that in a warming ocean, tropical copepods should become more widespread. Unfortunately, there are few time series of sufficient duration to demonstrate the increased range of tropical species, though there are tantalizing suggestions that this may be occurring.

Parvocalanus crassirostris is primarily a tropical and subtropical species [21], but its range may be extending. There are increasing numbers of reports of *P. crassirostris,* and it is unclear whether this is occurring because of real changes in distribution and abundance patterns, or is a result of the more widespread use of fine mesh nets. *P. crassirostris* is known to occur in the Mediterranean [21], but has been newly recorded in the Adriatic [83] where it occurs at densities up to 28,000 individuals m^{-3}. Oddly however, the peak abundance occurs in cool months (October–December). *P. crassirostris* is now dominant in the East China Sea [84], despite it not being recorded in the area in the 1960's [85,86]. Each of those studies recorded cyclopoid species of similar size as *P. crassirostris*, so this is not simply an effect of plankton net mesh size. The increased abundance of *Temora turbinata* in China may be evidence of a poleward extension of a warm-water species attributable to global warming [87].

In Australia, there is evidence that warm-water "signature" species are moving southward into Tasmania [88]. This conclusion was based on *Acartia danae*, *Pleuromamma gracilis*, *P. quadrangulata*, *Subeucalanus crassus*, *Temora turbinata*, *Sapphirina* spp., and *Corycaeus* spp. contributing to the discrimination of a cluster of samples in an ordination analysis belonging to warm years. However, all these species have distributions well into temperate areas and even the subantarctic [21].

The phenomenon of "tropicalization" of temperate plankton communities has potentially important ecosystem consequences. Since warm water copepods have less food value than cold water species, food chains may be impacted, resulting in loss of "societal value" [82]. Range extension is directly attributable to climate-driven warming, and is distinct from other more directly anthropogenic effects such as invasions of alien species [89] and changes from large to small species as a result of eutrophication [90], each of which occur independently of ambient water temperature.

CONCLUSION

We commenced this review by remarking on how much our knowledge of not only the community ecology of pelagic copepods, but of their ecology in general, is shaped by the long tradition of research in temperate ecosystems. However, our increasing knowledge of subtropical and tropical zooplankton together with the more widespread use of fine mesh nets, leads to the inescapable conclusion that small copepods, together with larvaceans, are the most important metazooplankton in tropical waters. In all the environments considered in this review, the dominance of Oithonidae and Paracalanidae is striking. For instance, on the shelf of NW Australia 15 species of Paracalanidae (31% of total copepod abundance) and 11 Oithonidae (50% of total copepod abundance) co-occur [33]. It remains a mystery how so many species of these families co-exist, and how their ecological niches are distinct.

Consequently, if we were to choose a model copepod representative of subtropical and tropical marine waters, rather than *Calanus* or *Acartia*, it should be a representative of the Oithonidae or Paracalanidae. The genus *Oithona* is ubiquitous and abundant in a wide range of habitats, but is represented by different species in each. In contrast, *Parvocalanus crassirostris* dominates numerically in many tropical coastal waters, though admittedly molecular taxonomy may yet demonstrate that this species is plural. If we were to choose a single species to be a model tropical copepod, then what better candidate than *P. crassirostris*? To date, however, there is not a single study on the biology of this species and we can only guess at the ecological role of *P. crassirostris* based on studies of other larger, temperate paracalanids. Realistically, however, conventional experimental methods used for copepod research become challenging when dealing with one of the smallest calanoid copepods , as *P. crassirostris* is only about 500 μm long as an adult.

Tropical pelagic copepods grow and thrive in environments characterized by warm temperature and low chlorophyll, and in which microbial food webs predominate. Temperature drives fast metabolic rates, and in most tropical coastal environments the particle field is dominated by detritus, whether in the form of marine snow in coral reefs and shelf environments, or as resuspended particulate material in mangrove systems and in turbulent macrotidal coastal environments. The same environmental factors influence the predators of copepods in these systems, i.e., chaetognaths, ctenophores, medusa and larval fish in the plankton, as well as a vast array of benthic suspension feeders. Copepods are caught in the crossfire between finding sufficient food to fuel fast metabolic rate, yet avoiding intense predation. We have much to learn to understand how these competing factors combine to shape community structure.

ACKNOWLEDGEMENTS

We wish to thank Peter Liston for permission to publish data from his 1990 thesis in Figures 4 and 5, and Eric Lawrey (AIMS) for reproducing Figure 4. We thank the Plymouth Marine Laboratory for access to the data from station L4 in the English Channel. Russ Shiel (University of Adelaide) provided advice on freshwater copepods, and Anthony Richardson (CSIRO) provided comments on an earlier draft of this manuscript.

REFERENCES

[1] Longhurst AR & Pauly D (1987) *Ecology of Tropical Oceans*. Academic Press, San Diego.

[2] Meiri S (2010) Bergmann's Rule–what's in a name? *Global Ecology and Biogeography*, 20, 203-207.

[3] Nishida S (1985) Taxonomy and distribution of the family Oithonidae (Copepoda, Cyclopoida) in the Pacific and Indian Oceans. *Bulletin of the Ocean Research Institute, University of Tokyo*, 20, 167 pp.

[4] Böttger-Schnack R (2001) Taxonomy of Oncaeidae (Copepoda, Poecilostomatoida) from the Red Sea. - II. Seven species of *Oncaea* s.str. *Bulletin of the Natural History Museum London (Zoology)*, 67, 25-84.

[5] Huys R & Boxshall GA (1991) *Copepod Evolution*. The Ray Society, London

[6] Boxshall GA, Halsey SH (2004) *An Introduction to Copepod Diversity*. The Ray Society, London.

[7] Cornils A, Blanco-Bercial L, Schnack-Schiel SB & Bucklin A (2011) Molecular and morphological phylogeny of Paracalanidae, Giesbrecht 1893 (Copepoda, Calanoida) *Poster presentation, 5th International Zooplankton Production Symposium*, Pucón, Chile March 14-18, 2011.

[8] Bradford JM (1978) *Paracalanus indicus* Wolfenden and *Corycaeus aucklandicus* Kraemer, two neritic pelagic copepods from New Zealand. *Journal of the Royal Society of New Zealand*, 8, 133-141.

[9] www.westernchannelobservatory.org.uk/l4_zooplankton.php [accessed October 17, 2012]

[10] McKinnon AD & Klumpp DW (1998) Mangrove zooplankton of North Queensland, Australia I. Plankton community structure and environment. *Hydrobiologia*, 362, 127-143.

[11] Geoscience Australia (2002) Australian Estuaries and Coastal Waterways. A Guide to the biogeochemistry of sediments and water. Canberra: CDROM, Geoscience Australia.

[12] Whitfield AK, Elliot M, Basset A et al. (2012) Paradigms in estuarine ecology - A review of the Remane diagram with a suggested revised model for estuaries. *Estuarine, Coastal and Shelf Science*, 97, 78-90.

[13] Bayly IAE & Williams WD (1973) *Inland Waters and Their Ecology*. Longman, Hawthorn, Victoria.

[14] Dussart BH, Fernando CH, Matsumura-Dundisi T & Shiel RJ (1984) A review of systematics, distribution and ecology of tropical freshwater zooplankton. *Hydrobiologia*, 113, 77-91.

[15] Duggan S, McKinnon AD & Carleton JH (2008) Zooplankton in an Australian tropical estuary. *Estuaries and Coasts*, 31, 455-467.

[16] Kimmerer WJ, McKinnon AD, Atkinson MJ & Kessell JA (1985) Spatial Distributions of Plankton in Shark Bay, Western Australia. *Australian Journal of Marine and Freshwater Research*, 36, 421-32.

[17] Chew LL & Chong VC (2010) Copepod community structure and abundance in a tropical mangrove estuary, with comparisons to coastal waters. *Hydrobiologia*, 666, 127-143.

[18] Almeida LR, Costa IS & Eskinazi-Sant'Anna EM (2012) Composition and abundance of zooplankton community of an impacted estuarine lagoon in Northeast Brazil. *Brazilian Journal of Biology*, 72, 13-24.

[19] Pillai PP (1971) Studies on the estuarine copepods of India. *Journal of the Marine Biological Association of India*, 13, 162-172.

[20] Tranter DJ & Abraham S (1971) Coexistence of species of Acartiidae (Copepoda) in the Cochin Backwater, a monsoonal estuarine lagoon. *Marine Biology*, 11, 222-241.

[21] Razouls C, de Bovée F, Kouwenberg J & Desreumaux N(2005 – 2012) *Diversity and Geographic Distribution of Marine Planktonic Copepods*. Available at http://copepodes.obs-banyuls.fr/en.

[22] Magris RA, Pereira JB & Loureiro Fernandes LF (2011) Interannual variability in tropical estuarine copepod assemblages off Northeastern Brazil. *Journal of Crustacean Biology*, 31, 260-269.

[23] Werry J & Lee SY (2005) Grapsid crabs mediate link between mangrove litter production and estuarine planktonic food chains. *Marine Ecology Progress Series*, 293, 165-176.

[24] Odum HT & Odum EP (1955) Trophic structure and productivity of a windward coral reef commúnty at Eniwetok Atoll. *Ecological Monographs*, 25, 291-320.

[25] Emery AR (1968) Preliminary observations on coral reef plankton. *Limnology and Oceanography*, 13, 293-303.

[26] Hamner WM, Jones MS, Carleton JH et al. (1988) Zooplankton, planktivorous fish, and water currents on a windward reef face: Great Barrier Reef, Australia. *Bulletin of Marine Science*, 42, 459-479.

[27] Yahel R, Yahel G & Genin A (2005) Near-bottom depletion of zooplankton over a coral reef I: diurnal dynamics and size distribution. *Coral Reefs*, 24, 75-85.

[28] Holzman R, Reidenbach MA, Monismith SG et al. (2005) Near-bottom depletion of zooplankton over a coral reef II: relationships with zooplankton swimming ability. *Coral Reefs*, 24, 87-94.

[29] Alldredge AL & King JM (2009) Near-surface enrichment of zooplankton over a shallow back reef: implications for coral reef food webs. *Coral Reefs*, 28, 895-908.

[30] Madhupratap M, Achuthankutty CT & Nair SRS (1991) Zooplankton of the lagoons of the Laccadives - diel patterns and emergence. *Journal of Plankton Research*, 13, 947-958.

[31] Ohlhorst SL (1982) Diel migration patterns of demersal reef zooplankton. *Journal of Experimental Marine Biology and Ecology*, 60, 1-15.

[32] McKinnon AD (1991) Community composition of reef associated copepods in the lagoon of Davies Reef, Great Barrier Reef, Australia. *Proceedings of the Fourth International Conference on Copepoda; Bulletin of the Plankton Society of Japan, Special Volume*, 467-478.

[33] McKinnon AD, Duggan S, Carleton JH & Böttger-Schnack R (2008) Summer Planktonic Copepod Communities of Australia's North West Cape (Indian Ocean) during the 1997-99 El Niño/La Niña. *Journal of Plankton Research*, 30, 839-855.

[34] Ambler JW (2002) Zooplankton swarms: characterisitics, proximal cues and proposed advantages. *Hydrobiologia*, 480, 155-164.

[35] Buskey EJ, Peterson JO & Ambler JW (1995) The role of photoreception in the swarming behavior of the copepod *Dioithona oculata. Marine and Freshwater Behaviour and Physiology*, 26, 273-285.

[36] Longhurst AR (1985) Relationship between diversity and the vertical structure of the upper ocean. *Deep Sea Research*, 32, 1535-1570.

[37] Tseng LC, Dahms HU, Hung JJ et al. (2011) Can different mesh sizes affect the results of copepod community studies? *Journal of Experimental Marine Biology and Ecology*, 398, 47-55.

[38] Hwang JS, Kumar R, Dahms HU et al. (2007) Mesh-size affects abundance estimates of *Oithona* spp. (Copepoda: Cyclopoida). *Crustaceana*, 80, 827-837.

[39] Paffenhofer GA (1993) On the ecology of marine cyclopoid copepods (Crustacea, Copepoda). *Journal of Plankton Research*, 15, 37-55.

[40] Gallienne CP & Robins DB (2001) Is *Oithona* the most important copepod in the world's oceans? *Journal of Plankton Research*, 23, 1421-1432.

[41] Böttger-Schnack R, Schnack D & Hagen W (2008) Microcopepod community structure in the Gulf of Aqaba and northern Red Sea, with special reference to Oncaeidae. *Journal of Plankton Research*, 30, 529-550.

[42] Rakhesh M, Raman AV, Kalavati C et al. (2008) Zooplankton community structure across an eddy-generated upwelling band close to a tropical bay-mangrove ecosystem. *Marine Biology*, 154, 953-972.

[43] Cornils A, Schulz J, Schmitt P et al. (2010) Mesozooplankton distribution in the Spermonde Archipelago (Indonesia, Sulawesi) with special reference to the Calanoida (Copepoda). *Deep-Sea Research II*, 57, 2076-2088.

[44] Webber MK & Roff JC (1995) Annual structure of the copepod community and its associated pelagic environment off Discovery Bay, Jamaica. *Marine Biology*, 123, 467-479.

[45] Rezai H, Yusoff FMd, Arshad A et al. (2004) Spatial and temporal distribution of copepods in the Straits of Malacca. *Zoological Studies*, 43, 486-497.

[46] McKinnon AD, Duggan S & De'ath G (2005) Mesozooplankton dynamics in inshore waters of the Great Barrier Reef. *Estuarine, Coastal and Shelf Science*, 63, 497-511.

[47] Liston PW (1990) *Spatial Variability and Covariability of Chlorophyll and Zooplankton on the Great Barrier Reef.* PhD thesis, James Cook University of North Queensland.

[48] Furnas MJ, Mitchell A, Skuza M & Brodie J (2005) In the other 90%: phytoplankton response to enhanced nutrient availability in the Great Barrier Reef Lagoon. *Marine Pollution Bulletin*, 51, 253-265.

[49] Farran GP (1936) Copepoda. *Scientific Report Great Barrier Reef Expedition*, 5, 73-142.

[50] Farran GP (1949) The seasonal and vertical distribution of the copepods. *Scientific Report Great Barrier Reef Expedition,* 2, 291-312.

[51] Williams DM, Dixon P & English S (1988) Cross-shelf distribution of copepods and fish larvae in the central Great Barrier Reef. *Marine Biology*, 99, 577-589.

[52] McKinnon AD & Thorrold SR (1993) Zooplankton community structure and copepod egg production in coastal waters of the central Great Barrier Reef lagoon. *Journal of Plankton Research*, 15, 1387-1411.

[53] Sammarco PW, Crenshaw H (1984) Plankton community dynamics of the central Great Barrier Reef Lagoon: Analysis of data from Ikeda et al. *Marine Biology*, 82, 167-180.

[54] Turner JT (2004) The importance of small planktonic copepods and their roles in pelagic marine food webs. *Zoological Studies*, 43, 255-266.

[55] Buskey EJ, O'Peterson J & Ambler JW (1996) The swarming behavior of the copepod *Dioithona oculata*: In situ and laboratory studies. *Limnology and Oceanography,* 41, 513-521.

[56] Hamner WM & Carleton JH (1979) Copepod swarms: Attributes and role in coral reef ecosystems. *Limnology and Oceanography*, 24, 1-14.

[57] Duggan S, Undu MC & McKinnon AD (2011) Community composition of zooplankton near fish cages in Lampung and South Sulawesi, Indonesia. *Asian Fisheries Science,* 24, 288-303.

[58] Longhurst AR (1985) The structure and evolution of plankton communities. *Progress in Oceanography*, 15, 1-35.

[59] Zalkina AV (1970) Vertical distribution and diurnal migration of some Cyclopoida (Copepoda) in the tropical region of the Pacific Ocean. *Marine Biology,* 5, 275-282.

[60] McKinnon AD, Duggan S, Böttger-Schnack R et al. (2010) Depth structuring of pelagic copepod biodiversity in waters adjacent to an Eastern Indian Ocean coral reef. *Journal of Natural History*, 47, 639-655.

[61] Jacoby CA & Greenwood JG (1991) Species-specific variations in emergence of coexisting *Stephos* and *Pseudodiaptomus* (Copepoda: Calanoida). *Proceedings of the Fourth International Conference on Copepoda; Bulletin of the Plankton Society of Japan, Special Volume*, 405-418.

[62] Jacoby CA & Greenwood JG (1988) Spatial, temporal and behavioural changes in emergence of zooplankton in the lagoon of Heron Reef, Great Barrier Reef, Australia. *Marine Biology*, 97, 309-328.

[63] Rios-Jara E & González JG (2000) Effects of lunar periodicity on the emergence behavior of the demersal copepod *Pseudodiaptomus cokeri* in Phosphorescent Bay, Puerto Rico. *Bulletin of Marine Science*, 67, 887-901.

[64] Melo PAM, Silva TA, Neumann-Leitão S et al. (2010) Demersal zooplankton communities from tropical habitats in the southwestern Atlantic. *Marine Biology Research*, 6, 530-541.

[65] Alldredge AL & King JM (1980) Effects of moonlight on the vertical migration patterns of demersal zooplankton. *Journal of Experimental Marine Biology and. Ecology*, 44,133-156.

[66] Richardson TL & Jackson, GA (2007) Small phytoplankton and carbon export from the surface ocean. *Science*, 315, 838-840.

[67] Calbet A & Landry MR (2004) Phytoplankton growth, microzooplankton grazing, and carbon cycling in marine systems. *Limnology and Oceanography*, 49, 51-57.

[68] Ohtsuka S, Böttger-Schnack R, Okada M & Onbé T (1996) In situ feeding habits of *Oncaea* (Copepoda: Poecilostomatoida) from the upper 250 m of the central Red Sea, with special reference to consumption of appendicularian houses. *Bulletin of the Plankton Society of Japan*, 43, 89-105.

[69] Nakamura Y & Turner JT (1997) Predation and respiration by the small cyclopoid copepod *Oithona similis*: How important is feeding on ciliates and heterotrophic flagellates? *Journal of Plankton Research*, 19, 1275-1288.

[70] Turner J (1986) Zooplankton feeding ecology: contents of fecal pellets of the cyclpopid copepods *Oncaea venustra*, *Corycaeus amazonicus*, *Oithona plumifera*, and *O. simplex* from the northern Gulf of Mexico. *Marine Ecology*, 7, 289-302.

[71] Lampitt RS (1978) Carnivorous feeding by a small marine copepod. *Limnology and Oceanography*, 23, 1228-1231.

[72] Ambler JW, Alcala-Herrera J & Burke R (1994) Trophic roles of particle feeders and detritus in a mangrove island prop root ecosystem. *Hydrobiologia*, 292/293, 437- 446.

[73] Gonzalez HE & Smetacek V (1994) The possible role of the cyclopoid copepod *Oithona* in retarding vertical flux of zooplankton faecal material. *Marine Ecology Progress Series*, 113, 233-246.

[74] Paffenhofer GA (1984) Food ingestion by the marine planktonic copepod *Paracalanus* in relation to abundance and size distribution of food. *Marine Biology*, 80, 323-333.

[75] Suzuki K, Nakamura Y & Hiromi J (1999) Feeding by the small calanoid copepod *Paracalanus* sp. on heterotrophic dinoflagellates and ciliates. *Aquatic Microbial Ecology*, 17, 99-103.

[76] O'Neil JM, Metzler PM & Glibert PM (1996) Ingestion of $^{15}N_2$-labelled *Trichodesmium* spp. and ammonium regeneration by the harpacticoid copepod *Macrosetella gracilis*. *Marine Biology*, 125, 89-96.

[77] Furnas MJ (1992) Pelagic *Trichodesmium* (= *Oscillatoria*) in the Great Barrier Reef region. *In*: Carpenter EJ, Capone DG & Reuter J (eds.), *Marine Pelagic Cyanobacteria: Trichodesmium and other Diazotrophs*. Kluwer Academic Publishers, 265-272.

[78] Turner J, Tester P & Conley W (1984) Zooplankton feeding ecology: predation by the marine cyclopoid copepod *Corycaeus amazonicus* F. Dahl upon natural prey. *Journal of Experimental Marine Biology and Ecology*, 84, 191-202.

[79] Landry M, Lehner-Fournier J & Fagerness V (1985) Predatory feeding behavior of the marine cyclopoid copepod *Corycaeus anglicus*. *Marine Biology*, 85, 163-169.

[80] Hays GC, Richardson AJ & Robinson C (2005) Climate change and marine plankton. *Trends in Ecology and Evolution*, 20, 337-344.

[81] Bonnet D, Richardson A, Harris R et al. (2005) An overview of *Calanus helgolandicus* ecology in European waters. *Progress in Oceanography*, 65, 1-53.

[82] Batten SD & Walne AW (2011) Variability in northwards extension of warm water copepods in the NE Pacific. *Journal of Plankton Research*, 33, 1643-1653.

[83] Eleftheriou A, Anagnostopoulou-Visilia E, Anastasopoulou E et al. (2011) New Mediterranean Biodiversity Records (December 2011). *Mediterranean Marine Science*, 12, 491-508.

[84] Tseng L-C, Souissi S, Dahms H-U et al. (2008) Copepod communities related to water masses in the southwest East China Sea. *Helgoländer Marine Research*, 62, 153-165.

[85] Mori T (1964) *The pelagic Copepoda from the Neighbouring Waters of Japan*. The Soyo Company Inc., Tokyo, Japan.

[86] Chen Q & Zhang Sh (1965) The planktonic copepods of the Yellow Sea and the East China Sea. I. Calanoida. *Studia Marina Sinica*, 7, 20–131.

[87] Zhang GT, Sun S, Xu ZL & Zhang QL (2010) Unexpected dominance of the subtropical copepod *Temora turbinata* in the temperate Changjiang River Estuary and its possible causes. *Zoological Studies*, 49, 492-503.
[88] Johnson CR, Banks SC, Barrett NS et al. (2011) Climate change cascades: Shifts in oceanography, species' ranges and subtidal marine community dynamics in eastern Tasmania. *Journal of Experimental Marine Biology and Ecology*, 400, 17-32.
[89] Bollens SM, Cordell JR, Avent S & Hooff R (2002) Zooplankton invasions: a brief review, plus two case studies from the northeast Pacific Ocean. *Hydrobiologia*, 480, 87-110.
[90] Uye S (1994) Replacement of large copepods by small ones with eutrophication of embayments: cause and consequence. *Hydrobiologia*, 292/293, 513-519.

In: Copepods: Diversity, Habitat and Behavior
Editor: Laurent Seuront

ISBN: 978-1-63117-846-7

Chapter 3

HARPACTICOIDA (COPEPODA) UNDER HYPOXIA CONDITIONS IN THE BLACK SEA

Elena A. Kolesnikova[1], Victor E. Zaika[2], Nelli G. Sergeeva[3,*] and Anna Lichtschlag[4]

[1,2,3]Institute of the Biology of the Southern Seas NAS Ukraine, Sevastopol, Ukraine
[4]Max-Planck-Institute for Marine Microbiology, Bremen, Germany
Current address: National Oceanography Center, Southampton, United Kingdom

ABSTRACT

The purpose of these investigations was (i) to understand the response of different meiobenthos taxa, in particular harpacticoides, to oxygen depletion and (ii) to identify Species that may be indicators of hypoxia in the shallow and deep-water areas of the Black Sea. The abundance and species composition of Harpacticoida under conditions of hypoxia were studied in shallow offshore waters off the north-western Crimea peninsula (cape Tarkhankut and Omega Bay of the Sevastopol region). The distribution of Harpacticoida in the permanent hypoxic conditions was also investigated in the deep-water zone of the Bosporus region (in the Istanbul Strait's outlet area of the Black Sea). Sources of hypoxia in bottom sediments of coastal areas of Crimea had various natures. A site of periodic hypoxia during summer time due to restricted vertical mixing of water was in the Omega Bay. Tarkhankut is the most western cape of the Crimea peninsular in the Black Sea. Cape Tarkhankut is site with local temporal hypoxia due to methane seeps. Near the Tarkhankut cape in the depth range between 1 and 12 m were the streams of bubbles of free-gas methane from bottom sediment. The significant seasonal changes were noticed in the Harpacticoida fauna in soft bottom sediments in the Crimea coast, connected with the changing oxygen content in the bottom sediments. The most number of species observed under increase of oxygen concentrations in porewaters. The dominant and subdominant species were changing. A new species *Darcythompsonia fairliensis* was recorded for the Black Sea; its populations were numerous only in hypoxia conditions in shallow water areas. In the Bosporus outlet of the Black Sea altogether 40 species of Harpacticoida were found at depths between 80 to 300 m. Species were not uniformly

* E-mail address: nserg05@mail.ru

distributed with water depth with *Haloschizopera pontarchis* present at almost all studied depths and reaching the highest abundance at 150 m. At the same depth a large number of *Amphiascoides subdebilis* was found, whereas *Bulbamphiascus imus* reached high abundance at 200 m. *Cletodes tenuipes* was often represented in different depths. Each of these species was a typical inhabitant of soft bottom sediments.

INTRODUCTION

The Black Sea is a deep marginal meromictic sea. It has an oxygenated surface water layer overlying an anoxic deep water layer. The current regime of the Black Sea induces a dome-like shape of the upper layer, with a thickness of the oxygenated surface layer of about 70-120 m in the open sea and of about 150-170 m near the shelf and continental slope [1]. Below, the water (up to 2200 m) is always anoxic and sulfidic.

The hypoxic layer of the Black Sea is situated in the lower part of the oxycline, near the hydrogen sulfide zone. In this layer the oxygen deficiency increases with depth. Correspondingly, the composition of benthic communities changes where hypoxic waters impinge on the seafloor and create a gradient of increasing hypoxia. Approaching the upper boundary of the hydrogen sulfide zone, the species composition of macrofauna becomes poorer and large forms occur only occasionally and finally vanish.

At larger depth metazoan meiofauna begin to dominate, and together with protozoans, create a distinctive benthic community. For example, in the northern part of the Black Sea at 150 m depth Nematoda dominate the meiobenthos community along with the deep-water polychaetes *Vigtorniella zaikai* [2] and *Protodrilus* sp., the largest of which reaching a length of up to 1–1.5 mm [3]. Representatives of Hydrozoa and protozoans (Gromiida, Foraminifera and Ciliophora) also constitute part of the meiobenthic community. As the meiobenthic community is characterized by this particular species of polychaetes living under permanent oxygen deficiency, it is named *V. zaikai* community [4].

In addition to the permanent hypoxia found all over the Black Sea, local and seasonal hypoxic zones are also present in shallower areas of the shelf. They are usually associated with increased organic matter input due to natural and anthropogenic eutrophication, whereas some of them are obviously also connected to local methane seepage. In this context, this paper aims at assessing the richness and distribution of harpacticoida species in the different hypoxic habitats of the Black Sea, in particular in

1) seasonal coastal near-bottom hypoxia associated with organic matter enrichment,
2) seasonal detrital-microbial benthic mat, formed around small active methane seeps, and
3) deeper water layer with permanent hypoxia (Figure 1).

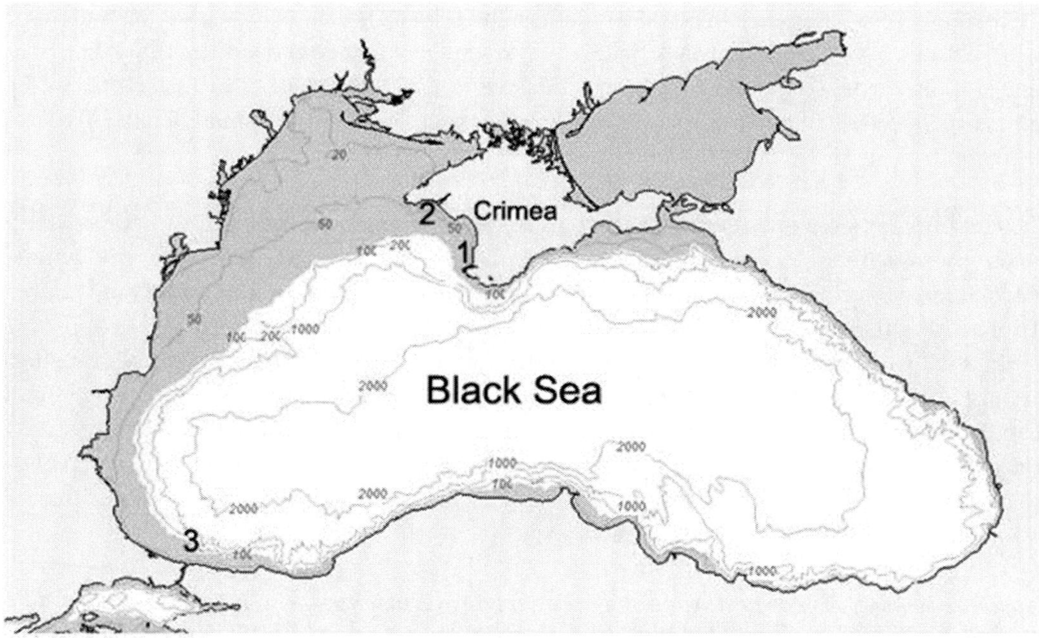

Figure 1. Researched sites in the Black Sea: 1. Omega Bay (seasonal coastal near-bottom hypoxia); 2. Tarkhankut area (seasonal detrital-microbial benthic mat, formed around small shallow active methane seeps); 3. Bosporus Strait (deeper water layer with permanent hypoxia).

MATERIALS AND METHODS

Study Sites: Different Hypoxic Habitats

First of all, in general terms, describe the types of hypoxic habitats in the Black Sea. Coastal bottom hypoxic areas are common in different regions of the Black Sea. In the north-western part of the sea they are reported frequently, but we'll just focus the studied cases of hypoxia in the bays of Sevastopol. Hypoxia is commonly observed in the summer, with the strong water stratification and the absence of storms. Propagation occurs in the summer in organisms from different parts of the ecosystem, which often leads to increased accumulation of organic matter on the bottom. These conditions favor the occurrence of hypoxia, which occurs frequently in the stagnant bottom lows.

Factor contributing to the appearance of the bottom spots of hypoxia were methane seeps from the bottom. Taking into account the peculiarity of the environmental situation, which is created around the methane seeps, we identify hypoxia associated with seepage, as a separate type. Active methane seeps in recent years discovered in the Black Sea at different depths. Harpacticoida studied in the seep, located at a shallow depth.

Spatially most extensive hypoxic habitat is a thick layer, which is located at the bottom of an oxycline and covers all the Black Sea. All layers in the water column of the Black Sea have a dome-shape, so the hypoxic layer is located in the central part of the sea closer to the surface. This layer at the periphery of the sea is in contact with the bottom at depths of 100 m and extends to about 150 meters where it borders with anoxic zone.

For this reason, most attention has been paid to Harpacticoida at depths of over 100 m. There is information about the penetration of oxygen at greater depths in the Bosporus Strait, which is downstream to the Black Sea water enriched with oxygen. Therefore, in this area Harpacticoida distribution was studied at depths of 200-250 and even 300 m, i.e., in the area of anoxic waters.

In shallow hypoxic habitats transition from normoxic to hypoxic water passes on short distance. It is difficult to study the preference of animals in relation to levels of oxygen in a spatially compressed redox-cline. In a permanent layer of hypoxia on the Black Sea stretched many meters along the bottom, in which each meter of the bottom has some specific level of hypoxia, there are the best conditions for studying the relationship between the degree of hypoxia and the distribution of meiobenthos.

Seasonal Coastal Hypoxia Associated with Organic Enrichment, Omega Bay

Omega Bay (44°36.184'-33°26.571') is 1.4 km long, maximally 90 m wide, and at its entrance 17 m deep. The water salinity in the bay is in the range 17.5 - 18.1 ‰ and the temperature at the water surface varies from 7.8 °C in February to 27° C in August [5].

To cover different water depths, samples were taken along the west coast instead of the shallower, central part of the bay. The sediment at the sampling site was sandy, mixed with pebbles overgrown by the seagrass Zostera marina. During research dives in summer 2009 small, oxygen-depleted areas were found on top of the sediment, indicated by dead grass and sulfide [6,7]. The main reason for the periodic near-bottom hypoxia is the intensive use of the bay as a recreational area and the discharge of waste water.

In Omega Bay meiobenthos samples were taken at 5 stations, at depths ranging from 2 to 11 m in July, October and December 2009 and March 2010 (Figure 2) and cores for pore water measurements and benthic samples were taken in parallel only on station 5.

At St. 5 the chemical composition of pore water was studied by co-workers from the Marine Hydrophysical Institute and the vertical distribution (1 mm resolution) of oxygen and hydrogen sulfide was measured in the upper 75 - 210 mm of the sediment. Voltametric microelectrodes provided accurate measurements of O_2 (5 μmol l^{-1}) and H_2S (0.5 μmol l^{-1}). Similar measurements were conducted in Omega Bay in 2010 [18].

St. 5 of the sampling was visually different from the other stations as more or less square spots consisting of reduced sediments were present at the sediment surface. Vertical distribution of oxygen and sulfide in the sediment core of St. 5 from the Omega Bay is shown in Figure 3.

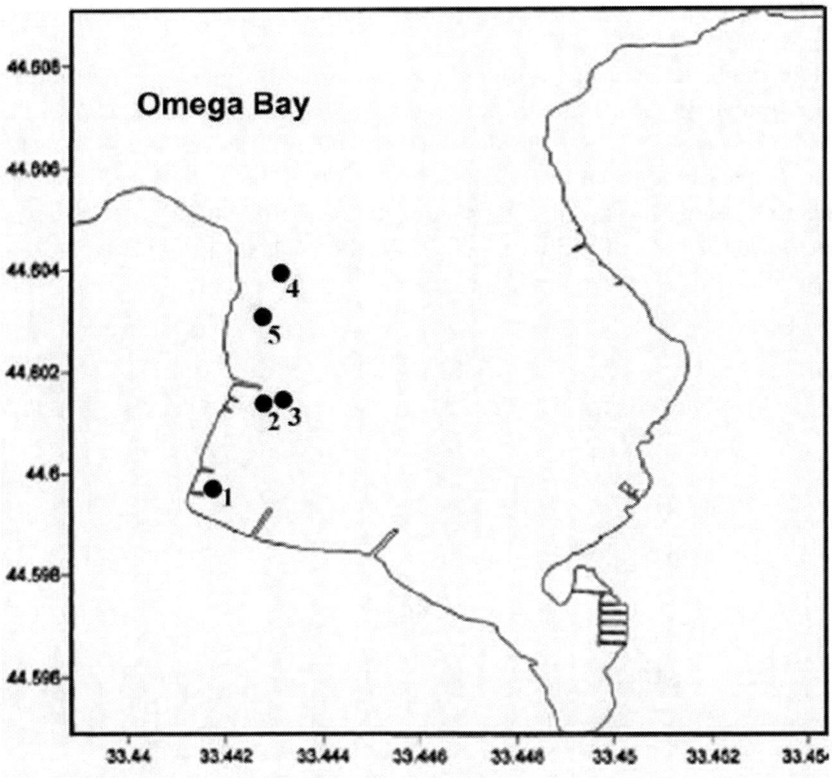

Figure 2. Meiobenthos sampling stations in Omega Bay. Station depths: 1. 1.5 m; 2. 4 m; 3. 3.6 m; 4. 11 m; 5. 8.5 m.

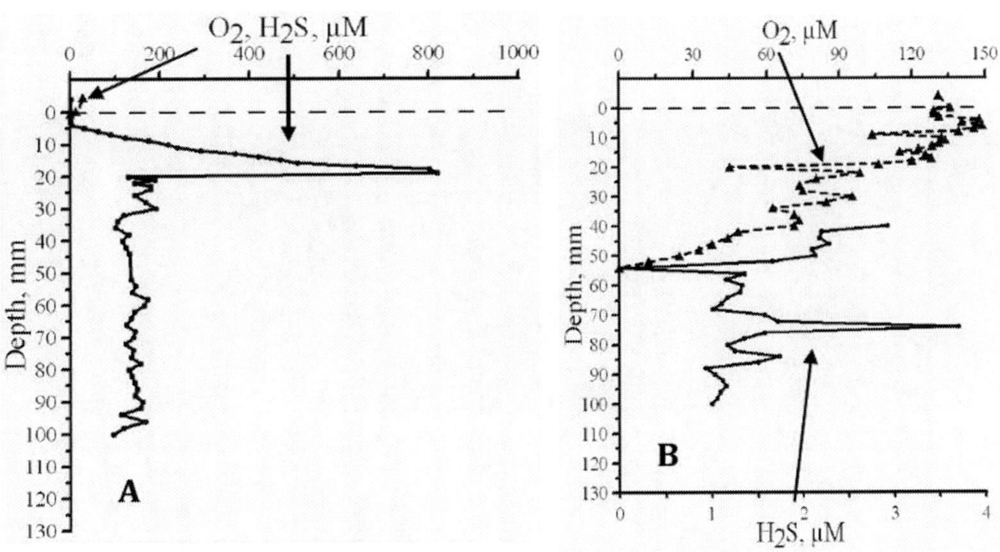

Figure 3. Vertical distribution of oxygen (dashed line) and sulfide (solid line) in the sediment core at St. 5 from Omega Bay recovered in July of 2009 (A) and in September of 2009 (B). Modified from [8].

Seasonal Detrital-Microbial Benthic Mats around Active Methane Seep, Cape Tarkhankut Area (45°21', 32°44')

Seepage of methane from the seafloor is known from different areas of the Black Sea, e.g., in Cape Tarkhankut area. Here, the strongest seepage is observed in a water depth of 3 - 7 m, approximately 100 - 170 m away from the shore (Figure 4) [13]. Gas seeps are associated with massive detrital-bacterial mat up to 12-17 cm thick and with an organic carbon content exceeding the background sediments by up to 7-50 times. These mats are high in hydrogen sulfide (Eh from -350 to - 400 mV) [9-11] and most likely contain sulfide-oxidizing bacteria.

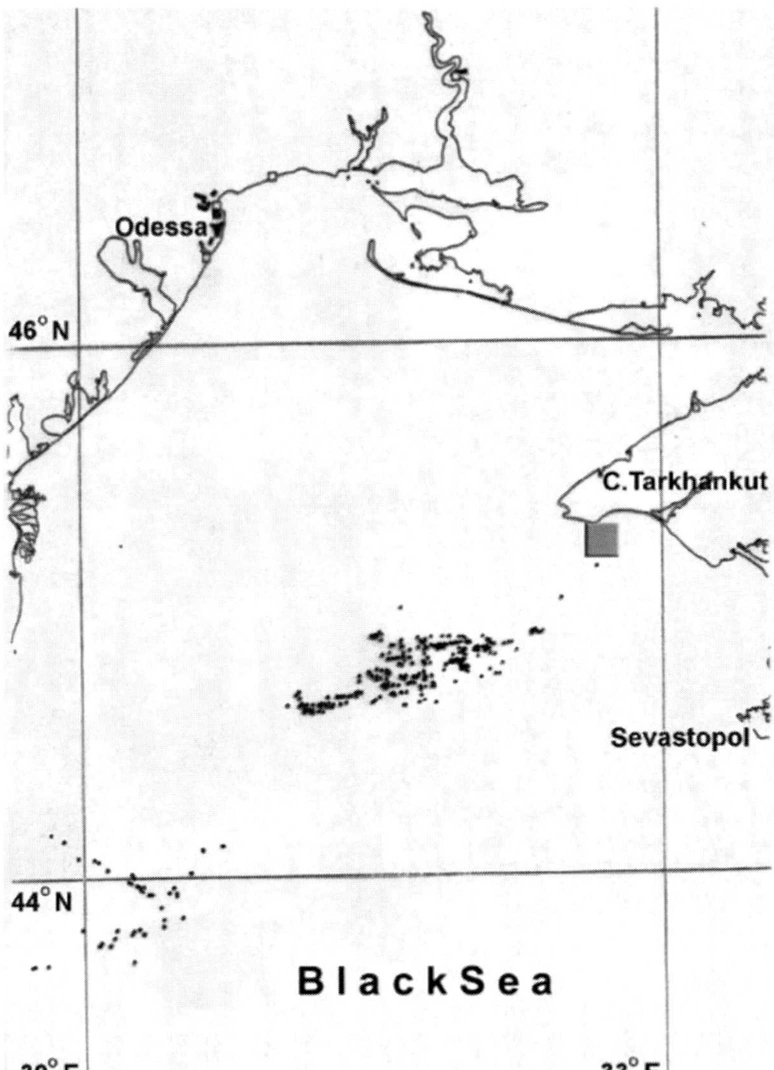

Figure 4. Meiobenthos sampling in the area of Tarkhankut cape (grey square) including a scheme of methane seeps locations (black dots). Modified from [13].

In early September 2009 the water temperature was 22°C at a depth of 3.5 m at the sampling site. In the bottom waters, O_2 concentration was 8.06 mg l^{-1} (*ca.* 360 μmol l^{-1}), but directly under the mat on the surface of the sediment it decreased to 0.6 - 0.0 mg l^{-1} (*ca.* 26 - 0 μmol l^{-1}) [11, 12].

Deep layer with Permanent Hypoxia, The Istanbul Strait's (Bosporus) Outlet Area of The Black Sea

Water exchange in the Black Sea occurs through the Bosporus, with Black Sea water flowing as an upper current through the Bosporus, while a lower current transports saline water from the Sea of Marmara to the Black Sea. This dense, saline water sinks at the continental slope and drives convection. Intermittent filaments of intrusive waters at depths of 100-500 m cause active ventilation of the anoxic waters and a vertical shift of the hypoxic layer [14].

SAMPLE COLLECTION AND GEOCHEMICAL MEASUREMENTS

In Omega Bay and Tarkhankut area divers collected undisturbed sediment cores with plastic liners for meiobenthos studies and voltammetric profiling. After recovery, cores were immediately processed to avoid changes in sediment conditions. Voltammetric profiling [8,15] to a depth of 100 to 400 mm with a sub-millimeter vertical resolution was performed on the sediment cores according to Luther et al. [16]. Data were usually acquired within 2 hours after sampling. A voltammetric analyzer (Analytical Instrument Systems Inc. Model DLK-60) with an Au/Hg amalgam solid-state working microelectrode was used for all measurements [17]. The calibration remained within 5% over the period of profiling. Scans were normally run from −0.1 to −1.7 V in a linear sweep or cyclic mode at the rate of 1.0 or 2.0 V/s. [11, 12].

Methane concentrations were measured with a gas chromatograph (Crystal 2000). A SE-30 column with a flame ionization detector heated to 65°C was used for analyses [11]. Nitrogen was used as a carrier gas with a flow rate of 30 ml min^{-1}. To study the response and penetration depth of fauna in sulfidic sediments, sediments were subsampled with syringes in the upper 10 cm layer in a 10 cm^2 area.

Sampling was conducted by divers with core liners (area of 18.1 cm^2). Undisturbed sediment samples were recovered, which allowed us to study the vertical distribution of fauna. Three replicates of the sediment cores were sampled at the following intervals: 0-1, 1-2, 2-3, 3-5, 5-7, 7-8 cm. All sediment sections were preserved in 75% alcohol. The sediments were washed, sieved from <1mm to > 63 μm, and stained with Rose Bengal for further counts of organisms with an optic microscope. In summer 2010 direct microscopic counts of non-fixed samples of bottom sediments were conducted.

In the Istanbul Strait's (Bosporus) outlet area of the Black Sea, samples were obtained during two expeditions: in November 2009 during a cruise with the R/V '*Arar*' of the Istanbul Technical University and in April 2010 during the cruise 15/1 of the R/V '*Maria S. Merian*' (Figure 5). The meiobenthos on the Black Sea shelf and the upper slope area off the Bosporus Strait outlet was studied along vertical profiles from 75 to 300 m. In this area the less saline surface waters of the Black Sea interact with the saline Mediterranean waters, creating a special ecological system and rapid transitions from oxic, hypoxic and anoxic water conditions can occur [14,27].

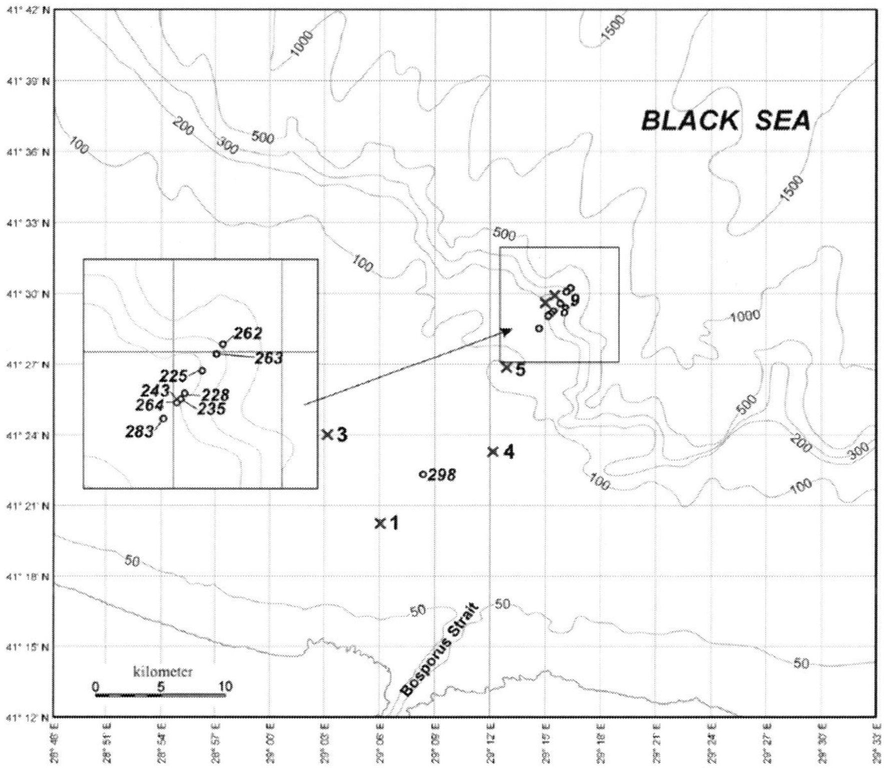

Figure 5. Meiobenthic stations with recorded Harpacticoida in the Istanbul Strait's (Bosporus) outlet area of the Black Sea (RV '*Arar*' cruise, November 2009, St. 1-9, and RV '*Maria S. Merian*' cruise, April 2010, St. 225-298).

Samples for biological studies were collected using a multicorer and a gravity corer, devices that obtain virtually undisturbed sediment samples. The sediment cores were sliced into 1 cm-thick layers down to a depth of 5 to 10 cm in order to study the vertical distribution of the fauna. Each layer was washed through two sieves, the upper one with a mesh size of 1 mm, the lower one with a mesh size of 63 μm. All organisms found in the samples were picked out and recorded. All sediment sections were preserved in 75% alcohol, which is known to preserve morphological structures of fauna without distortion. The sediments were stained with Rose Bengal before being sorted in water under a microscope for *living* (i.e., stained) organisms. Only those specimens were extracted that stained intensely with Rose Bengal and showed no signs of morphological damage. All of the isolated organisms were counted and identified to higher taxa. Harpacticoides were identified to the species level.

To follow the main inflow of the oxic water body through the Bosporus into the Black Sea, water column oxygen concentrations were measured with a CTD equipped with an oxygen sensor (SBE43) during the R/V '*Maria S. Merian*' cruise 15/1. CTD casts were performed in parallel to benthic sampling at benthic sampling station 225, 228, 235, 243, 262,

280 and 298 (Table 1). In addition vertical distribution and concentration of pore water sulfide in sediments was measured along a transect from 100 to 250 m water depth. For this, pore water was extracted with Rhizons from the upper 30 cm of sediments sampled with a multicorer as described in Lichtschlag et al. [19]. Pore water samples were fixed in ZnAc and measured on shore spectrophotometrically according to the method of Cline [20]. Detection limit was 7 μmol l^{-1}.

Table 1. Water depths and coordinates of meiobenthos stations in the Istanbul Strait's (Bosporus) outlet area of the Black Sea (RV '*Arar*' cruise, November 2009 and RV '*Maria S. Merian*' April 2010)

Station	Water depth, m	Latitude, N	Longitude, E
1	75	41°3388	29°1005
3	82	41°4003	29°0535
4	88	41°3882	29°2040
5	103	41°4477	29°2158
6	122	41°4780	29°2468
7	160	41°4832	29°2523
8	190	41°4893	29°2588
9	250	41°4988	29°2687
10	300	41°5023	29°2723
224	200	41°4932	29°2623
228	173	41°4847	29°2538
235	159	$41^0$4828	$29^0$2520
243	152	$41^0$4820	$29^0$2510
262	296	$41^0$5022	$29^0$2723
263	248	$41^0$4987	$29^0$2687
264	150	$41^0$4818	$29^0$2512
283	117	$41^0$4757	$29^0$2438
298	80	$41^0$3718	$29^0$1387

RESULTS AND DISCUSSION

Seasonal Hypoxia in Coastal Sediments (Omega Bay, Sevastopol)

A total number of 17 samples were collected at this sites and every sample contained 2 to 15 Harpacticoida species (Table 2).

In total 43 Harpacticoida species were found in Omega bay sediments (see Annex) with the largest number of species detected in July and March, especially in St. 2 and St. 5 (13 and 15 species, Table 2). Least numbers of species (2) were found in October in St. 3 and St. 5.

In the Black Sea zone occupied by populations of different benthic species are replaced with changes in depth; there is a change belts. The most densely animals of each species are located at depths with the most favorable conditions for the species. Usually this place ("core") is located in the central part of a horizontal belt of species settlement. Closer to the edges of the belt animals become more and more rare and can be considered '*random*'.

Table 2. Number of Harpacticoida species detected in samples from Omega Bay

Year / Month	Station No/Depth (m)				
	1/2	2/4	3/6	4/11	5/8.5
2009					
July	11	13	-	10	12
October	7	9	2	4	2
December	7	5	10	-	6
2010					
March	-	13	-	7	15

To exclude rare and random forms from the analysis, we selected 14 '*common*' species out of the 40 species, which

1) were defined to the species levels (with 3 exclusions, owing high abundance),
2) had a density of more than 1,000 ind m^2, and
3) were detected at least in three samples (Table 3).

Table 3. List of 14 "common" Harpacticoida species for all stations of the Omega Bay

Species
Ameira parvula (Claus, 1866)
Enhydrosoma propinquum (Brady & Robertson in Brady, 1880)
Darcythompsonia fairliensis (Scott T., 1899)
Ectinosoma melaniceps Boeck, 1865
Halectinosoma herdmani (Scott T. & A., 1894)
Harpacticus flexus Brady & Robertson, 1873
Asellopsis bacescui Por, 1959
Amphiascus longirostris (Claus, 1863) Huys, 2010
Amphiascus sp.
Amphiascoides subdebilis (Willey, 1935)
Amphiascoides brevifurca (Czerniavsky, 1868)
Amphiascoides sp.
Haloschizopera pontarchis Por, 1959
Tisbe sp.

Of these 14 species some occurred in large numbers only at depths of up to 4 m (*Halectinosoma herdmani*, *Harpacticus flexus*), where they were present from July, but peak abundances were found in October. Other species were found at different depths, with maximum numbers in July and a lower abundance in March (*Amphiascus longirostris*) or October (*Asellopsis bacescui*). Some species occurred from September to December (Normanella mucronata), others from December to March (*Enhydrosoma propinquum*), or from October to March (*Ectinosoma melaniceps*, *Amphiascoides subdebilis*).

In addition to water depth and seasonal succession of species, we obtained data on bottom water hypoxia dynamics as well as on the chemical composition of the pore water in the upper centimeters of the sediment, which allows us to discuss the impact of oxygen fluctuations on the Harpacticoida species composition. Since the chemical studies were only performed at St. 5, we compare changes in Harpacticoida species composition at this sampling location at a specific sampling occasion. To analyze changes in abundance of macro- and meiobenthos, different criteria were applied. For instance, the depth of hydrogen sulfide onset in pore waters of the sediments was used [8, 21]. Concentration of oxygen in bottom water was also considered. Both criteria indicated that the most severe hypoxic events at St. 5 were observed in July and December 2009. Indeed, at the time H2S was present at the sediment-water interface, and even escaped into the bottom water (Table 4).

Table 4. Depth of H_2S and total abundance (ind.· m^{-2}) of Harpacticoida at St.5 of the Omega Bay (Samples are in order of presence of H_2S in the sediment)

Depth of H_2S (mm)	Date	Harpacticoida abundance
0	July 09	10,304
0	Dec 09	3,680
34	Mar 10	22,080
36	Sep 09	28,888
60	May 10	116,472

When analyzing the dynamics of Harpacticoida abundance, an overall downward trend in the total number of Harpacticoida was observed during the presence of H_2S close to the sediment surface, however, concurrently numbers of *Darcythompsonia fairliensis* were increased (Figure 6).

With these data on the species composition of Harpacticoida, it is now possible to compare the responses of the different species to the onset of H_2S in the sediment. Altogether, 12 species of Harpacticoida were found in July 2009 and six species in December 2009 at St. 5. In July, only *D. fairliensis* was abundant (6956 ind/m2). Other species abundances were very low (*Laophonte* sp., *Dactylopusia tisboides*, *N. mucronata*).

In contrast, in December 2009, a very different species composition was found at St. 5: *Asellopsis bacescui* was present with only 1160 ind.· m^{-2}, whereas other, previously less dominant species, (*Ameira parvula*, *E. propinquum*, *E. melaniceps* and *Amphiascus* sp.) had higher abundance (100 to 750 ind.· m^{-2}). Comparing the species composition of Harpacticoida in July (when the water temperature was still high) and December at the same location, but under comparable hypoxic conditions, leads to the conclusion that in addition to oxygen concentrations also other factors have a large influence.

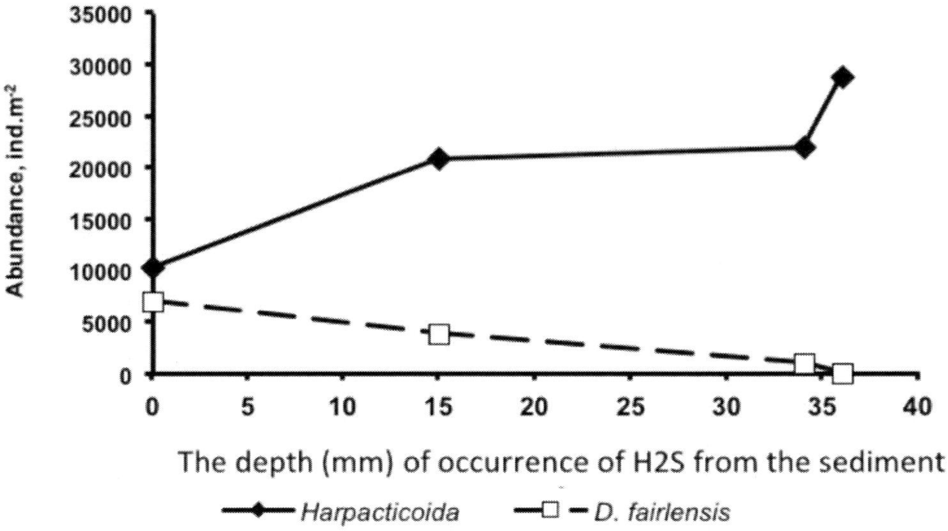

Figure 6. Abundance of Harpacticoida (black rhombi) and *Darcythompsonia fairliensis* (open squares) as a function of depth of occurrence of H_2S in the sediment of Omega Bay.

For our comparison study *D. fairliensis* (T. Scott, 1899) is most important, not only because of its high abundance, but primarily as it was previously found in sediments subject to severe hypoxia in the cape Tarkhankut area [11, 22]. At St. 5 this species was detected during three sampling campaigns: in July 2009 the number was 6954 ind m^2, in October the same year the abundance dropped to 3846 ind.· m^{-2}, however, the lowest number (984 ind.· m^{-2}) of the species was found in March 2010. This leads us to the conclusion that the abundance of *D. fairliensis* is clearly linked to the onset of H2S in the sediment: The closer to the surface H_2S is present the higher is the abundance of this species (Figure 7). Thus, response of *D. fairliensis* to increase of hypoxic conditions is opposite to the response of most other Harpacticoida species.

As the depth of H_2S onset is closely related to the oxygen concentration in the water, we get similar result when we compared *D. fairliensis* abundances with the oxygen concentration at the sediment-water interface (Figure 7).

The observed behavior of *D. fairliensis* during increasing hypoxia indicates that this species might represent an ideal indicator species for hypoxia. Interestingly, in Omega Bay *D. fairliensis* was found at St. 2 (July, 1570 ind m^2) and St. 3 (October, 7118 ind.· m^{-2}). In the latter case the highest number of *D. fairliensis* was found and together with this species only a small number of *E. melaniceps* was detected. This suggests that also at these sites hypoxic condition occurred. On the other hand, *D. fairliensis* was not always present during severe hypoxia, as for example in December 2009 at St. 5.

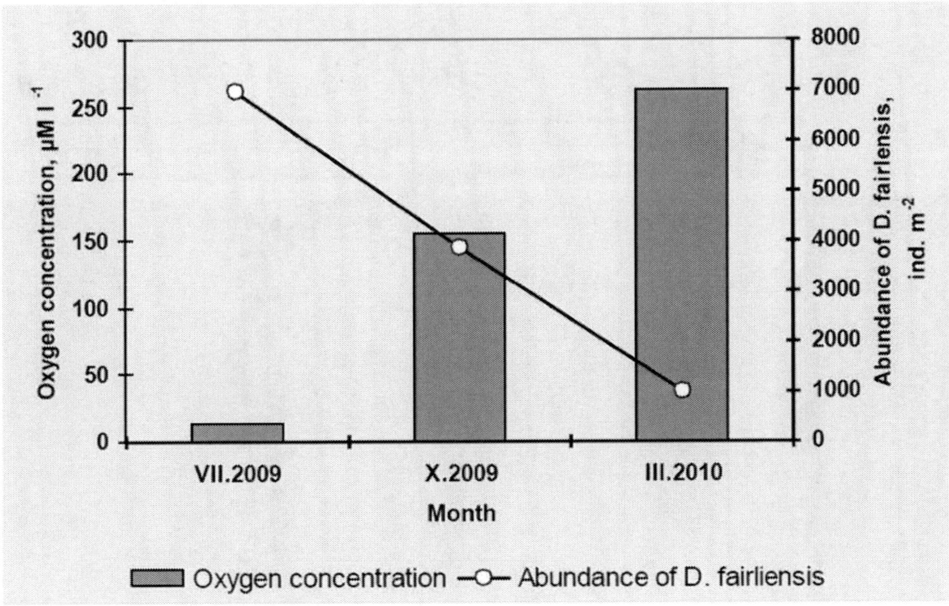

Figure 7. Abundance of *Darcythompsonia fairliensis* in relation to oxygen concentration in pore water at the bottom sediment of Omega Bay.

Species composition and Distribution of Harpacticoida in Methane Seep Area (Cape Tarkhankut)

The flow of methane from the sediments to the water column ranges from 1.7 ml m^{-2} d^{-1} (1.2 mg m^{-2} d^{-1}) for well-aerated sites with strong oxidizing conditions to 8170 ml m^{-2} d^{-1} (5800 mg m^{-2} d^{-1}) for sediment with high rates of methanogenesis [11].

Voltammetric profiling of all sediment cores retrieved during different seasons showed presence of sulfide within the bacterial mats [11]. Generally, vertical profile showed an increase of sulfide from zero or minimal values at the sediment surface to maximal concentrations of 1300 µmol l^{-1} at a depth of 40 to 70 mm. Interestingly, sulfide decreases below a depth of 100 mm. This can be due to the lack of electron acceptor for methane oxidation (oxygen, oxidized forms of iron, or sulphate [11] caused by to strong advection rates, limiting a deep penetration of any oxidant for methane supplied from the water column [23].

Only during winter storms sulfide was absent in the sediments. Thus, during an average annual cycle minimal or no sulfide was present in the sediments during the winter storms and the highest concentration of sulfide was found by the end of the summer (Figure 8). Unfortunately, up to now we were not able to sample methane seep sediments after a storm event, as weather conditions didn't allow diving. The minimal concentrations of sulfide were about 50 µmol l^{-1} at a depth of 25 to 60 mm with no or extremely low concentration of sulfide in the upper 15 mm of sediments and concentration of oxygen was 250 µmol l^{-1} in the overlying waters (2 to 3 mm above the sediment surface). Thus, conditions were most likely very favourable for meiobenthos during this time, as they were exposed to plenty of oxygen from the water column and plenty of fresh organic matter in the sediments.

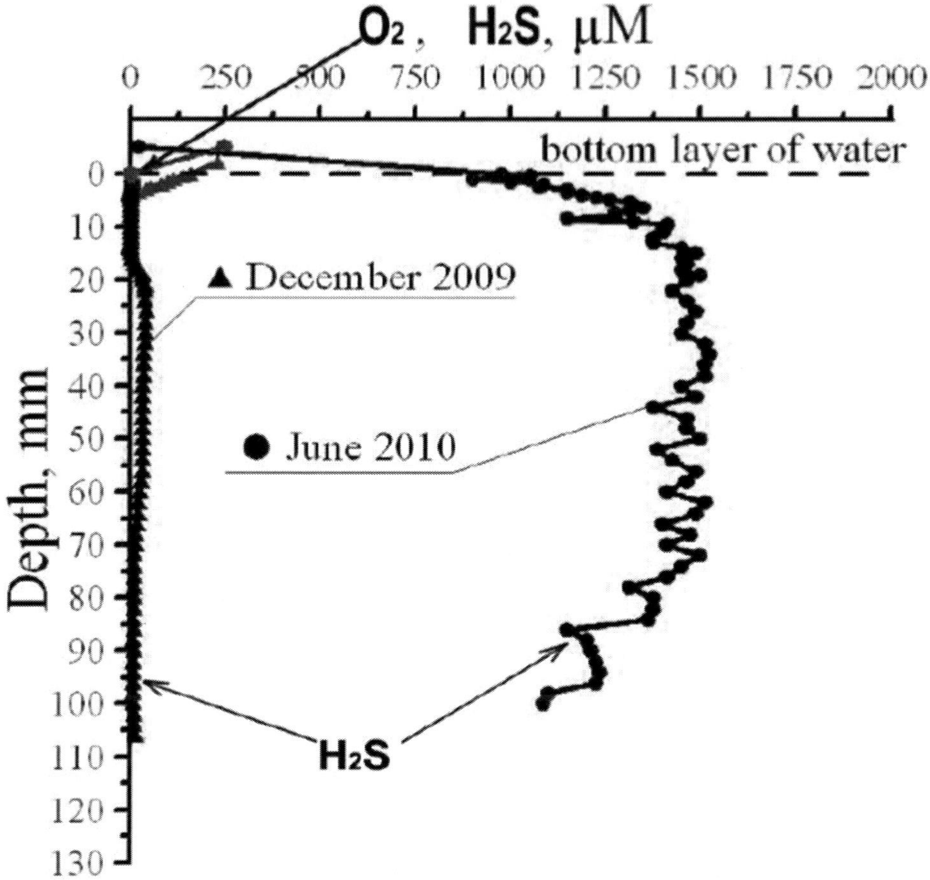

Figure 8. Seasonal variations in the vertical distribution of oxygen and hydrogen sulfide in pore water in the bottom sediment (Tarkhankut area).. Modified from [11].

During the time of maximum sulfide concentrations at the end of the summer, concentrations easily reached 1500 µmol l^{-1} with maxima of almost 3000 µmol l^{-1} found in September 2009. Sulfide concentrations at the sediment surface were as well very high (1000 µmol l^{-1}), revealing extremely high vertical gradients of sulfide and oxygen [11]. Oxygen only penetrated the top layer (0-4 mm) of the detrital-microbial mat [9].

The meiobenthos density in the upper 5 cm of the sediments varied strongly with redox conditions (Figure 9). Moreover, the population density varied for each of the sampled sites. For example, it was 20×10^3 to 44×10^3 ind.· m^{-2} at one site and 115×10^3 to 120×10^3 ind.· m^{-2} at another site, both sampled in September 2009. Most likely this is due to limited oxygen penetration, as at both sites organic matter content was high. However, the abundance of meiobenthos was higher and varied from 97×10^3 to 632×10^3 ind.· m^{-2} and from 243×10^3 to 800×10^3 ind.· m^{-2} at two background sites sampled at the same day [11].

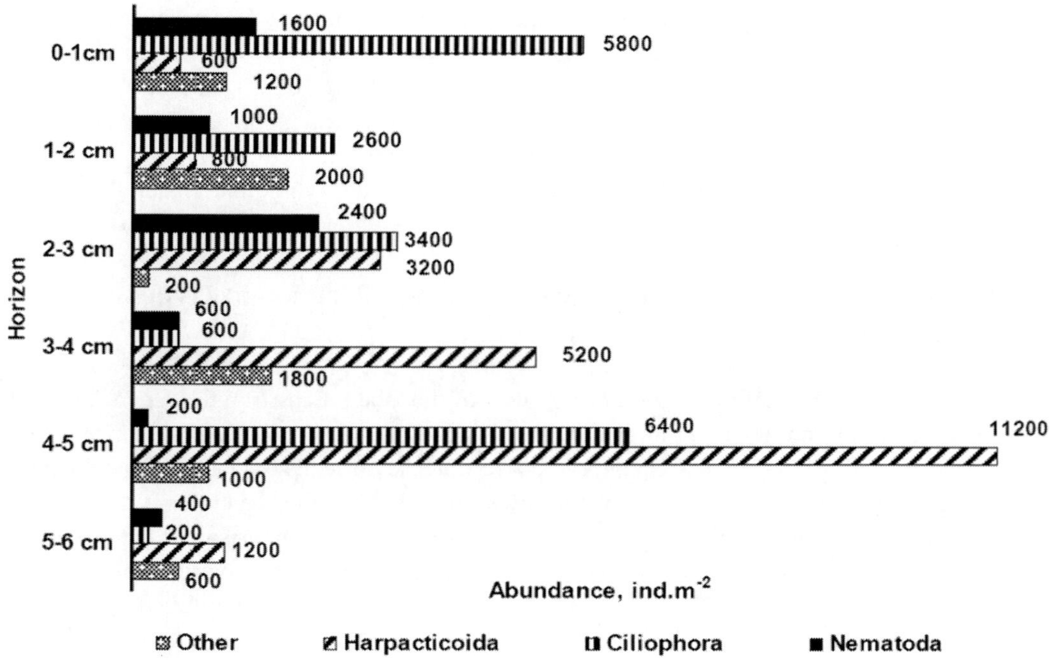

Figure 9. Vertical distribution of the main meiobenthos taxa found in bottom sediment (Tarkhankut area).

An expected (classical) distribution of the abundance and diversity of meiofauna has been observed at the background site, where these characteristics are the highest in the upper layer. Contrary to this, the meiofauna abundance in the sulfidic sediments is higher in lower layers where the hydrogen sulfide concentration reaches its maximum. At the same time, taxonomic richness is similar in all layers.

Figure 9 displays the response of the main taxa and the depth of fauna penetration in anoxic/sulfidic sediments in the Cape Tarkhankut area. These results suggest that some benthic fauna can adapt to H_2S, although it is thought to be toxic for many higher organism.

In sediments with high pore water hydrogen sulfide concentration some meiobenthic groups were found from 3 cm to be dominated by harpacticodes (Figure 9). However, only *Darcythompsonia fairlensis* was found and other species of Harpacticoida were not detected in these samples. Analysis of the vertical distribution of *D. fairlensis* in sediments from the surface to a depth of 7 cm (1cm intervals) showed the highest concentration at 4-5cm, where oxygen was absent and hydrogen sulfide concentrations were highest. In this sediments layer, adult females, males, and copepodites at different stages were found.

In July 2010, where samples were not fixed, but immediately analyzed, the results were similar. We found alive and actively moving individuals of *D. fairlensis*, which proved their tolerance to these environmental conditions.

D. fairlensis has been detected in various samples from the methane seeps collected from July to September. Its numbers reached up to 140×10^3 ind.·m^{-2}. Other Harpacticoida species, *Haloshizopera junodi*, and *Dactylopodia* sp. were found only occasionally. Perhaps all these species are able to tolerate hypoxia, however, their rare occurrence does not allow further conclusions. On the other hand, we can conclude that *D. fairlensis* does not only tolerate hypoxia, but our results suggest that it is confined to this condition, as *D. fairlensis* occurs in lower abundance in normoxic condition, as at the background site with normoxia, where Harpacticoida species composition was more diverse, but *D. fairlensis* was absent.

D. fairlensis is one of the few species of the order Harpacticoida, which has been previously shown to inhabit hypoxic environments in marine sediments [24]. It is an endobenthic species, and it usually lives in shallow waters on decaying plant material.

Distribution of Harpacticoida in Deeper Layers of Permanent Hypoxia (Bosporus Area)

The distribution of benthos in the deeper shelf of the Black Sea strongly depends on the oxygen gradient in the water column. In the Bosporus area the inflow of oxygen-rich Mediterranean waters into the permanently hypoxic water layer leads to changes in the degree of hypoxia and causes a vertical shift in the lower boundary of the oxycline. It has a great effect on the penetration of macro- and meiofauna, as shown by a comparison of the Bosporus area with the districts in the northern part of the sea. During a study of the vertical distribution of benthic fauna conducted with the RV '*Arar*' [25] macrobenthic Annelides (Oligochaeta and Polychaeta) were only found in the Bosporus area at depths greater than 125 m with abundances of 10920 and 6760 ind.·m^{-2} at 162 m and 250 m water depth, respectively. Meiobenthic polychaetes, belonging to the species *Vigtorniella zaikai* Kisseleva predominated at a depth of 250 m, whereas in other parts of the Black Sea these polychaetes have their maxima of abundance at a shallower depth (150 - 170 m) [25].

Figure 10 shows the changes in the total number of meiobenthos sampled during the R/V '*Arar'* cruise. After a gradual decrease with increasing water depth, meiobenthos abundance was increased again at 250 m depth, so near the border of the hypoxic/anoxic zone Nematodes are the most important contributors to the meiofauna, followed by Harpacticoida, Gromiida and Ostracoda [27].

Two surveys carried out in the same area allows us to compare the distribution of Harpacticoida in a water depth from 75 to 300 m (Figure 11). Altogether 40 species of Harpacticoida were found in this depth interval at the Bosporus area of the Black Sea (see Annex). Distribution with depth and total numbers of Harpacticoida in specific depth differed between the two cruises. In the RV '*Arar*' data, the abundance of Harpacticoida rapidly decreased from > 200 000 ind.· m^{-2} at a depth of 75 m to 520 ind.· m^{-2} at a depth of 190 m. In contrast, in the data obtained during the RV '*Maria S. Merian*' cruise peaks of abundances were found at 150-160 m and 200 m depths (Figure 11), and abundance were low (7000 to 9000 ind.· m^{-2}) at 80-117 m. However, in both datasets the number of Harpacticoida was close to 1000 ind.· m^{-2} at 250 m depth.

Altogether, the number of Harpacticoida species found per sample decreased with depth: 4 to 27 (usually 7-10) species at 75-100 m, 6-12 species at 150-170 m, and 1-3 species at 200-300 m.

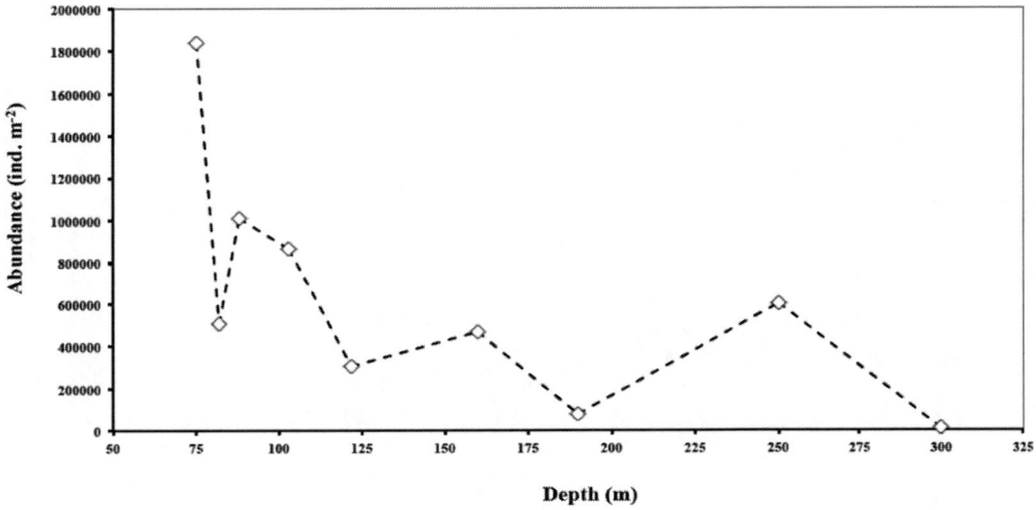

Figure 10. Vertical distribution of total meiobenthos in the Bosporus area (cruise RV '*Arar*').

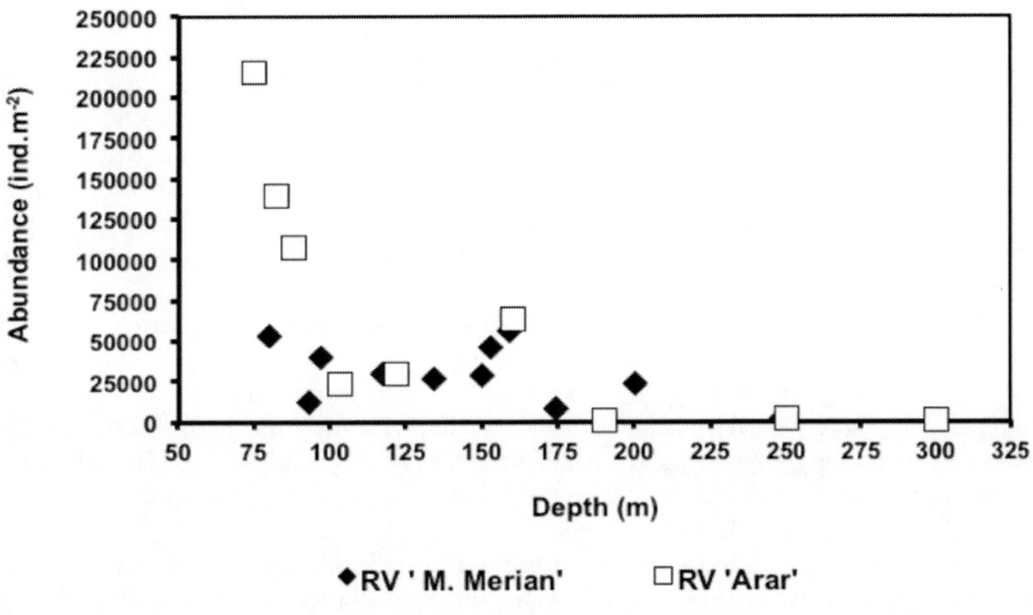

Figure 11. Vertical distribution of Harpacticoida (without naupliar stages) as a function of depth in the Bosporus Strait area of the Black Sea found during the RV'*Arar*' cruise (black triangles; November, 2009) and the RV '*Maria S. Merian*' (white; April, 2010).

Changes in the number of Harpacticoida species with depth found during the R/V '*Maria S. Merian*' cruise are shown in Figure 12. Here the greatest number of species (12) in one sample was detected at 152 m depth, while during the R/V '*Arar*' cruise the abundance maximum (27 species) was at 82 m (Figure 12).

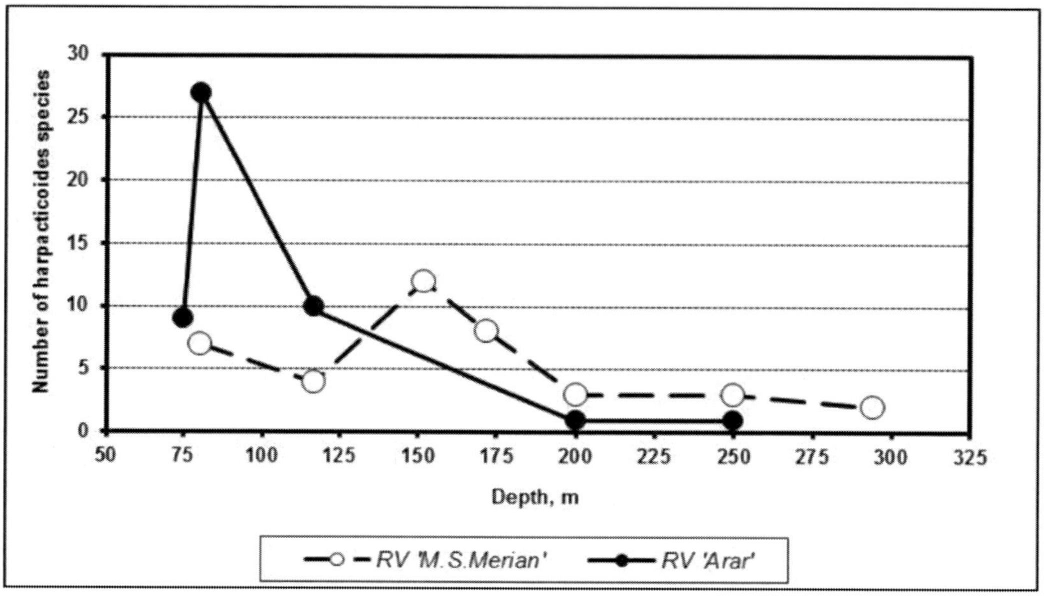

Figure 12. Vertical distribution of the total number of Harpacticoida species at different depth (RV '*Arar*', November 2009 and RV '*Maria S. Merian*' May 2010).

The highest abundance of Harpacticoida (27,400 ind m^{-2}) during the RV '*Maria S. Merian*' cruise were detected at 152 m depth, i.e., eight times lower than the highest total abundance the RV '*Arar*' cruise. At a depth of 152 m three species dominated: *H. pontarchis* (10,400 ind.· m^{-2}), *A. subdebilis* (5,350 ind.· m^{-2}), *E.longifurcatum* (4,930 ind.· m^{-2}). In comparison, different species dominated at 200 m during the RV '*Maria S. Merian*' cruise with a total abundance of 34,440 ind.· m^{-2} (essentially *E. longicaudum*, *B. imus* and *Mesochra* sp.).

Thus, the species composition at similar depths varied considerably between the two cruises. To identify the most common trends in species composition we again will focus on species that occur frequently or are more abundant. For instance during both cruises, *C. tenuipes* was found with no more than 1600 - 1700 ind.· m^{-2}. This species was found in almost all stations in a broad range of depths from 80 m to 250 m (Figure 13). The abundance of this species was highest at 152 m depth and then decreased with increasing depth. Deepest depth of occurrence was 250 m (240 ind.· m^{-2}).

During the RV '*Arar*' cruise the abundance of Harpacticoida was 213,700 ind.· m^{-2} at 75 m, with *E. melaniceps* (116,600 ind.· m^{-2}) being the dominant species. At a depth of 100 m, 11 species were found, of which a relatively high number (4,000 to 18,000 ind.· m^{-2}) were *E. caeni*, *E. sarsi* and *N. mucronata*; the other species were less abundant than 500 ind.· m^{-2}. Most of these species were also found at 75 m or 82 m.

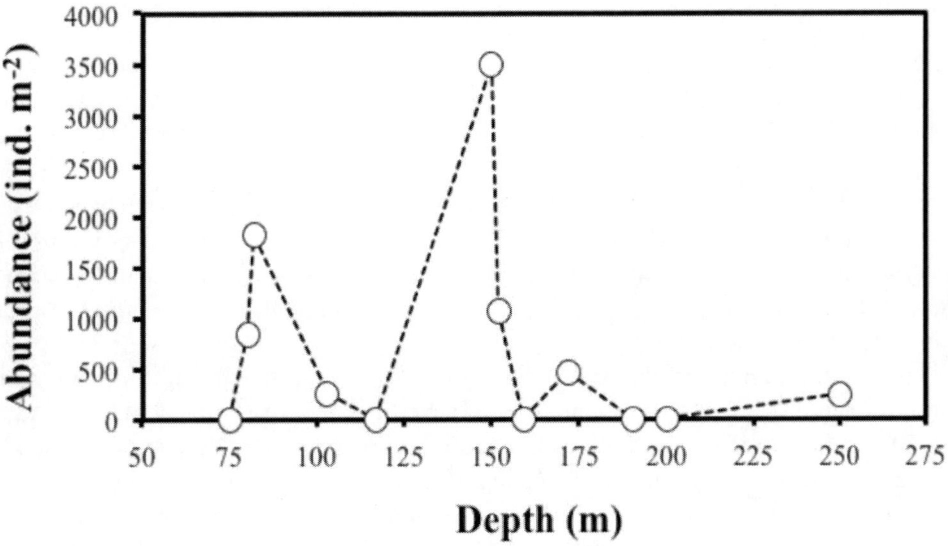

Figure 13. Vertical distribution of *C. tenuipes* (cruises RV's '*Arar*' and '*M. S. Merian*').

For this study the depth distribution of the three species (*H. pontarchis*, *E. longifurcatum* and *B. imus*, was compared that occur near the anoxic zone (Figure 14).

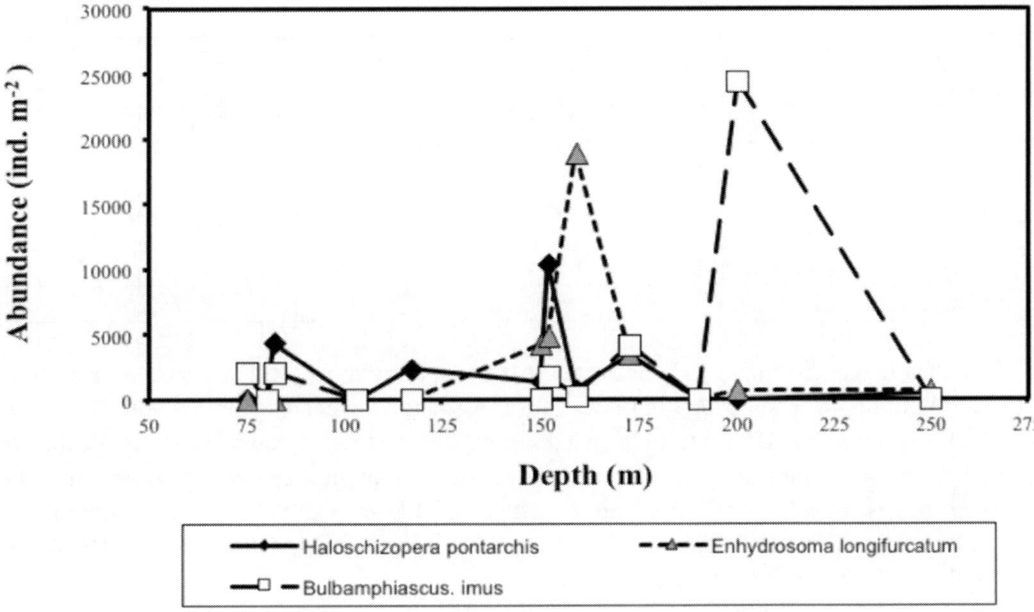

Figure 14. Vertical distribution of *H. pontarchis*, *E. longifurcatum* and *B. imus* at different depths in the Bosporus area (RVs '*Arar*' and '*M.S. Merian*').

The maximum abundance of *H. pontarchis* (10,400 ind.· m^{-2}) was observed at a depth of 152 m, *E. longifurcatum* abundance peaked at 159 m depth (19,000 ind.· m^{-2}), and *B. imus* (24,400 ind.· m^{-2}) at 200 m depth. At a depth of 172 m, all three species had a similar abundance (4,000 to 5,000 ind.· m^{-2}). However, at 250 m depth only the first two species were present in low abundances (400 to 700 ind.· m^{-2}). According to the RV '*Arar*' data, *B. imus* abundance ranged between 11,000 and 13,000 ind.· m^{-2} at a depth of 75-80 m. Changing the depths at which the species is plentiful, can be explained by seasonal differences as well as differences in power in the downstream Bosporus Strait, through which flows into the Black Sea oxygenated water. At 172 m, 8 species of Harpacticoida were found. Apart from the types discussed above, there were *H. herdmani*, *C. tenuipes*, *Typhlamphiascus confusus*, *Laophonte brevifurca* and *Amphiascus* sp.

Thus, in the depths of the Black Sea, previously considered 'lifeless' (in terms of eukaryotic fauna), we found representatives of several genera of Harpacticoida. We attribute this to Marmara water jets that fall into the top of the anoxic zone. Previously, some species of Harpacticoida were only reported from depths ranging from 50 to 150 m, near the southern coast of Crimea [26]. In the Bosporus area we found that meiofauna present not only at a depth of 172 m, but it is much deeper, and let us give you the data obtained at depths of 200, 250 and 300 m (Table 5).

Table 5. Harpacticoida recorded in the Black Sea at depths of 200 to 300 m

Species	Depth (m)		
	200	250	300
Bulbamphiascus imus (Brady, 1872)	24,366	0	0
Mesochra sp.	9,348	0	0
Enhydrosoma longifurcatum Sars G.O., 1909	726	707	0
Enhydrosoma gariensis Gurney, 1930	0	1,040	0
Haloschizopera pontarchis Por, 1959	0	471	0
Cletodes tenuipes Scott T., 1896	0	236	0
Eurycletodes (Oligocletodes) latus (T Scott, 1892)	0	0	376
Laophonte sp.	0	0	188

The data presented in Table 5 indicates that the total number of different Harpacticoida decreases with depth. The change in depth causes a change in species composition. However, at a depth of 250 m four Harpacticoida species were still present. Apparently, these depths are close to the limit because of anoxia to Harpacticoida, even in an area near the Bosporus. The discovery of two members of the group at a depth of 300 m was not further considered, as they only occurred in small abundances and the occurrence might be accidental. However, this depth might be worth investigating in future studies.

The finding of two Enhydrosoma species at 250 m depth and the very large number of *B. imus* at the 200 m depth, leads us to the conclusion that conditions at this depth are quite suitable for some Harpacticoida, adapted to sever hypoxia. According to the available data, the oxygen concentration was 0.12-0.16 µmol l^{-1} at depth of 90-120 m during the RV '*Arar*' cruise [27]. Unfortunately, the measurements of oxygen were carried out only at some depths. However, vertical distribution of oxygen was quite different during the '*Maria S. Merian*' cruise, as the inflow of oxic Mediterranean waters seems to have ceased during the sampling period. Oxygen data indicated the onset of the anoxic zone at around 140-160 m water depth and fully oxic conditions were present only above 116 m water depth (Table 6).

Table 6. Bottom water oxygen concentration measured with the CTD

Station	Parallel benthic sampling station	Water depth (m)	Bottom water oxygen concentration
295	298	82	< 220 µmol L^{-1}
280	283	116	< 107 µmol L^{-1}
242	243	152	< 13 µmol L^{-1}
232	235	159	< 10 µmol L^{-1}
227	228	172-174	anoxic below 140 m
223	225	197-200	anoxic below 140 m
197	263	248-260	anoxic below 140 m
190	262	296-301	anoxic below 140 m

The limited information on the chemical environmental variables at each depth limits the possibilities of further analysis of the reaction of different types of Harpacticoida to hypoxia. In addition, in the area of the Bosporus there are significant fluctuations in environmental conditions, as evidenced by the change in the depth of the peak settlement of the polychaetes Vigtorniella zaikai [28] and the water column oxygen concentrations.

The data presented in the Table 5 show that total number of different Harpacticoida is reduced with increasing depth. Obviously, the change in depth causes a change in species composition, however, at a depth of 250 m four Harpacticoida species were still present.

The discovery of two members of the group at a depth of 300 m was not further considered, as they only occurred in small abundances and the occurrence might be random, however, this depth might be worth examining during future studies. The finding of two Enhydrosoma species at 250 m depth and the very large number of *B. imus* at the 200 m depth, leads us to the conclusion that conditions at this depth are quite suitable for some Harpacticoida, adapted to sever hypoxia. According to the available data, the oxygen concentration was 0.12 - 0.16 µmol l^{-1} at depths of 90 to 120 m during the RV '*Arar*' cruise. [27].

However, vertical distribution of oxygen was quite different during the '*Maria S. Merian*' cruise, as the inflow of oxic Mediterranean waters seems to have ceased during the sampling period. Oxygen data indicated the onset of the anoxic zone at around 140-160 m water depth and fully oxic conditions were present only above 116 m water depth (Table 6). No sulfide was detected in sediments sampled in 100 and 150 m water depth. At 200 m water depth sulfide was present in the sediment below 5-10 cm, however, at 250 m the upper 30 cm of the sediment contained sulfide. Concentrations were 400 µmol l^{-1} at the sediment water interface and sulfide will have been seeping into the water column.

The limited information on the chemical environmental variables at each depth limits the possibilities for further analysis of the reaction of different types of Harpacticoida to hypoxia. In addition, in the area of the Bosporus there are significant fluctuations in environmental conditions, as evidenced by the change in the depth of the peak settlement of the polychaetes *Vigtorniella zaikai* [28] and the water column oxygen concentrations.

Conclusion

A decrease in oxygen in the environment led to a reduction in both the abundance and total number of Harpacticoida species in soft bottom sediment. In the Black Sea during study of hypoxic habitats, various species of Harpacticoida have been found, that have adapted to life in hypoxic environments. Some of these species prefer a depth of 20 m, while others live near the anoxic zone at a depth of *ca.* 150 m.

Each species of meiofauna can choose their optimum level of hypoxia. Typically this level is difficult to study in a compressed vertically redox-cline at the sediment surface. In a permanent layer of hypoxia on the Black Sea stretched many meters along the bottom, there are the best conditions for studying the relationship between the degree of hypoxia and the distribution of meiobenthos. In hypoxic/anoxic areas significant abundance of species such as *H. pontarchis, A. subdebilis, E. longicaudum, Mesochra* sp. and *B. imus*, were recorded, the latter inhabiting the surface layer of the sediments as well as the first 1-2 cm. All of these species form in hypoxic habitats population having all stages of development from eggs to adults (not identified to species only stage of eggs and nauplii). *C. tenuipes* was found at almost all stations in a range of depths ranging from 80 to 250 m. Previously, these species of Harpacticoida were reported from depths ranging from 50 to 150 m near the southern coast of Crimea [26], where until now only *A. longirostris* was abundant in significant numbers at 120–160 m depth and was recorded at 170 m [29].

A study by Por [30] also indicate that the peculiar deep-water harpacticoids such as *T. confusus, E. latus, E. parasimilis, Heteropsyllus dimorphus and Hemimesochra atargatis* as well as *A. longirostris, H. pontarchis, N. mucronata* and *Stylicletodes longicaudatus* inhabit in 'the Modiolus muds' (deep Black Sea; the limit of the macrobenthic life is reached within this zone). Following this author, *A. longirostris*, at 140 m is the deepest recorded Harpacticoids of the Black Sea, while *E. latus* has been found at 124 m [30].

These data show that the presence of a gradient of hypoxia causes a change in species composition of Harpacticoida and, species present are tolerant to hypoxic conditions. Interestingly, in shallow hypoxic habitats different species dominate than in the deep hypoxic conditions.

Only one harpacticoids, *D. fairlensis*, can be suggested as indicator species of hypoxic conditions in shallow offshore areas. In the hypoxic zone numerous well-developed populations of the *D. fairlensis* were present. *D. fairlensis* is one of the few species of the order Harpacticoida, which is known to inhabit anoxic-, hypoxic-conditions in other seas as well [24]. This endobenthos species usually occurs in shallow water areas on decaying plant material. Interestingly, this species probably has a history of origin. Analyzing its distribution Gomez [31] concluded that, "*the ancestral stock of Darcythompsonia could have originated in the equatorial Thetys Sea at sometime during the Late Triasic - Early Jurassic periods*'.

ACKNOWLEDGMENTS

This work was partly financed by the EC 7th FP project HYPOX 226213(In situ monitoring of oxygen depletion in hypoxic ecosystems of coastal and open seas, and land - locked water bodies) and the project TÜBITAK-NASU (Comparative studies of meiobenthic community structure off Crimea (Ukraine) and Sinop (Turkey) with respect to identification of Marine Protected Areas (MPAs) in Turkish Black Sea). We thank the anonymous reviewers for detailed comments that helped to improve the paper considerably. We are grateful thank the captains, cruise leaders, crew and shipboard scientists of the RV '*Arar*' cruise and the RV '*MSM*' 15/1 cruise for their excellent work at sea. We thank to Sergey Konovalov and Nataly Orekhova for help with the O_2 and H_2S data and Maksim Gulin, Vitaly Timofeev, Tetiana Kosheleva, Oksana Anykeeva, Ludmila Lukyanova for sample collection in the coastal Crimean areas.

REFERENCES

[1] Vinogradov ME, Shushkina EA, Flint MV & Tumantseva NI (1986) Distribution of plankton in the lower part of oxygen zone in the Black Sea. *Oceanology*, 26, 300-309.

[2] Kisseleva MI (1992) New genus and species of the family Chrysopetalidae Polychaeta from the Black Sea. *Russian Journal of Zoology,* 71, 128–132. (*In Russian)*

[3] Kisseleva MI (1998) Peculiarity of vertical distribution of polychaetes family Protodrilidae and Nerilidae in the Black Sea. *Russian Journal of Zoology*, 77, 533–539. (*In Russian*)

[4] Zaika VE , Sergeeva NG & Gulin MB (2009) The population structure of polychaeta *Vigtorniella zaikai* (Kisseleva, 1992) in the Black Sea and the characteristics of a community in which it is the dominating species. *Marine Ecological Journal*, 8, 59-66. (*In Russian*)

[5] Kuftarkova EA, Rodionova NYu, Gubanov VI, and Bobko NI (2008) Hydrochemical description of some marine inlets in Sevastopol area. *Proceedings of YugNIRO*, 46, 110-116. (*in Russian*).

[6] Sergeeva NG, Gulin MB & Ivanova EA (2008). Meiobenthos of the coast redox-zone (Kruglaya Bay, Black Sea). *In*: Gilbert A & Yanko-Hombach V (eds.), *Proceedings of "IGCP 521-INQUA 0501 Fourth Plenary Meeting and Field Trip"*, Bucharest, Romania-Varna, Bulgaria, 160-161.

[7] Sergeeva NG & Ivanova EA (2008) Benthic fauna associated with the active methane seeps at the Crimean coastal zone (Black Sea). *In*: Gilbert A & Hallock P (eds.), *Proceedings of the Fifth International Conference "Environmental Micropaleontology, Microbiology", EMMM'2008*, Chennai, India, 275-278.

[8] Zaika VE, Konovalov SK & Sergeeva NG (2011). The events of local and seasonal hypoxia at the bottom of the Sevastopol bays and their influence on macrobenthos. *Marine Ecological Journal,* 10, 15 -25.

[9] Gulin MB, Timofeev VA & Bondarenko LV (2010). Zoobenthos in the microbiotopes of methane seeps of the shelf zone of the Crimean coast. *In: Systems for*

Environmental Monitoring. Ecosi-Gidrofisika, Sevastopol, Ukraine, 225-229 *(In Russian)*

[10] Zaika VE & Gulin MB (2011). The maximal depths of fish inhabitation in the Black Sea and features of their trophic strategy nearby of oxic/anoxic interface. *Marine Ecological Journal*, 10, 39-47. (*In Russian*)

[11] Sergeeva NG, Konovalov SK, Kolesnikova EA & Chekalov VP (2011). Response of the meiobenthos communities to hypoxia in the coastal zone (Tarhankut, Crimea, the Black Sea). *In*: Gilbert A & Yanko-Hombach V (eds.), *Proceedings of "IGCP 521-INQUA 0501 Seventh Plenary Meeting and Field Trip"*, Odessa, Ukraine, 151-154.

[12] Orekhova NA, Sergeeva NG, Gulin MB & Konovalov SK (2010). Events of hypoxia and anoxia in the Crimean coastal waters. *In*: *Proceedings of "IGCP 521-INQUA 0501 Sixth Plenary Meeting and Field Trip",* Rhodes, Greece, 153-154.

[13] Gulin MB (2004) Dependens of location and intensity of methane cold seeps on geophysical factors: the Black Sea near-shore shallow-water seeps, underwater videomaterials. 2004. 1st General Assemly of the European Geosciences Union: Nice (France) 25-30 April 2004. Id Nr: EGU04-A-00976 BG 25.

[14] Özsoy E & Ünlüata Ü (1997) Oceanography of the Black Sea: a review of some recent results. *Earth-Science Reviews*, 42, 231-272.

[15] Orekhova NA & Konovalov SK (2009) Voltammetric studies of the bottom sediments of the Sevastopol Bay. *Marine Hydrophysical Journal,* Ecosi-Gidrofisika, Sevastopol, Ukraine, 52-66

[16] Luther III GW, Brendel PJ, Lewis BL et al. (1998) Simultaneous measurement of O2, Mn, Fe, I−, and S(−II) in marine pore waters with a solid-state voltammetric microelectrode. *Limnology and Oceanography*, 43, 325-333.

[17] Brendel PJ & Luther III GW (1995) Development of a gold amalgam voltammetric microelectrode for the determination of dissolved Fe, Mn, O2, and S(−II) in porewaters of marine and freshwater sediments. *Environmental Science and Technology*, 29, 751-761.

[18] Orekhova NA (2010) Hypoxia and anoxia in the coastal sediments of Crimean. *In*: Oleinik YB (ed.), *Proceedings of the "Theoretical and applied edition "Geography and tourism" Conference,* Kiev, Alterpress, 4, 146-252.

[19] Lichtschlag A, Felden J, Bruechert V et al. (2010) Geochemical processes and chemosynthetic primary production in different thiotrophic mats of the Haakon Mosby Mud Volcano (Barents Sea). *Limnology and Oceanography*, 55, 931-949.

[20] Cline JD (1969) Spectrophotometric determination of hydrogen sulfide in natural waters. *Limnology and Oceanography,* 14, 454-458

[21] Zaika VE, Ivanova EA & Sergeeva NG (2012). Seasonal changes of meiobenthos of the Sevastopol Bays with the analysis of influence of bottom hypoxia. *Marine Ecology Journal*, Special Issue, 2, 29-36 (*in Russian*).

[22] Kolesnikova EA & Sergeeva NG (2011) First record of *Sarsameira parva* (Boeck, 1872) and *Tachidiella minuta* GO Sars, 1909 (Copepoda, Harpacticoida) in the Black Sea. *Marine Ecology Journal*, 10, 72.

[23] de Beer D, Sauter E, Niemann H et al. (2006) *In situ* fluxes and zonation of microbial activity in surface sediments of the Hakon Mosby Mud Volcano. *Limnology and Oceanography*, 51, 1315-1331.

[24] Cristoni C, Colangelo MA & Ceccherelli VU (2004) Spatial scale and meiobenthic copepod recolonisation: testing the effect of disturbance size in a seagrass habitat *Journal of Experimental Marine Biology and Ecology*, 298, 49-70.

[25] Sergeeva NG, Zaika VE & Bondarev IP (2010). The lowest zoobenthos border in the Black Sea Near-Bosporus region. *Marine Ecological Journal,* Ecosi-Gidrofisika, Sevastopol, Ukraine, 10, 65-72

[26] Kolesnikova EA (1983) Harpacticoida in Communities of Soft Bottom Sediments near the Crimean Southern Coast. *Ekologiya Morya*, 15, 20-26. (*In Russian*)

[27] Sergeeva NG, Mazlumyan SA, Çağatay N & Lichtschlag A (2013) Hypoxic meiobenthic communities of the Istanbul Strait's (Bosporus) outlet area of the Black Sea. *Turkish Journal of Fisheries and Aquatic Sciences*, 13, 33-41.

[28] Zaika VE & Sergeeva NG (2012) Deep-water benthic polychaetes (*Vigtorniella zaikai* and *Protodrilus* sp.) in the Black Sea as indicators of the hydrogen sulfide zone Boundary *Vestnik Zoology*, 46, 19-27.

[29] Sergeeva NG, Gooday AJ Mazlumyan SA et al. (2011) Meiobenthos of the Oxic/Anoxic Interface in the Southwestern Region of the Black Sea: Abundance and Taxonomic composition. *In*: Altenbach AV, Bernhard JM & Seckbach J. (eds.), *Anoxia: Paleontological Strategies and Evidence for Eukaryote Survival*, 369-401.

[30] Por FD (1964) A study of the Levantine and Pontic Harpacticoida (Crustacea, Copepoda). *Zoologische Verhandelingen*, 64, 1-128.

[31] Gomez S (2000) A new genus, a new species, and a new record of the family Darcythompsoniidae Lang, 1936 (Copepoda, Harpacticoida) from the Gulf of California, Mexico. *Zoological Journal of the Linnean Society*, 129, 515-536.

In: Copepods: Diversity, Habitat and Behavior
Editor: Laurent Seuront
ISBN: 978-1-63117-846-7

Chapter 4

ANTARCTIC LACUSTRINE COPEPODS: SCATTERED REMNANTS OF AN ANCIENT FAUNA AND RECENT MARINE INVADERS

Kerrie M. Swadling* and John A. E. Gibson
Institute for Marine and Antarctic Studies, University of Tasmania, Hobart, Australia

ABSTRACT

The Antarctic continent is home to myriad lakes, ranging from freshwater to hypersaline environments. These lakes provide habitat for low trophic level organisms and are often dominated by bacteria and protists. However, in some of the lakes copepods and other invertebrates have managed to thrive. At least eight species of copepods have been identified, with the possibility of several more being present. While the number of copepod species is not high, their ability to adapt to the Antarctic environment is reflected in their behavior, life history strategies and physiology. Calanoid copepods have been observed in both marine-derived and freshwater lakes, while cyclopoids are presently identified only from freshwater lakes and harpacticoids from saline lakes. Evidence compiled from both continental and peninsular Antarctica suggests that some of the copepods, notably *Gladioferens antarcticus* and *Boeckella poppei,* which inhabit freshwater lakes, could well have been *in situ* prior to the Last Glacial Maximum and possibly for much longer. This evidence has led to the conclusion that, with respect to Antarctic freshwater zoogeography, "*vicariance rather than dispersal may yet prove to be the most parsimonious and reasonable explanation*". However, it is also possible for the freshwater habitats that dispersal has played a significant role, particularly in the colonization of lakes along the Antarctic Peninsula by copepods from South America during the Holocene. Populations of marine-derived copepods are found living in saline lakes, particularly in oases along the margins of East Antarctica. The populations have been present for many thousands of years, following isolation of the lakes from the ocean *via* isostatic rebound approximately 5000 years BP. The animals were either passively trapped in newly-forming lakes that arose as a result of a decrease in the local sea level,

* E-mail address: Kerrie.Swadling@utas.edu.au.

or they were transported via sea spray and/or by animal vectors. Physiological traits such as broad salinity and/or temperature tolerance have promoted the establishment of copepod populations in Antarctic lakes. The family Centropagidae has been particularly successful in colonizing the freshwater lakes, with three species currently known. Members of this family are often common in coastal, estuarine and freshwaters of other continents and are known for their euryhalinity. Similarly, the most successful saline species, *Paralabidocera antarctica*, is a member of the Acartiidae, a family that is often numerically dominant in estuarine and coastal waters in temperate and tropical regions throughout the world. It is not yet understood how these populations will respond to relatively rapid environmental change.

INTRODUCTION

The Antarctic terrestrial environment appears to be hostile to life, with over 99% of the continent permanently covered by ice, long periods of near or total darkness (and conversely, 24-hour sunshine), frequently strong winds, air temperatures that can plummet to -80 °C in winter, and limited availability of liquid water. While these ice-covered areas may harbor biological activity [1], there are areas of the continent where far greater activity occurs. Ice-free 'oases', ranging in area of exposed rock up to 4000 km^2, are scattered around the continent's margins and provide an environment suitable for terrestrial life [2]. In particular, the availability of liquid water, in the form of myriad lakes, ponds, streams and soaks, is far greater in these areas than on the ice cap. While no vertebrates are associated exclusively with these habitats, a range of invertebrates thrives. Arthropods have been the most successful colonisers: mites, chironomids and springtails inhabit the moss beds [3], while the lakes contain at least 18 species of crustacea, including copepods, cladocerans, ostracods and anostracans [4]. Copepods are the most biodiverse group, with four species currently recorded from saline lakes and a further four, at least, from freshwater lakes. Calanoid, cyclopoid and harpacticoid copepods all have representatives in Antarctic lakes, with calanoids present in both saline and freshwater environments [4].

The origin of Antarctic lacustrine copepods is under debate, especially for those species living in the freshwater lakes. Two competing theories have been proposed: colonization from extra-continental sources after the Last Glacial Maximum (LGM) approximately 18000 years ago [5], and long-term survival in suitable habitat on the continent, possibly pre-dating the onset of complete glaciation of the continent approximately 35 million years ago. The first theory is based on the argument that the expanded LGM ice-sheet must have covered the entire continent, in particular the low-lying oases where most of the lakes occur. Notwithstanding the possibility of survival within the ice biome [1], the coverage of the continent would most likely have precluded copepod survival on the continent. This point of view was apparent in the copepod literature as early as the late 1950s [6], and dispersal *via* vertebrate vectors or wind was accepted as the default explanation for the presence of freshwater lacustrine invertebrates in Antarctica for many decades [5].

The suggestion that long-term survival on the continent may have played a role in the development of the current biota stemmed initially from the description of a series of mites from an endemic Antarctic family and genus, *Maudheimia*, associated with mountain ranges on opposite sides of Antarctica. These species clearly had a pre-LGM, and probably ancient, pre-glaciation origin, and both the distribution on the continent and the presence of more than one species suggested a role for vicariance, though whether speciation occurred prior to the onset of continental glaciation is not certain [7]. The next step was the development of an understanding that many of the ice-free areas that currently occur along the Antarctic coastline were not completely ice-covered at the LGM, and therefore the habitats in which copepods live may have been present at least through the last glacial cycle [8]. The conclusion to be drawn from these observations is that it is possible that the freshwater fauna present in Antarctic lakes may have had at least a pre-LGM origin [4].

The origins of the copepods inhabiting saline lakes are clearer, as these lakes have in most cases post-LGM marine origins [9] and the species present are marine species. Thus the copepods would appear to be recent colonizers of Antarctic lakes. However, their presence does beg the question: can these populations develop into more permanent members of the Antarctic fauna?

In this chapter we will discuss the habitat, diversity and behavior of the copepod species found in Antarctic lakes, with emphasis on their origins and the ecological and physiological traits that have promoted their successful survival. While there is not always detailed information about the lakes and the copepods that inhabit them, there is sufficient knowledge to make some generalizations about this intriguing group of animals. We have restricted our discussion to those copepods that are found in continental lakes or in lakes on nearby offshore islands (e.g., the South Shetland group adjacent to the Antarctic Peninsula). However, in some instances there is relevant information about the biology of one or more species that has been gleaned from populations living in lakes on Subantarctic islands or on the South American and Australian continents.

Antarctic Lakes As Habitat

Antarctic oases are areas that are naturally free of permanent snow or ice. The larger Antarctic oases include the McMurdo Dry Valleys, areas on the Antarctic Peninsula, including Alexander Island, Vestfold Hills, Bunger Hills, Larsemann Hills, areas of the Prince Charles Mountains, the Syowa Coast, and the Schirmacher Oasis (see Figure 1). All of these regions contain numerous lakes and ponds (Figure 3). For example, the Vestfold Hills has more than 350 lakes measuring over 1 ha in area in its 410 km^2. Although these regions are cold, sufficient solar energy is absorbed by the exposed dark rock and soil to melt the little snow that does fall, or else the snow is scoured by katabatic winds. Allochthonous inputs to the lakes are generally limited, but can include carbon- and nutrient-rich runoff from moss-beds and penguin rookeries. Soils are generally poorly-developed and are largely mineral in nature. Autochthonous (i.e., within-lake) processes dominate in Antarctic lakes [10].

The lakes of the oases vary in size from small, shallow ponds up to large lakes greater than 10 km^2 in area. The deepest Antarctic lake, Radok Lake, is over 300 m deep. Salinity ranges from ultra-fresh for lakes whose water mainly comes from melt of the polar ice-cap, to highly saline (> 250 g L^{-1}) for some marine-derived lakes. Common features include low water temperatures, low productivity and simple food webs. Most of the lakes develop an annual ice cover, with open water present for a short period in the summer months. There are, however, many deviations from this pattern. Many lakes in the McMurdo Dry Valleys and the Prince Charles Mountains are effectively permanently ice-covered, but with often tens of metres of unfrozen water beneath the ice. Other lakes remain ice-free even during winter.

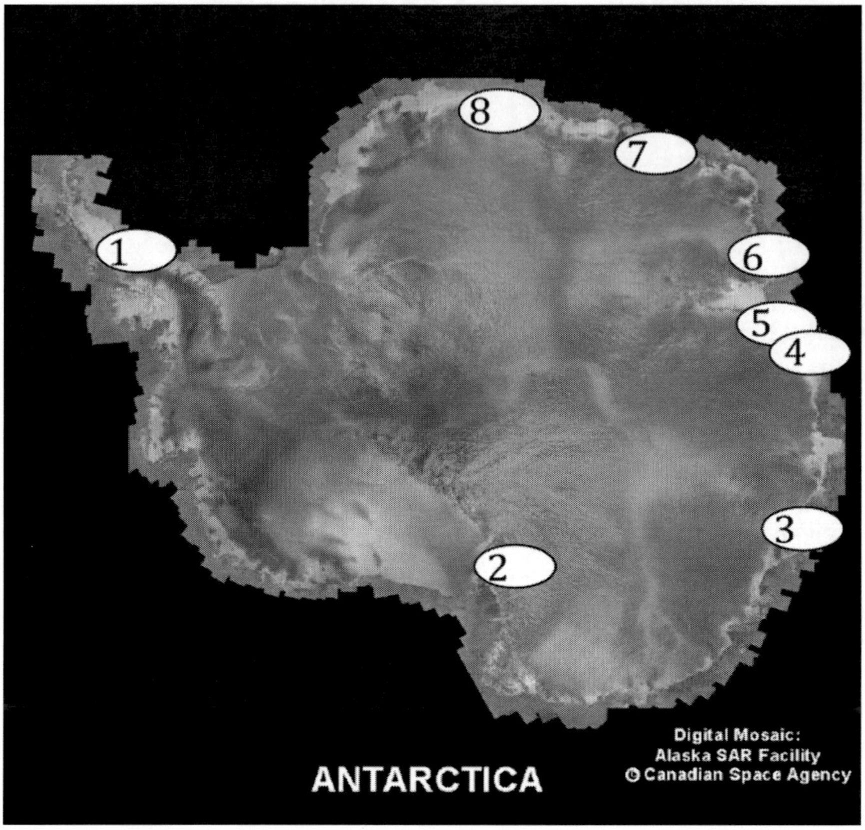

Figure 1. Antarctic Oases mentioned in the text. 1. Antarctic Peninsula; 2. McMurdo Dry Valleys; 3. Bunger Hills; 4. Vestfold Hills; 5. Larsemann Hills; 6. Amery Oasis at the northern end of the Prince Charles Mountains; 7. Syowa Oasis; 8. Schirmacher Oasis.

The extreme cases appear to be Lake Vida, in the McMurdo Dry Valleys, in which a 20 m ice cover overlays a small pocket of highly saline water, and hypersaline Deep Lake in the Vestfold Hills, the water of which has a nominal freezing point of *ca.* -27 °C, but which drops to only *ca.* -18 °C in winter, so there is no ice cover at all throughout the year. Water temperatures in the lakes are not always close to freezing and can reach as high as 25 °C in Lake Vanda in the McMurdo Dry Valleys and 15 °C in Ekho Lake in the Vestfold Hills [11]. Productivity ranges from oligotrophic in large, deep clear-water lakes to eutrophic in shallow ponds and lakes along the Antarctic Peninsula. The unique combination of depth, sediment type, location, height above sea level, water source, bedrock chemistry and water column processes means that each lake has evolved with its own distinctive character. Freshwater, brackish and hypersaline lakes can therefore all occur within close proximity.

Freshwater Lakes

Freshwater lakes, defined as having salinity < 3, are the most common lake type in Antarctica. They form when glacial meltwater runoff is trapped in depressions, which can either be completely rock-based or part rock and part ice (e.g., Lake Hoare in the McMurdo Dry Valleys). Large, deep (> 50 m) freshwater lakes occur across the continent in the McMurdo Dry Valleys, Bunger Hills, Vestfold Hills, and the Prince Charles Mountains. The best studied of these are the lakes of the McMurdo Dry Valleys. The ice cover of these lakes is permanent, and is typically 3-5 m thick, which means that turbulent mixing by wind or by convective overturning due to loss of heat to the atmosphere is not a general feature of these lakes. This has led to the conclusion that they are highly stable environments, which has allowed the organisms that live in them to thrive [12].

Smaller freshwater lakes occur in the Vestfold Hills, Bunger Hills, Larsemann Hills, the Shirmacher Oasis, the Antarctic Peninsula and other, smaller oases, and number in the thousands. These lakes generally are shallow (< 20 m) and have only seasonal ice cover. The lakes typically have relatively low diversity and abundance of primary producers within the water column, but in contrast have a rich benthic community of cyanobacteria, diatoms, other algae and protists that develop into coherent microbial mats. The ice-free period often results in higher productivity through greater light availability, especially when accompanied by a source of nutrients. An example is Lake Boeckella, located at Hope Bay at the northern tip of the Antarctic Peninsula. This productive lake receives meltwater from nearby glaciers and also gets runoff from an Adélie penguin rookery that is present in the catchment [13].

Saline Lakes

Saline lakes are generally defined as having salinity >3 [14]. Many Antarctic saline lakes were formed when pockets of seawater were trapped as a result of isostatic uplift of the land that occurred as the ice cap retreated after the LGM [9]. Most of the lakes have since undergone complex cycles of flushing by freshwater and/or decrease of their water volume by evaporation to reach their current state. In many cases salinity of the marine-derived saline lakes is no longer close to that of seawater, although the major ion ratios are typically similar to seawater. Saline lakes can also form through the slow accumulation of solutes that reach the lake either through atmospheric deposition or weathering processes within the lake's drainage basin. These lakes often have strikingly different major ion ratios.

The biota of the saline lakes depends strongly on the salinity. The most saline lakes, such as Deep Lake in the Vestfold Hills, have a very limited biota, but biodiversity increases with decreasing salinity. As found for the freshwater lakes, the greatest productivity occurs in benthic microbial mats rather than in the water column.

Many of the Antarctic saline lakes, particularly in the McMurdo Dry Valleys, Vestfold Hills and along the Syowa Coast are meromictic (i.e., permanently stratified) [11], and form when fresher water flows onto the lake surface and is prevented from mixing. Stratification is then maintained by a strong density gradient that results in an increase in salinity with depth. Most meromictic lakes are anoxic at depth. Meromixis also leads to strong stratification in the biology of the lake, both in terms of productivity and the species present. An important consequence of stratification is that different species occupy different vertically-separated niches within the water column,

One example of a saline, meromictic lake is Ace Lake, in the Vestfold Hills. It has a maximum depth of 25 m and an area of 18 hectares, and drains a small catchment that is approximately three times the size of the lake. There is very little allochthonous input to the lake as the catchment is free of any permanent bird colonies, and the only plant life present are small and sparse patches of lichens and mosses. The ice cover on the lake persists for at least 11 months each year and so the period of open water is short. During winter, small snow banks develop in the catchment. These banks melt during summer, leading to some meltwater input. The oxycline in the lake is at approximately 11 m and below this the water is anoxic.

The biota of the lake is highly stratified, with primary production by algae taking place in the oxygenated upper waters, sulphide oxidation in a dense, near axenic plate of photosynthetic sulphur bacteria occurring at the top of the anoxic zone [15], and fermentation, sulphate reduction and archaeally-mediated methanogenesis in the deeper water [16]. Benthic algal mats are found in the oxic waters around the edges of the lake.

Copepods in Freshwater Lakes

Boeckella Poppei (Mrázek)

The calanoid copepod *Boeckella poppei* was first recorded in Antarctic lakes during the British Graham Land Expedition 1934-37 as *Pseudoboeckella poppei* [17]. The genus *Pseudoboeckella* has been fused with *Boeckella* [18], and *Boeckella poppei* is the currently accepted name. The species is predominantly associated with Patagonia and the Antarctic Peninsula, and also occurs in lakes on the islands in the Drake Passage (where it has at times been misidentified as *Pseudoboeckella silvestri* [19,20]). On the Antarctic Peninsula it occurs south to Alexander Island (71°S) [21,22]. However, what is most intriguing about this species is that it occurs in an isolated population in at least three lakes of the Amery Oasis in the Prince Charles Mountains, East Antarctica [23,24], many thousands of kilometres from the Antarctic Peninsula. There is also a possible unconfirmed record in another geographically disparate location, Lake Hoare in the McMurdo Dry Valleys, though the single animal present could not be completely identified [25].

The isolated populations in the lakes of the Amery Oasis appear to be a dwarf form of the species, being up to 50% smaller than the South American populations [23]. There are also some morphological differences between the Amery Oasis populations and those from Patagonia, notably in the endopod of the fifth leg of the males [24]. However, the importance of these differences as a taxonomic marker has been questioned, as the species is morphologically plastic in its 'typical' range. The question of the origin of these geographically disparate populations was addressed by Pugh et al. [5], who concluded that the Antarctic Peninsula populations arose through Holocene colonisation, presumably from South America. The Amery Oasis population was considered to be '*anomalous*' and '*likely a recent alien import from the Maritime or sub-Antarctic*' [5].

The question of the true origin of Antarctic populations, and in particular those isolated from the core distribution, can be approached using a number of methods. Molecular genetic studies are underway, but as yet no conclusions can be drawn. A more direct method is to study microfossil remains preserved in sediments of lakes in which the species occur. *Boeckella poppei* leaves a number of identifiable microfossils, including hatched egg casings and spermatophores, and the distribution of these microfossils has been studied in three Antarctic lakes. On Signy Island in Drake Passage the occurrence of empty copepod egg casings in the basal sediments of a lake indicated that the species has been present from immediately after the deglaciation of the island and formation of the lake, which is estimated to have occurred approximately 5500 yr before present (BP). The sediments of Lake Boeckella at the far northern end of the Antarctic Peninsula have revealed a similar situation, with the species present since the formation of this lake, also about 5500 yr ago [13]. The third lake studied was Lake Terrasovoje, one of the lakes with an isolated population in the Amery Oasis. The sediment record in this lake extended back to at least 12000 years BP, and the egg casings and spermatophores were present throughout the sediment [26]. This indicated that the population in the lake was not the result of a recent introduction, but rather the copepod was a long-term inhabitant of the lake. Furthermore, the Amery Oasis population therefore pre-dated the deglaciation of the Antarctic Peninsula.

If *B. poppei* is a Holocene immigrant from South America, its presence in the basal sediments of newly forming lakes implies that dispersal during the early to mid-Holocene was very efficient. However, in general, freshwater calanoid copepods are poor dispersers, as highlighted by the distinct faunas of eastern and western Australia [27], which have been unable to recolonize areas of the continent from which they were separated as long ago as the early Cretaceous (115–110 Ma BP). The disjunct distributional patterns of several South American species [22] also indicate poor dispersal abilities. A possible mechanism for *B. poppei* becoming established in lakes in the early to mid-Holocene is that they had already reached the Antarctic Peninsula region prior to significant deglaciation and had colonised small proglacial lakes or ponds, a habitat which it colonizes today [28]. The larger lakes that formed following deglaciation were then seeded from these short-lived lakes.

Another possibility is that *B. poppei* are relict populations that survived the LGM in continental refugia. *Boeckella poppei* is an adaptable species, occurring in a wide range of physical habitats from small, oligotrophic–mesotrophic ponds to large, deep, ultra-oligotrophic lakes [21,24,29]. It has also been recorded from two epishelf lakes, which are bodies of fresh water dammed by an ice shelf and floating on denser seawater [18,24,30]. This adaptability means that *B. poppei* might have been able to survive in small, proglacial lakes, and perhaps epishelf lakes that were dammed by more extensive late Pleistocene ice shelves, which have since collapsed.

While colonization of the Amery Oasis from South America or perhaps relict populations on the Antarctic Peninsula is possible, it is more likely that *B. poppei* survived the LGM in a local refugium. Recent cosmogenic isotope data has shown that parts of the oasis have been ice-free for up to 2 Ma [31], indicating that *B. poppei* could have survived not only the most recent glaciation but also many previous glaciations in the area. While it still cannot be concluded that *B. poppei*, or its ancestors, have been present on the continent since the onset of glaciation, it is likely that the Amery Oasis populations are ancient. Molecular genetic analysis of individuals from the Amery Oasis, Antarctic Peninsula and South America will help to confirm or refute this conclusion.

Boeckella poppei often represents the highest trophic level in the lakes, though sometimes it co-occurs with the predatory copepod *Parabroteas sarsi*. Early studies of the morphology of its feeding appendages indicated that *B. poppei* was a benthic feeder, with mouthparts adapted to stir up bottom debris into a suspension that could then be filtered [19]. Later work that examined gut contents of *B. poppei* found a correlation between algae in the guts and those found in the lake and suggested that it fed unselectively on phytoplankton [32]. More recent experimental studies applying bottle clearance methods to determine grazing rates [33,34] have shown that *B. poppei* is an omnivorous feeder that can take a range of prey sizes and is able to consume both phytoplankton and periphyton, depending on the lake environment and the prey species present. One well-studied lake, Sombre Lake on the South Orkney Islands, has changed from an oligotrophic state prior to the mid-1980s [35] to its current eutrophic state [36]. The eutrophy has been the result of increases in the local fur seal population, which has led to nutrient enrichment in the lake and its catchment area [34]. Therefore, early suggestions that *B. poppei* was a benthic browser in Sombre Lake [19,20] might not be wrong, but rather reflect a change in diet as the trophic status of the lake has changed. Large numbers of phototrophic flagellates now live in the pelagos and have become an important part of the copepod's diet [34]. This dietary flexibility is reflected in the life cycle plasticity. For example, in Boeckella Lake, where there is a rich source of phytoplankton, the copepods have a multivoltine life cycle, while in oligotrophic Chico Lake (Hope Bay) where food is scarce they exhibit single annual recruitment (univoltine) with a small clutch size [29].

***Gladioferens Antarcticus* Bayly**

Gladioferens antarcticus is a calanoid copepod only known from four epishelf lakes in the Bunger Hills [24], although this distribution may partly reflect the limited sampling of epishelf lakes in this area. Epishelf lakes are an equivalent of estuaries in polar areas, and contain freshwater derived from snow and ice melt on land floating on denser seawater and dammed by ice, but with hydraulic connection to the ocean [30]. If sea level changes an epishelf lake will raise or lower, much as an estuary responds to such changes. They therefore contain a relatively stable environment that will remain even if there is significant sea level or other climate change. Comprehensive sampling of rock-based freshwater lakes in the Bunger Hills indicated that *G. antarcticus* is not present in other lake types of the area [37].

The genus *Gladioferens* has a Gondwanan distribution and, along with its presence in lakes of the Bunger Hills, it is found in lakes and estuaries along the eastern and southern coasts of Australia and around New Zealand; the genus is apparently absent from the sub-Antarctic islands. While *Gladioferens* can be prominent in estuarine waters, species in this genus are much rarer in neritic waters (e.g., *Gladioferens inermis*) and have not been recorded from the open ocean [38].

As highlighted by Bayly et al. [24], the discovery of *Gladioferens* in Antarctica resulted in a strikingly disjoint distribution for the genus. While noting that *G. antarcticus* is endemic, it was postulated that this species, along with other continental copepods, were either trapped from seawater or arrived via anthropogenic introduction [5]. However it has been shown that *G. antarcticus* has significant morphological differences to the other members of the genus, and that it is the most primitive species in its subgenus [38]. Therefore it was argued that *G. antarcticus,* as a basal species, could not have colonized Antarctica from an extra-continental site [24]. The genus *Gladioferens* has never been recorded from the Southern Ocean, in particular amongst the coastal marine zooplankton [39, K. Swadling, personal observation]. Furthermore, *G. antarcticus* seems to require fresh or low salinity waters (0.02–1.0) [24], which makes it unlikely that it has either recently invaded the freshwater lakes from the marine environment or that it has been dispersed across more than 2,000 km of open ocean. Reconstructions of East Gondwana (Figure 2) indicate that the region encompassing the present day Bunger Hills would have lain opposite south-western Western Australia, a region where another species of *Gladioferens*, *G. imparipes*, now occurs. Therefore, the most parsimonious conclusion is that *G. antarcticus* (or its forbears) has inhabited the coastline of Antarctica since the separation of Australia and Antarctica approximately 65 million yr ago. Bayly et al. [24] claim that '*If ever there was a species that bespeaks vicariance it is G. antarcticus*'.

While its ancestors probably inhabited poikilosaline waters, *G. antarcticus* is now adapted to the stable freshwater habitats provided by the epishelf lakes.

Figure 2. Partial reconstruction of East Gondwana, showing the close physical relationship between the only known location of *Gladioferens antarcticus* (Bunger Hills; filled square) and the region now inhabited by *Gladioferens imparipes* (open square). Used by permission from Gibson and Bayly (2007).

Figure 3. Examples of lake types in Antarctica. A. Lake Boeckella, Hope Bay, Antarctic Peninsula (Google Earth image 2/5/2011). B. Ace Lake, in the Vestfold Hills (John Gibson); C. Lake Joyce in the McMurdo Dry Valleys (Dale Anderson); D. Preparing to sample through Lake Terrasovoje, Prince Charles Mountains (Louise Newman).

The continued survival of *G. antarcticus* on the Antarctic continent argues for the existence of epishelf lakes, or similar habitats in which freshwater from summer melt accumulated in lakes, through successive glacial maxima. That part of the Bunger Hills was ice free at the LGM [40] provides evidence, albeit indirect, for the existence of such habitats (at least for the most recent glacial period) in the immediate area of the modern distribution of *G. antarcticus*.

Little is known about the behaviour of *G. antarcticus*. When collected from White Smoke Lake in March 1992 adults were present, and the copepods were found to be most common near the base of the lake at 90 m. The water from the lake is particularly clear, with very low chlorophyll levels (0.37 to 0.54 µg l^{-1}). The lakes have been permanently covered by ice at least from the 1940s, being up to 4 m thick. Some microbial mat was recovered when the lake was re-sampled in 1999/2000, suggesting that this may be the food source for the animals [37]. All four lakes in the Bunger Hills where *G. antarcticus* has been collected are deep (68 – 122 m), cold (<1.2 °C) and fresh (salinity of 0.1 to 1). It has never been described as common in the lakes, indicating that it has either been undersampled or is just innately scarce [37].

Parabroteus Sarsi (Mrázek)

The calanoid copepod *Parabroteus sarsi* was first recorded from an Antarctic lake during the British Graham Land Expedition 1934-37 [17]. This species has also been recorded as *Limnocalanus sarsi*, and both names appear to be accepted. However, the third identification as *Giantella sarsi* [41] is not in use [42]. *Parabroteus sarsi* has been found in lakes of Graham Land and Palmer Land on the Antarctic Peninsula, but has not been recorded in either the Dry Valleys lakes or those of Eastern Antarctica. The copepod is very large for a freshwater species (5 to 7 mm) and is a brilliant orange-red in lakes where fish are absent in southern Argentina, but colourless in lakes when fish are present [43,44]. There appear to be no records of its colour in Antarctic lakes, but it is possible that it is colored as there are no fish present. Furthermore, the lakes are often shallow and when the ice cover has melted the pigments in the exoskeletons of the copepods could help them avoid UV damage. Indeed *P. sarsi* has been shown to be tolerant of UV exposure [45], which is probably assisted by its colouration. The main habitats of *P. sarsi* are shallow permanent or ephemeral ponds [46], though they have been found in larger, deeper lakes on the Antarctic Peninsula [5].

Much of what is known about *P. sarsi* is derived from studies in Argentinian and Chilean lakes, and those of the Subantarctic islands (e.g., South Georgia), while there is almost no information about the Antarctic continental populations other than records of their presence. It appears that in Antarctic lakes the populations are sparse, reaching only 1-2 animals m^{-3} [46]. *Parabroteus sarsi* is carnivorous, at least at the adult stage, and this is highlighted by its well-developed maxilla and maxillipeds [19]. There is some evidence from South American lakes that *P. sarsi* undergoes dietary shifts as it develops through the copepodid stages. It seems that CI are largely herbivorous, then the stages become increasingly omnivorous, gradually adding more animal prey to the diet [43]. Once it reaches adulthood it appears to be primarily predatory. The adults grip their prey with the well-developed setae of the feeding limbs. They bite off chunks and devour the entire animal in the case of copepod prey, but are less able to devour entire cladocerans, possibly due to the presence of tail spines on these animals [47]. *Parabroteus sarsi* is likely to prey on *Boeckella poppei* when they occur in the same lakes; the predatory pressure that *P. sarsi* exerts is small, with less than 1% of the copepods removed per day in lakes on South Georgia [48].

The origins of Antarctic populations of *P. sarsi* have not been investigated in any detail, though its similar distribution (at least in Patagonia and on the Antarctic Peninsula) to *Boeckella poppei* may suggest a similar history.

Diacyclops Mirnyi (Borutsky and Vinogradov) and Related Species

This cyclopoid copepod was first identified *as Acanthocyclops mirnyi* by Russian scientists [6] in samples collected from Algae Lake in the Bunger Hills. It has also been recorded as *Diacyclops mirnyi* and it seems that the *Diacyclops* designation is correct [49]. The copepod has appeared sporadically in the literature under both names [5,38]. It has also been recorded from lakes in the Vestfold and Larsemann Hills, well to the west of the Bunger Hills [50]. Up until recently there had been no systematic study of this species group. The recent discovery of a further population of Antarctic *Diacyclops* in Lake Joyce in the McMurdo Dry Valleys (Figure 4), however, has instigated a review of the group that has shown that there may be at least four species present [50]. Unfortunately the type material for Borutsky and Vinogradov's [6] description of *Acanthocyclops mirnyi* has been lost, so at present there isstill enigma concerning this group. However, it appears that animals from the Vestfold Hills and Lake Joyce represent new species, while there may be two species within the Bunger Hills.

Representatives of the genus *Diacyclops* are found in South America and Australia, in freshwater lakes. All of the four forms found in Antarctica are in the *Diacyclops michelseni* group, which otherwise occurs in the islands of Drake Passage, other Subantarctic Islands and Tasmania and south-eastern Australia. The current thinking about this group is that it possibly evolved in Antarctica, and migrated to South America and Australia during glacial periods when the icesheets were larger [50]. As yet there has been no critical examination of affinities between these species, and therefore possible routes of ancient or post-glacial colonisation. However, the recognition that there are multiple species in Antarctica supports an ancient and possibly vicariant origin for these species.

Little is known of the life history of *Diacyclops* species in Antarctica. All stages have been collected from the water columns of the Antarctic lakes, where it is restricted to lakes of low salinity (< 4). Nauplii are usually found in higher abundance in net samples, leading to the suggestion that only juvenile animals inhabit the water column, and that adults might be strongly associated with benthic mats as adults. Korotkevic [52] reported capturing only juvenile stages in Algae Lake, Bunger Hills, in January 1957, though no further details were given. Roberts et al. [53] recorded only juveniles in net-hauls from Lake Joyce, while adults had to be collected by divers from the surface of the microbial mat. An ovigerous adult female of one of the putative Bunger Hills species, captured in March, was pictured in Barker [54], suggesting that the species wintered at an early stage (possibly even as diapause eggs) before developing to adults in summer. In the Vestfold Hills, ovigerous females were present in late February (D. Grace, unpublished notes, 1973).

At least one of the species in the Bunger Hills is bright red-orange, while the Vestfold Hills and Lake Joyce species are colourless. Two possible mechanisms might explain this observation: protection against UV and protection against predation. However, with the lake ice covers lasting for most of the year and often a snow cover on this, the importance of detrimental exposure to UV is uncertain.

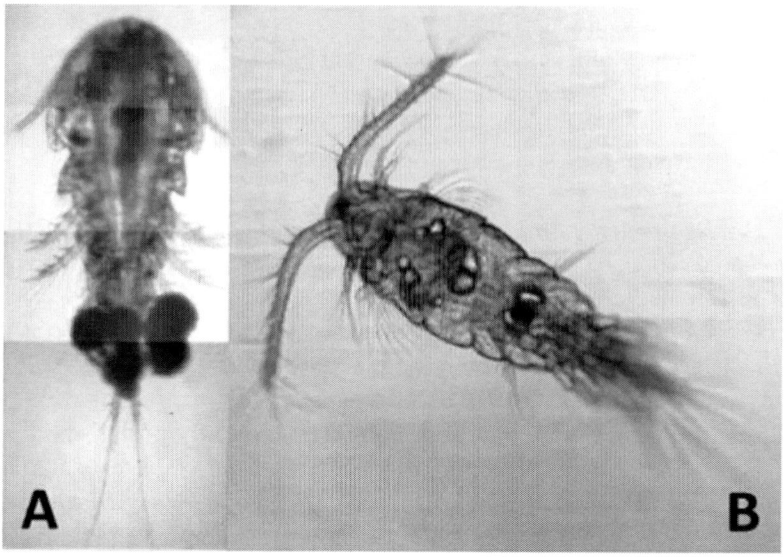

Figure 4. A. Female freshwater cyclopoid copepod (*Diacyclops* sp.) from Lake Joyce (Dale Anderson). B. Copepodite of the calanoid copepod *Paralabidocera antarctica* that is found in marine-derived Ace Lake (Sarah Payne).

Alternatively, the colourless specimens in the Vestfold Hills are possibly a response to predation, however, as this species represents the highest trophic level this seems unlikely.

Colonisation of Lakes by Antarctic Freshwater Copepods

While the number of copepod species that are currently known to inhabit freshwater lakes is not large, it does highlight that the Antarctic terrestrial environment is not completely hostile to life and that some invertebrate species exhibit characteristics that have allowed them to settle and thrive there. One striking feature of the Antarctic freshwater calanoid copepods is that they are all members of the family Centropagidae. It is very likely that the three species have had different colonisation histories in Antarctica: *Parabroteus sarsi* is possibly a Holocene immigrant that has become established on the Antarctic Peninsula following dispersal from South America; *Boeckella poppei* has possibly arrived via dispersal from South America and has also possibly survived as relict populations in glacial refuges in East Antarctica; and *Gladioferens antarcticus* is an ancient Gondwanan species that is endemic to the continent. While the methods of their colonisation might differ the '*remarkable degree of evolutionary euryhalinity*' that Centropagids display [24], will have assisted the copepods in colonising areas that experience variable conditions. For instance, a population of *B. poppei* survives in a shallow (1.38 m) lake that freezes to the bottom for four months each year and is also periodically breached by violent storms and flooded with marine water [55]. When salinity in the lake increased from 0 to 20 following a marine incursion there was high mortality in *B. poppei* but the population was not wiped out and managed to recover in subsequent months. Further, the three lakes in the Amery Oasis that contain *B. poppei* differ in their salinity by a factor of over 100 [24]. *Gladioferens antarcticus* appears to be much more restricted in its salinity tolerance, based on its current known distribution in four freshwater lakes [24], which might explain why its distribution does seem limited.

Many Centropagids produce resting eggs [56], which are a method of avoiding unfavourable environmental conditions. Resting eggs in the Antarctic environment would be an especially useful trait for those species inhabiting shallow ponds that might periodically freeze to the bottom, or those that are impacted by salt spray, e.g., along the Antarctic Peninsula or on the South Shetland Islands. *Boeckella* produces resting eggs in lakes on the South Shetland Islands and they have been shown to be viable for up to a mean age of 46 years and a maximum of 200 years [57].

The second group of copepods that has been successful in colonising Antarctic freshwater lakes are the cyclopoids in the genus *Diacyclops*. This is the richest genus within the family Cyclopidae, with more than 100 known species [58]. *Diacyclops* spp. can colonise all kinds of freshwater habitats, including temporary pools, lakes, springs, marshes, streams, and alluvial and karstic groundwaters. This genus is capable of producing resting stages and is known to be tolerant of harsh environmental conditions [58]. Interestingly, conspecifics do not usually co-occur. There is a lot more to be learnt about the *Diacyclops* species living in Antarctica. To date nearly all observations have identified the specimens as *Diacyclops* (or *Acanthocyclops*) *mirnyi*. Yet it is clear from detailed taxonomic descriptions that are currently being undertaken [50] that at least 3 more species are present in Antarctic lakes. Once the taxonomy has been improved it will be worthwhile undertaking more ecological studies of this group.

Copepods in Saline Lakes

Paralabidocera Antarctica (I. C. Thompson)

During the summer of 1974-75 a population of copepods was discovered in saline Ace Lake in the Vestfold Hills. The species present was identified to be a lacustrine form of the marine calanoid copepod *Paralabidocera antarctica* (Figure 4) [59]. *Paralabidocera antarctica* is a member of the Family Acartiidae, and was first described from marine plankton collections near the South Shetland Islands [60]. Bayly [58] compared *P. antarctica* to genera within the Acartiidae and concluded that it is '*somewhat more primitive and less reduced*' (in segmentation). The lacustrine population differs from the marine population by being considerably smaller (about 60 % of the total length of the oceanic form), and by several morphological differences between the fifth legs of the males.

Following its discovery in Ace Lake, populations of *P. antarctica* were also found to be thriving in two other isolated saline lakes in the Vestfold Hills – Pendant Lake and Lake Abraxas [61] – and Burton Lake, a saline lagoon that retains intermittent connection to the sea [62]. There are no other records of *P. antarctica* from lakes elsewhere in Antarctica, though there is intriguing evidence that this may have occurred previously, and this specific issue is discussed further below.

The lacustrine populations of *P. antarctica* have been present for at least the last 6000 years, when they became isolated from the sea following isostatic uplift of the Vestfold Hills as a result of retreat of the polar ice cap [9]. The history of Ace Lake, in particular, has been followed through study of microfossils observed in the sediments [63]. Burton Lake is currently going through the same process. A drop in water level of approximately 1 m due to continued isostatic rebound will result in the isolation of the basin as a lake, presumably with its copepod population *in situ*.

A year-long investigation of the life history strategies of the Ace Lake and nearby marine populations of *P. antarctica* revealed some similarities and several striking differences. Similarities included long overwintering phases (0.5 months) by late-stage nauplii, rapid development through the copepodid stages, and a short adult life span of 2–3 weeks. Adults appeared in late spring or early summer and spawned and died soon after [64]. The most notable difference between the populations was that *P. antarctica* living in the lake were almost never found within the ice cover. In Antarctic coastal marine waters this species is strongly associated with the sea ice, where it spends much of the year living within the extensive brine channel system [65]. Gut contents analyses have shown that the marine copepods are feeding on algae that either grows on the walls of the brine channels, or which forms dense mats at the ice-water interface [66]. The animals are also preyed on by the ice fish *Pagothenia borchgrevinski* that live in inverted depressions and crevices at the underside of the ice [67]. Therefore, the association of marine *P. antarctica* with the sea ice is a strategy that affords the copepods both protection from predation and access to a food source during winter when phytoplankton in the water column is absent. In contrast, the ice cover of the lakes is much harder, with few, if any, brine channels forming during its growth. Chlorophyll *a* concentrations in the lake ice do not exceed 1.0 mg m^{-2} (c.f. 300 mg m^{-2} in coastal sea ice; [64]), indicating that the algal biomass is low. At the same time, there are measurable quantities of protists in the oxic waters of the lake throughout the year [68], providing a dietary source for the copepods. As *P. antarctica* represents the highest trophic level in Ace Lake they are also released from the predation pressures that constrain the marine population.

It is likely that this combination of factors, including ice hardness, a lack of predation threat, and a consistent food supply has freed the lacustrine population from the constraints imposed by living within the ice cover. This freedom has been manifested as diminishing synchronicity in the developmental cycle, where nearly all stages can be found in the lake at any one time. The presence of adults is still restricted to spring, though they appear in the lake up to one month earlier and survive for longer than in the marine habitat [64]. To cope with the possibility of competition for food that has arisen from many stages co-existing in the lake there is some indication that *P. antarctica* has adopted a strategy of vertical separation of life stages [68]. Subadult and adult males were found most commonly in a narrow layer just above the oxic-anoxic interface and coincided with a peak in the prasinophyte *Pyramimonas gelidcola*. Younger stages and females were found in a band just beneath the lake ice and corresponded with a high abundance of microflagellates [69]. A similar vertical separation of life stages was observed for Lake Abraxas, where males were again found just above the oxycline [61].

The grazing pressure of *P. antarctica* in Ace Lake was measured using a radioisotopic tracer and it was calculated that between 7 and 17% of primary production was grazed by the population, indicating that it could exert some top-down control on the protists [70]. Ingestion rates of all stages were higher when the ice cover was present than during the one month when there was open water in the lake [70]. As pointed out above, lacustrine *P. antarctica* reached maturity at a smaller size than the marine population. Two features of the lake might account for this difference. Firstly, the warmer water temperatures of meromictic Ace Lake have resulted in faster instantaneous growth rates of the copepods: 0.10 – 0.26 d^{-1} compared to the marine population of 0.04 – 0.14 d^{-1} [64]. The marine population would rarely experience water temperatures above 0 °C [71], while the Ace Lake population would be exposed to environmental temperatures reaching up to 10 °C [72]. Secondly, food quantity and quality probably play a role. The available prey in Ace Lake is comprised mostly of small flagellates, while larger species such as diatoms and dinoflagellates are rare. Ice diatoms are clearly the preferred food of the marine population [73], and at times these prey reach very high biomass (of > 300 mg m^{-2} in marine ice; [64]).

Harpacticoid Copepods

A small number of species of harpacticoid copepods has been observed in Antarctic lakes, and for most of them there is very little information available concerning any aspect of their ecology. A copepod that was identified as *Idomene scotti* Lang was found inhabiting the fringing algal mats in Ace Lake, Vestfold Hills [74], and this observation was later confirmed (J. Gibson, unpublished data, 1994). This species was most abundant in microbial mat samples, but nauplii also turned up occasionally in sediment trap samples. Nearby Lake Abraxas contains two harpacticoid copepod species: *Amphiascoides* sp. and an unidentified member of the family Laophontidae [61,75,]. The post-naupliar population of *Amphiascoides* sp. was found distributed throughout the oxygenated waters of Lake Abraxas, though its main distribution was centred close to the oxycline at a low oxygen concentration of only 2.2 mg l^{-1} [61]. Lake Abraxas is highly transparent, so it is possible that sufficient primary production can occur at around 17 m and provide enough sustenance for *Amphiascoides* sp. It is presumed that this species is herbivorous [61], but there are no data to actually support this idea. Although *Amphiascoides* sp. in Lake Abraxas is one of the few truly planktonic harpacticoids; its congeners are benthic and able to tolerate very low oxygen levels. Thus it appears that the planktonic population of *Amphiascoides* sp. has retained this adaptation to low oxygen and carried it into its planktonic life style in Lake Abraxas. In contrast, the unidentified Laophontid in the lake maintains a benthic habit typical of harpacticoid copepods, though nothing else is known about this population.

Other records of harpacticoid copepods occurring in Antarctic lakes include two species that were recovered from the water-sediment interface in Lake Nurume Ike, a saline, meromictic lake on the Sôya Coast close to Syowa Station [76]. The more abundant species was *Harpacticus furcatus* Lang, and adult female specimens with egg sacs attached were collected. The second species was smaller in size and was also morphologically similar to the genus *Harpacticus*. *Harpacticus furcatus* has also been observed in rock pools along the shores of the Antarctic Peninsula [77]. It is a marine species that is sometimes found living in association with the sea ice cover [78], a habitat which can function like an upside-down benthic environment as it has surfaces that the copepods can crawl across. Records of this species in small pools on the Antarctic Peninsula are probably marine interlopers [5], while its presence in Lake Nurume Ike suggests long-term survival (and perhaps speciation) in an isolated hypersaline, habitat.

Mechanisms for Colonisation of Saline Lakes by Copepods

Populations of marine-derived copepods that live in the saline lakes of the Vestfold Hills and the Syowa Oasis have survived for many thousands of years [61,64,76] following isolation of the lakes from the ocean via isostatic rebound approximately 5000 years BP [9].

To colonise these lakes marine species require a physical mode of transport of individuals to supralittoral pools through spray and/or by animal vectors. Alternatively, animals could be passively trapped in newly-forming lakes that arise as a result of a decrease in local sea level. The presence of *Harpacticus furcatus* in Antarctic Peninsula ponds is thought to occur via the first mechanism, and probably does not lead to permanent lacustrine populations [5]. The second mechanism, passive trapping, is probably responsible for the populations of *Paralabidocera antarctica* and harpacticoids that are found in lakes in Eastern Antarctica. However, to date, there has been little indication that marine species that are trapped in Antarctic lakes could survive glaciations and become part of a longer-term fauna. It was long accepted that the formation of the saline lakes in the Vestfold Hills arose as a result of a drop in sea level after a local mid-Holocene marine highstand [9]. Evidence certainly points towards modern Ace and Pendant lakes being isolated 2000–6000 yr BP [9,63,79].

The formation of Abraxas Lake, which is within 5 km from Ace and Pendant lakes, is more enigmatic, as the sill between it and the ocean was above the reach of the mid-Holocene marine highstand [9]. Pre-LGM sediment has been recovered from this lake (^{14}C dated to more than 25,000 cal yr BP), and the characteristics of the sediment indicate that the lake has existed continuously since that time [80]. It therefore appears probable that the lake fauna was isolated prior to the LGM. As highlighted previously, the copepods present in these lakes (*Paralabidocera antarctica* and the harpacticoid copepods) have undergone some behavioural changes since being trapped, indicating that they are able to adapt to the lacustrine environment.

Saline lakes are inherently variable environments, where changes in water balance can either lead to hypersaline or hyposaline conditions that are outside of the salinity tolerance of stenohaline marine species. Euryhaline organisms, those species which can tolerate a broad range of salinities, are more likely to survive. *Paralabidocera antarctica* is in the family Acartiidae, a group of copepods that are often dominant in estuarine and coastal environments and which are known to be both euryhaline and eurythermal. Likewise, many harpacticoids inhabit nearshore environments where both temperature and salinity fluctuates on daily, seasonal and annual bases.

Analysis of sediment cores has revealed that there was a general decline in biodiversity in Ace Lake after it was cut off from marine input that was probably due, in part, to the fluctuating salinity [63]. It might be expected that other saline lakes in the Vestfold Hills of similar salinities to Ace, Pendant and Abraxas lakes could contain copepod populations. Most of these lakes, however, appear to have gone through hypersaline phases [11], and if copepods were present they have probably died out.

Intriguing evidence for longer term survival of marine-derived species in Antarctic lakes was observed in the study of ancient copepod DNA extracted from the sediment of freshwater Lake Terrasovoje in the Amery Oasis [81]. An unidentified calanoid copepod with DNA that grouped very closely with that of *P. antarctica* was present from the formation of the lake until approximately 3000 yr BP. Furthermore, faecal pellets of very similar morphology to those of modern *P. antarctica* from Ace Lake [63] were observed through most of the core. While the unknown species was not necessarily *P. antarctica*, it was clearly an Acartiid that had adapted to a freshwater environment, contrasting with all other members of the family, which are estuarine/marine. This enigmatic species is neither present in modern Lake Terrasovoje nor in any other lakes of the area that have been sampled [24]. The probable explanation for the occurrence of an Acartiid in Lake Terrasovoje is that *P. antarctica*, or a related marine species, colonized a saline lake in the northern Prince Charles Mountains prior to the most recent glacial maximum and possibly much earlier [31]. This euryhaline acartiid adapted to existence at lower salinities, and was eventually able to become a part of the freshwater fauna of the region. It survived glaciations in local refugia (as did *B. poppei*), and was able to colonize Lake Terrasovoje soon after its late Pleistocene formation [82]. An as yet unknown environmental stress(es) resulted in its extinction (but not that of *B. poppei*) in Lake Terrasovoje.

It is clear that this originally marine species successfully colonized at least one continental lake and was able to survive the LGM in a local refugium. Along with the apparent pre-LGM origin for copepods in Lake Abraxas, this suggests that the other Crustacea found in saline lakes should not be viewed a priori as '*marine interlopers*' *sensu* [5] but rather the possible precursors of similar long-term populations.

CONCLUSION

It has become apparent that Antarctic lakes may be quite specialized habitats for a small number of copepod species. The copepods themselves, while exhibiting suitable physiological traits for colonization, appear to be relatively poor dispersers. It is not uncommon to find lakes of seemingly suitable habitat devoid of copepods, while neighboring lakes contain them. It is also apparent that, in East Antarctica at least, the lakes have been environmentally stable for long periods and the copepods have a long association with them. Further palaeolimnological and DNA-based phylogenetic studies are needed to address questions of longevity and colonization pathways. While the existence of copepod remnants in sediments has shown that some species were present prior to the LGM, their known distributions are so disjoint that it is not yet possible to determine their routes of dispersal and settlement.

We still have incomplete knowledge of how many species of copepods inhabit the lakes. On the one hand, as outlined previously, sediments from Lake Terresovoje in the Amery Oasis show clear evidence of a reduction in the number of copepod species present in the last 3000 years [81]. Examination of ancient DNA from other sediments may well reveal similar declines. On the other hand, detailed morphological studies, as are currently being undertaken for *Diacyclops* [50], could result in an expanded number of species than the current eight that are recognized. It has to be remembered that only a small number of the lakes has been sampled comprehensively.

It is unclear how environmental change, such as climate change, could impact the copepod populations. Of course, to understand the impacts of future change it is helpful to comprehend how climate change proceeded in the past. Lake sediment cores provide an invaluable tool for addressing this issue, e.g., [63,83]. Environmental stressors on the lakes include rising UV-B, increases in airborne contaminants, changes associated with increased tourism and rising air temperatures [84]. The low species diversity of the lakes, along with the absence of many functional groups, means that the loss or addition of even one species through environmental change could have a major impact on the productivity and food web integrity of a lake [84]. Further, the highly transparent nature of many of the lakes means that they do not obtain significant UV-B screening *via* dissolved organic carbon. This makes them vulnerable to rising UV-B, particularly along the Antarctic Peninsula where air temperatures have increased by up to 3.5 °C, and so lake ice covers are likely to be thinner and persist for shorter times each year. Changes in the trophic status are also apparent, again along the Antarctic Peninsula where increases in fur seal colonies have led to eutrophication of some lakes. To ensure that these unique environments suffer as little impact as possible, it is important that we gain more information about the past and present distributions of species, their genetic diversity, their primary means of dispersal and their ecology, including feeding preferences, reproductive patterns and environmental tolerances. This information will assist in the management of Antarctic lakes and their inhabitants into the future.

ACKNOWLEDGMENTS

This work has been supported by Australian Antarctic Science Grants (projects 691 and 2387) and the Australian Research Council (DP0342815). We thank Dr. Tomislav Karanovic for discussions about *Diacyclops*. We also acknowledge all the Antarctic expeditioners over the years who have helped us sample Antarctic lacustrine crustaceans.

REFERENCES

[1] Anesio AM & Laybourn-Parry J (2012) Glaciers and ice sheets as a biome. *Trends in Ecology and Evolution*, 27, 219-225.

[2] Terauds A, Chown SL, Morgan F et al. (2012) Conservation biogeography of the Antarctic. *Diversity and Distributions*, 18, 726-741.

[3] Pugh PJA, Convey P (2008) Surviving out in the cold: Antarctic endemic invertebrates and their refugia. *Journal of Biogeography*, 35, 2176-2186.

[4] Gibson JAE & Bayly IAE (2007) New insights into the origins of crustaceans of Antarctic lakes. *Antarctic Science*, 19, 157-164.

[5] Pugh PJA, Dartnell HJG & McInnes SJ (2002) The non-marine crustacea of Antarctica and the islands of the Southern Ocean: biodiversity and biogeography. *Journal of Natural History*, 36, 1047-1103.

[6] Borutzky EV & Vinogradov ME (1957) Occurrence of cyclopidae (*Acanthocyclops mirnyi* sp. n.) on the Antarctic continent. *Zoologicheskiy Zhurnal*, 36, 199-202 [In Russian with English summary].

[7] Marshall DJ & Coetzee L (2000) Historical biogeography and ecology of a Continental Antarctic mite genus, *Maudheimia* (Acari, Oribatida): evidence for a Gondwanan origin and Pliocene–Pleistocene speciation. *Zoological Journal of the Linnean Society*, 129, 111-128.

[8] Hodgson DA, Verleyen E, Sabbe K et al. (2005) Late Quaternary climate-driven environmental change in the Larsemann Hills, East Antarctica, multi-proxy evidence from a lake sediment core. *Quaternary Research*, 64, 83-99.

[9] Zwartz D, Bird M, Stone J & Lambeck K (1998) Holocene sea-level change and ice-sheet history in the Vestfold Hills, East Antarctica. *Earth and Planetary Science Letters*, 155, 131-145.

[10] Vincent WF, Hobbie JE & Laybourn-Parry J (2008) Introduction to the limnology of high-latitude lake and river ecosystems. *In*: Vincent WF & Laybourn-Parry J (eds.), *Polar Lakes and Rivers: Limnology of Arctic and Antarctic Aquatic Ecosystems.* Oxford University Press, Oxford, 1-23.

[11] Gibson JAE (1999) The meromictic lakes and stratified marine basins of the Vestfold Hills, East Antarctica. *Antarctic Science*, 11, 175-192.

[12] Spigel RH & Priscu JC (1998) Physical limnology of the McMurdo Dry Valley lakes. *In*: Priscu JC (ed.), *Ecosystem Dynamics in a Polar Desert: The McMurdo Dry Valleys. Antarctic Research Series*, 72, 153-187.

[13] Gibson JAE & Zale R (2006) Holocene development of the fauna of Lake Boeckella, northern Antarctic Peninsula. *The Holocene*, 16, 625-634.

[14] Burton HR (1981) Marine lakes of Antarctica. Antarctic Symposium, ANZAAS, University of Tasmania, 51-57.

[15] Ng C, DeMaere MZ, Williams TJ et al. (2010) Metaproteogenomic analysis of a dominant green sulfur bacterium from Ace Lake, Antarctica. *The ISME Journal*, doi:10.1038/ismej.2010.28.

[16] Lauro FM, DeMaere MZ, Yau S et al. (2010) An integrative study of a meromictic lake ecosystem in Antarctica. *The ISME Journal*, doi:10.1038/ismej.2010.185.

[17] Harding JP (1941) Lower Crustacea. *Scientific Reports of the British Graham Land Expedition, 1934 – 1937*, 1, 319-322.

[18] Bayly IAE (1992) Fusion of the genera *Boeckella* and *Pseudoboeckella* (Copepoda) and revision of their species from South America and sub-Antarctic islands. *Revista Chilena de Historia Natural*, 65, 17-63.

[19] Heywood RB (1970) The mouth parts and feeding habits of *Parabroteus sarsi* (Daday) and *Pseudoboeckella silvestri* (Daday) (Copepoda, Calanoida). *In*: Holdgate MS (ed.), *Antarctic Ecology*. Academic Press, London, 639-650.

[20] Heywood RB (1970) Ecology of the fresh-water lakes of Signy Island, South Orkney Islands: III. Biology of the copepod *Pseudoboeckella silvestri* Daday (Calanoida Centropagidae). *British Antarctic Survey Bulletin*, 23, 1-17.

[21] Heywood RB (1977) A limnological survey of the Ablation Point area, Alexander Island, Antarctica. *Philosophical Transactions of the Royal Society of London B: Biological Sciences*, 279, 39-54.

[22] Menu-Marque S, Morrone JJ & De Mitrovitch CL (2000) Distributional patterns of the South American species of *Boeckella* (Copepoda: Centropagidae): a track analysis. *Journal of Crustacean Biology*, 20, 262-272.

[23] Bayly IAE & Burton HR (1993) Beaver Lake, Antarctica, and its population of *Boeckella poppei* (Mrázek) (Copepoda: Calanoida). *Verhandlungen, Internationale Vereinigungen für Limnologie*, 25, 975-978.

[24] Bayly IAE, Gibson JAE, Wagner B & Swadling KM (2003) Taxonomy, ecology and zoogeography of two Antarctic freshwater calanoid copepod species: *Boeckella poppei* and *Gladioferens antarcticus*. *Antarctic Science*, 15, 439-448.

[25] Hansson L-A, Hylander S, Dartnall HJG et al. (2012) High zooplankton diversity in the extreme environments of the McMurdo Dry Valley lakes, Antarctica. *Antarctic Science*, 24, 131-138.

[26] Cromer L (2008) *Palaeobiogeography and Palaeoecology of the Antarctic Freshwater Fauna*. PhD thesis, University of Tasmania.

[27] Maly EJ & Bayly IAE (1991) Factors influencing biogeographic patterns of Australian centropagid copepods. *Journal of Biogeography*, 18, 455-461.

[28] Jones VJ, Hodgson DA & Chepstow-Lusty A (2000) Palaeolimnological evidence for marked Holocene environmental changes on Signy Island, Antarctica. *The Holocene*, 10, 43-60.

[29] Izaguirre I, Allende L & Marinone MC (2003) Comparative study of the planktonic communities of three lakes of contrasting trophic status at Hope Bay (Antarctic Peninsula). *Journal of Plankton Research*, 25, 1079-1097.

[30] Gibson JAE & Andersen DT (2002) Physical structure of epishelf lakes of the southern Bunger Hills, East Antarctica. *Antarctic Science*, 14, 253-262.

[31] Fink D, McKelvey B, Hambrey MJ et al. (2006) Pleistocene deglaciation chronology of the Amery Oasis and Radok Lake, northern Prince Charles Mountains, Antarctica. *Earth and Planetary Science Letters*, 243, 229-243.

[32] Weller DLM (1977) Observations on the diet and development of *Pseudoboeckella poppei* (Calanoida, Centropagidae) from an Antarctic lake. *British Antarctic Survey Bulletin*, 45, 77-92.

[33] Almada P, Allende L, Tell G & Izaguirre I (2004) Experimental evidence of the grazing impact of *Boeckella poppei* on phytoplankton in a maritime Antarctic lake. *Polar Biology*, 28, 39-46.

[34] Butler H, Atkinson A & Gordon M (2005) Omnivory and predation impact of the calanoid copepod *Boeckella poppei* in a maritime Antarctic lake. *Polar Biology*, 28, 815-821.

[35] Hawes I (1985) Factors controlling phytoplankton populations in maritime Antarctic lakes. *In*: Siegfried WR, Condy PR & Laws RM (eds.), *Antarctic Nutrient Cycles and Food Webs*. Springer, New York, 245-252.

[36] Butler HG (1999) Temporal plankton dynamics in a maritime Antarctic lake. *Archiv für Hydrobiologie*, 146, 311-339.

[37] Gibson JAE (2000) *The Environment of the Bunger Hills*. Australian Antarctic Division, unpublished, 1-98.

[38] Bayly IAE (1994) *Gladioferens* Henry (Copepoda: Calanoida) discovered in Antarctica: *G. antarcticus* sp. nov. described from a lake in the Bunger Hills. *Polar Biology*, 14, 253-259.

[39] Razouls S, Razouls C & De Bovee F (2000) Biodiversity and biogeography of Antarctic copepods. *Antarctic Science*, 12, 343-362.

[40] Gore DB, Rhodes EJ, Augustinus PC et al. (2001) Bunger Hills, East Antarctica: ice free at the Last Glacial Maximum. *Geology*, 29, 1103-1106.

[41] Ekman S (1905) Cladoceran und Copepoden aus Antarktischen and subantarktischen Binnengewassern, gesammelt von der schwedichen antarktischen Expedition 1901–1903. *Wissenschaftliche Ergebnisse der schwedischen Sudpolarexpedition*, 5, 1-40, plts 1–3.

[42] Boxshall GA & Halsey SH (2004) *An Introduction to Copepod Diversity*. The Ray Society, London.

[43] Vega MP (1999) Life-stage differences in the diet of *Parabroteus sarsi* (Daday) (Copepoda, Calanoida): & field study. *Limnologica*, 29, 186-190.

[44] Reissig M, Modenutti B, Balseiro E & Queimaliños C (2004) The role of the predaceous copepod *Parabroteus sarsi* in the pelagic food web of a large deep Andean lake. *Hydrobiologia*, 524, 67-77.

[45] De los Rios P (2007) Short-term effects of exposition to artificial radiation on *Parabroteus sarsi* (Copepoda, Calanoida). *Biologia, Bratislava*, 62, 210-213.

[46] De los Rios P & Rivera R (2008) On the geographic distribution of *Parabroteas sarsi* (Mrázek, 1901) (Copepoda, Calanoida). *Anales Instituto Patagonia (Chile)*, 36, 75-78.

[47] Vega MAP (1995) Morphology and defensive structures in the predator-prey interaction: an experimental study of *Parabroteas sarsi* (Copepoda, Calanoida) with different cladoceran prey. *Hydrobiologia*, 299, 139-145.

[48] Christoffersen KS, Jeppesen E, Moorhead DL & Tranvik LJ (2008) Food-web relationships and community structures in high-latitude lakes. *In*: Vincent WF & Laybourn-Parry J (eds.), *Polar Lakes and Rivers. Limnology of Arctic and Antarctic Aquatic Ecosystems*. Oxford University Press, Oxford, 269-289.

[49] Morton DW (1977) *A Revision of the Australian Cyclopidae of the Subfamilies Cyclopinae and Eucyclopinae*. MSc thesis, Monash University.

[50] Karanovic, T, Gibson J, Hawes I et al. (2014) Three new species of *Diacyclops* (Copepoda: Cyclopoida) from continental Antarctica. *Antarctic Science*, in press.

[51] Borutzky EV (1962) New data on *Acanthocyclops mirnyi* Borutzky et M. Vinogradov from Antarctica. *Zoologicheskii Zhurnal*, 41, 1106-1107 [In Russian with English Abstract].

[52] Korotkevich VS (1964) Concerning the population of water bodies in the oases of East Antarctica. *Soviet Antarctic Expedition, Information Bulletin*, 1, 154-161.

[53] Roberts EC, Priscu JC, Wolf C et al. (2004) The distribution of microplankton in the McMurdo Dry Valley Lakes, Antarctica: response to ecosystem legacy or present-day climate controls? *Polar Biology*, 27, 238-249.

[54] Barker RJ (1977) A biological reconnaissance of the Bunger Hills, (March 1977). Australian Antarctic Division Technical Memorandum 67.

[55] Pociecha A & Dumont H (2008) Life cycle of *Boeckella poppei* Mrázek and *Branchinecta gaini* Daday (King George Island, South Shetlands). *Polar Biology*, 31, 245-248.

[56] Marcus NH & Boero F (1998) The importance of benthic-pelagic coupling and the forgotten role of life cycles in coastal aquatic systems. *Limnology and Oceanography*, 43, 763-768.

[57] Jiang X, Zhao S, Xu Z et al. (2012) Abundance and age of viable resting eggs of the calanoid copepod *Boeckella poppei* Mrázek in sediments: evidence of egg banks in two Antarctic maritime lakes. *Polar Biology*, 35, 1525-1531.

[58] Stoch F (2001) How many species of *Diacyclops*? New taxonomic characters and species richness in a freshwater cyclopoid genus (Copepoda, Cyclopoida). *Hydrobiologia*, 453/454, 525-531.

[59] Bayly IAE (1978) The occurrence of *Paralabidocera antarctica* (I.C. Thompson) (Copepoda: Calanoida: Acartiidae) in an Antarctic saline lake. *Australian Journal of Marine and Freshwater Research*, 29, 817-824.

[60] Thompson IC (1898) Report on a small collection of Antarctic plankton from the neighbourhood of the South Shetland Islands, collected by the staff of a Dundee Whaler in 1892-3. *Transactions of the Liverpool Biological Society*, 12, 291-297.

[61] Bayly IAE & Eslake D (1989) Vertical distributions of a planktonic harpacticoid and a calanoid (Copepoda) in a meromictic Antarctic lake. *Hydrobiologia*, 172, 207-214.

[62] Bayly IAE (1986) Ecology of the zooplankton of a meromictic antarctic lagoon with special reference to *Drepanopus bispinosus* (Copepoda: Calanoida). *Hydrobiologia*, 140, 199-231.

[63] Cromer L, Gibson JAE, Swadling KM & Ritz DA (2005) Faunal microfossils: Indicators for ecological change in an Antarctic saline lake. *Palaeogeography, Palaeoclimatology, Palaeoecology*, 221, 83-97.

[64] Swadling KM, McKinnon AD, De'Ath G & Gibson JAE (2004) Growth, production and life cycle plasticity of the copepod *Paralabidocera antarctica* in East Antarctica. *Limnology and Oceanography*, 49, 644-655.

[65] Tanimura A, Hoshiai T & Fukuchi M (1996) The life cycle strategy of the ice-associated copepod, *Paralabidocera antarctica* (Calanoida, Copepoda), at Syowa Station, Antarctica. *Antarctic Science*, 8, 257-266.

[66] Hoshiai T, Tanimura A & Watanabe K (1987) Ice algae as food of an Antarctic ice-associated copepod, *Paralabidocera antarctica* (I.C. Thompson). *Proceedings of the National Institute for Polar Research Symposium on Polar Biology*, 1, 105-111.

[67] Hoshiai T, Tanimura A, Fukuchi M & Watanabe K (1989) Feeding by the nototheniid fish, *Pagothenia borchgrevinki* on the ice-associated copepod, *Paralabidocera antarctica*. *Proceedings of the National Institute for Polar Research Symposium on Polar Biology*, 2, 61-64.

[68] Bell EM & Laybourn-Parry J (1999) Annual plankton dynamics in an Antarctic saline lake. *Freshwater Biology*, 41, 507-519.

[69] Bayly IAE & Burton HR (1987) Vertical distribution of *Paralabidocera antarctica* (Copepoda: Calanoida) in Ace Lake, Antarctica, in summer. *Australian Journal of Marine and Freshwater Research*, 38, 537-543.

[70] Swadling KM & Gibson JAE (2000) Grazing rates of a calanoid copepod (*Paralabidocera antarctica*) in a continental Antarctic lake. *Polar Biology,* 23, 301308.

[71] Gibson JAE & Trull TW (1999) Annual cycle of fCO_2 under seaice and in open water in Prydz Bay, East Antarctica. *Marine Chemistry*, 66, 187-200.

[72] Rankin LM, Gibson JAE, Franzmann PD & Burton HR (1999) The chemical stratification and microbial communities of Ace Lake, Antarctica: a review of the characteristics of a marine-derived lake. *Polarforschung*, 66, 35-52.

[73] Watanabe K, Satoh H, Hoshiai T (1990) Seasonal variation in ice algal assemblages in the fast ice near Syowa Station in 1983/84. *In*: Kerry KR & Hempel G (eds.), *Antarctic Ecosystems—Change and Conservation*. Springer, New York, 136-142.

[74] Dartnall HJG (2000) A limnological reconnaissance of the Vestfold Hills. *ANARE Reports*, 141, 1-57.

[75] Burton H & Hamond R (1981) Harpacticoid copepods from a saline lake in the Vestfold Hills, Antarctica. *Australian Journal of Marine and Freshwater Research*, 32, 465-467.

[76] Kudoh S, Tanabe Y & Takahashi KT (2008) Abundance of benthic copepods in a saline lake in East Antarctica. *Polar Biology*, 31, 1539-1542.

[77] Ito T & Fukuchi M (1978) *Harpacticus furcatus* Lang from the Antarctic Peninsula, with reference to the copepodid stages (Copepoda: Harpacticoida). *Antarctic Record*, 61, 40-64.

[78] Dahms HU & Schminke HK (1992). Sea ice inhabiting Harpacticoida (Crustacea, Copepoda) of the Weddell Sea (Antarctica). *Bulletin de l'Institute Royal des Sciences naturelles de Belgique*, 62, 91-123.

[79] Coolen MJL, Hopmans EC, Rijpstra WIC et al. (2004) Evolution of the methane cycle in Ace Lake (Antarctica) during the Holocene: response of methanogens and methanotrophs to environmental changes. *Organic Geochemistry*, 35, 1151-1167.

[80] Gibson JAE, Paterson KS, White CA & Swadling KM (2009) Evidence for the continued existence of Abraxas Lake, Vestfold Hills, East Antarctica during the last Glacial Maximum. *Antarctic Science*, 21, 269-278.

[81] Bissett A, Gibson JAE, Jarman SN et al. (2005) Isolation, amplification and identification of ancient copepod DNA from lake sediments. *Limnology and Oceanography Methods*, 3, 533-542.

[82] Wagner B, Cremer H, Hultzsch N et al. (2004) Late Pleistocene and Holocene history of Lake Terrasovoje, Amery Oasis, East Antarctica, and its climatic and environmental implications. *Journal of Paleolimnology*, 32, 321-339.

[83] Swadling KM, Dartnall HJG, Gibson JAE et al. (2001) Fossil rotifers and the early colonization of an Antarctic lake. *Quaternary Research*, 55, 380-384.

[84] Walton DH, Vincent WF, Timperley MH et al. (1997) Synthesis: polar deserts as indicators of change. *In*: Lyons WB, Howard-Williams C & Hawes I (eds.), *Ecosystem Processes in Antarctic Ice-free Landscapes*. Balkema, Rotterdam, 275-279.

In: Copepods: Diversity, Habitat and Behavior
Editor: Laurent Seuront
ISBN: 978-1-63117-846-7

Chapter 5

MARINE COPEPODS AND SOLAR RADIATION

Rodrigo J. Gonçalves [1,2,*] ***and Samuel Hylander*** [2,3]
[1]Estación de Fotobiología Playa Unión (EFPU) and Consejo Nacional de Investigaciones Científicas y Técnicas (CONICET), Argentina
[2]Centre for Ocean Life, Technical University of Denmark (DTU Aqua), Denmark
[3]Centre for Ecology and Evolution in Microbial model Systems – EEMiS, Linnaeus University, Kalmar, Sweden

ABSTRACT

Ultraviolet radiation (UVR) has an overall negative effect on zooplankton. The direct effects (i.e., when the individual is exposed to UVR) include mortality due to DNA damage, and changes in feeding, reproduction and respiration. The effects can also be indirect, for example when UVR affects food quality of phytoplankton. However much of our knowledge comes from freshwater zooplankton (copepods but especially cladocerans), whereas marine studies are comparatively scarce. Here we summarize the current knowledge in marine copepods and try to give an overview of knowledge gaps and interesting areas for further research. The indirect effects of UVR are likely to be more important (and less predictable due to interactions with other factors across trophic levels) than the direct effects.

INTRODUCTION

Planktonic organisms living in surface layers of the ocean are exposed to solar radiation, which is used for photosynthesis but can also be detrimental depending on the wavelength composition (see below). Zooplankton can be directly or indirectly affected by solar radiation: direct effects may be observed when zooplankters receive sunlight, whereas indirect effects involve other trophic levels (e.g., when phytoplankton is affected first). Here we give an overview of known effects of ultraviolet radiation (UVR, radiation with wavelengths between 280 and 400 nm) on marine copepods and identify some knowledge

* E-mail address: rodrigo@efpu.org.ar.

gaps. We focus on marine pelagic copepods, even though the UVR research has been traditionally skewed towards freshwater zooplankton as we want to highlight differences and similarities in UVR responses between these systems. In this context, fundamental differences between freshwater and marine environments are the lack of altitude gradients in the sea (a lake at 1000 m receives *ca.* 19% more UVR than a lake at sea-level, see [1,2] and the intrinsic longitudinal and latitudinal limits in the movement of copepods imposed by lake boundaries. However, many of the relevant mechanisms in plankton (e.g., UVR-induced mutations, DNA repair, avoidance strategies) work in a similar manner in both freshwater and saline systems, so we mention them to highlight potential similarities and differences where applicable.

One could a priori argue that marine systems are globally more important than lakes in terms of secondary production and size. However a bibliographic survey shows that the research conducted on copepod UVR photobiology has mainly focused on freshwater species. This is most likely due to an operational convenience, although it can be speculated that UVR exposure in lakes, ponds and shallow waters may be more evident. Regardless of the cause, research on zooplankton photobiology in freshwater environments has been a rich source of knowledge about zooplankton responses and adaptations under UVR exposure. In this perspective, we focus on what we believe are critical gaps in UV-related studies in marine copepods, identify potential areas where more research is needed, and give an overall perspective of the importance of UVR.

LIGHT IN THE OCEAN

The energy from the Sun that reaches the ocean surface is mostly a combination of infrared radiation (IR, heat), photosynthetically active radiation (PAR, basically visible light which provides energy for photosynthesis) and ultraviolet radiation (UVR [1]) (Table 1). Solar infrared, visible and UV radiation have different shares of energy in wavelength spectrum, hence within the Sun's spectral distribution of energy there are wavelengths with different (sometimes opposite) effects. For example, a given intensity of PAR would generally favor photosynthesis whereas UVR would tend to inhibit it, and heat would increase metabolic rates in the upper mixed layer whereas too much PAR could also produce photoinhibition. The ratio of these three components of solar emission is of great importance for the ecosystem. This ratio changes considerably as different wavelength are absorbed and scattered through the atmosphere (Figure 1) and the ocean (Figure 2). Therefore, changes in composition of the atmosphere and seawater water will change these ratios.

In broad terms, visible light enhances photosynthesis, heat increases metabolism, and UVR would be mostly deleterious [3]. Another beneficial effect of PAR is the activation of the repair activity of UVR-induced damage in the cell (through a process called photorepair, which also uses the range of longer UVA wavelengths, see more details below). This means that if the UVR:PAR ratio increases, the overall biological effects of solar radiation will also change. UVR reaching the ocean surface actually comprises a continuum of UVR wavelengths which include ultraviolet B (UVB) and ultraviolet A (UVA) radiation (Table 1).

---

[1] UVR is probably the less intuitive component to think of, partially because it doesn't produce an immediate perception in our experience (as visible light or heat do).

Although co-occurring, UVA and UVB are not absorbed in the same way by a given compound.

Table 1. Solar radiation reaching the surface of the water column

Incoming Radiation	Wavelength range [+](nm)
Ultraviolet total (RUV)	280 - 400
- Ultraviolet B (UVB)	280 - 315
- Ultraviolet A (UVA)	315 - 400
Visible, photosynthetic available light (PAR)	400 - 700
Infrared (IR)	700 nm - 1000

[+] (Commission Internationale d'Eclairage,CIE)

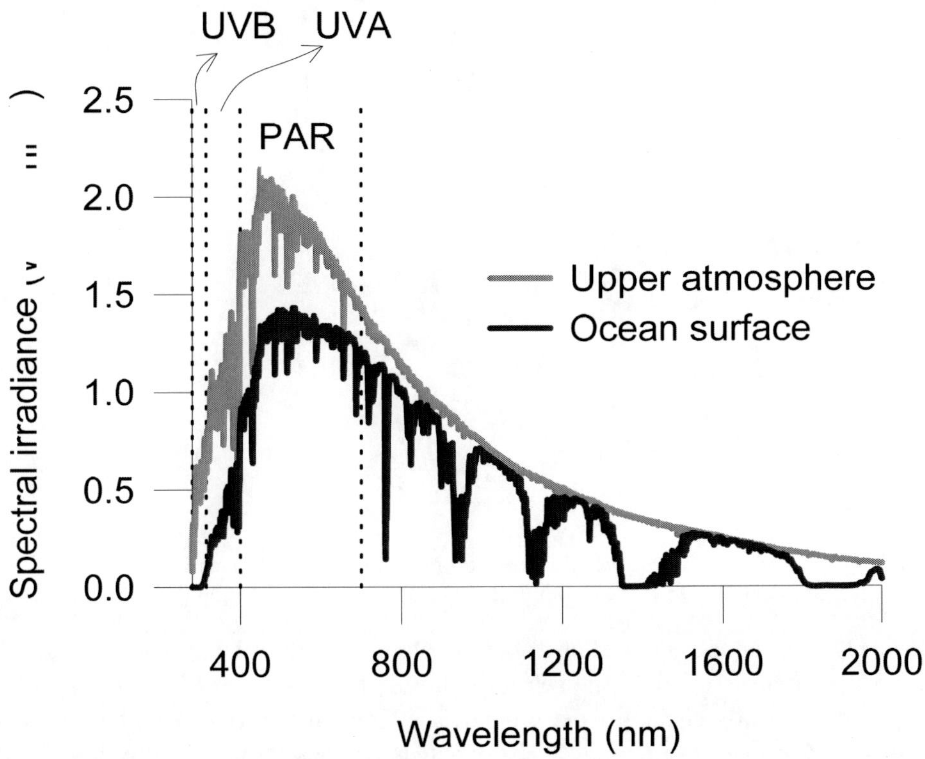

Figure 1. Spectral distribution of solar radiation: irradiance before (top line) and after (bottom line) passing through the atmosphere. Limits of PAR, UVA and UVB ranges are marked with dotted lines. Data from Standard ASTM G173, Standard Tables for Reference Solar Spectral Irradiances: Direct Normal and Hemispherical on 37° Tilted Surface, Amer. Society for Testing Matls., West Conshocken, PA, USA.

For example, UVB is strongly absorbed by atmospheric ozone [4], but not UVA (see below). Therefore the UVB:UVA ratio will be determined by the quality and quantity of different absorbing substances in the path of Sun rays from the upper atmosphere down to the biomolecules inside planktonic organisms. In addition, UVA and UVB can have distinct effects on biomolecules, i.e., shorter wavelengths have stronger interactions with molecules such as DNA.

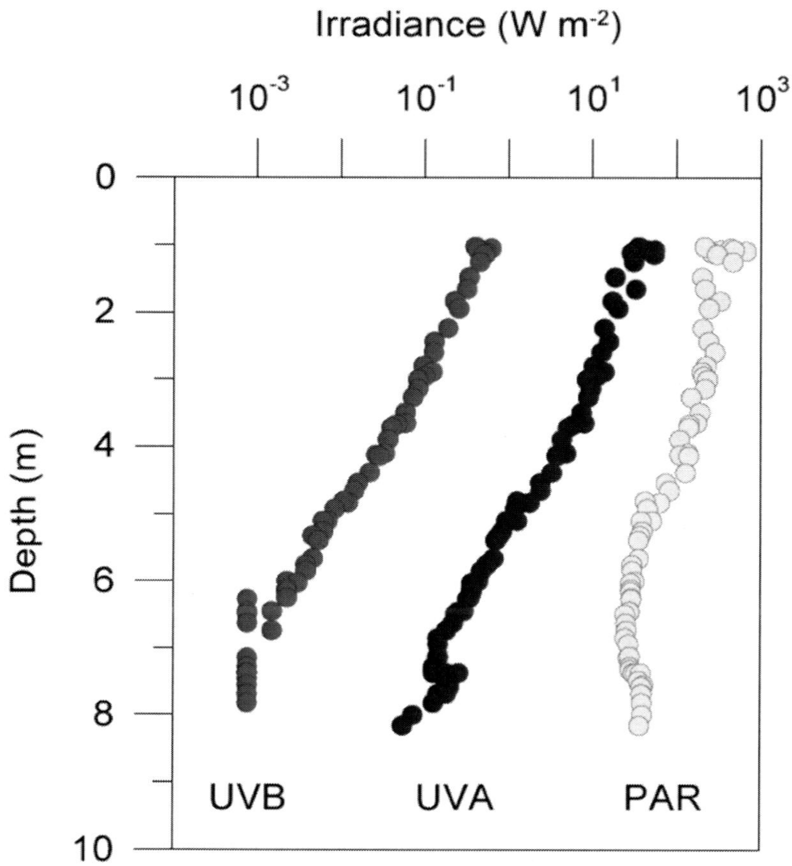

Figure 2. Common attenuation patterns of underwater radiation in a coastal station in Atlantic Patagonia. Data provided by Dr. E.W. Helbling from Estación de Fotobiología Playa Unión (Argentina)

Thus, changes in the spectral distribution of energy of solar radiation (e.g., due to differential absorbance by the atmosphere/hydrosphere in different wavelengths) can be expected to dynamically shape the net biological effects on the pelagic realm. For example nauplii of *C. finmarchicus* respond to UVB, whereas females respond also to UVA [5]. This reinforces the notion that the contribution of each spectral range (in this case UVB:UVA ratio) is important for the net biological effect.

Around 90% of PAR (100% being the extraterrestrial solar spectrum) passes through the atmosphere to reach the Earth surface, but UVR is highly attenuated and only *ca.* 25% and 7% of the UVA and UVB (respectively) incoming energy reaches the surface (percentages estimated for central Europe in midsummer, see [4]). However these small percentages are enough to produce significant effects in many organisms.

But why is UVR important and how can it affect copepods? The main negative effect of UVB is its capacity to interact with biomolecules and induce mutations in the genetic material [6-8], but it can also cause damage in proteins, membranes and other cellular components

[9-13]. On the other hand, UVA is known to induce the production of reactive oxygen species (ROS), which causes oxidative stress in the cell [14-16] so copepods need to cope with it using different endogenous enzymatic reactions [17,18].

In the last decades, an increase in UVR research was triggered mainly by the discovery of the Antarctic 'ozone hole' [19]. Stratospheric ozone acts as a filter in the atmosphere by strongly absorbing in the UVB wavelengths range. After initial focus on the local phenomenon of the ozone hole, it was discovered that the thinning of the stratospheric ozone layer is a global phenomenon, also related to a global increase of UVB radiation reaching the Earth surface [20,21]. Reductions in ozone concentration were subsequently also described in the North Hemisphere [22-24] and according to atmospheric models, the ozone layer is not expected to recover anytime soon [25-28]. Increases in UVB due to ozone reduction are negative for most species [3]. However, even at normal ozone concentrations, UVR is an important factor in many aquatic ecosystems and it has been shown that other factors may be more important than ozone concentration in determining the net UVR exposure, such as clouds and attenuation in the water column [29-31]. UVR may be even more important when considered together with with climate change [32] because its increase may act in synergy with other phenomena such as global warming [33].

Solar ultraviolet radiation is exponentially attenuated in the water column. This is measured with underwater radiometers to make a profile at a given station (Figure 2). The specific values of irradiance and ratios (e.g., PAR:UV) will change substantially as water transparency is affected by many different components. The underwater attenuation of solar radiation is due to absorption of radiant energy by dissolved organic carbon (DOC) and its optically active component, chromophoric dissolved organic matter (CDOM), as well as particulates such as phytoplankton cells [34]. The attenuation by CDOM is selective towards shorter wavelengths, i.e., even a small increase in DOC would lead to a substantially large PAR:UVR ratio [35]. Therefore, changes in these absorbing components will result in different UVB and UVA penetration into the surface layers of the pelagic realm. Specifically, Tedetti & Sempéré [36] reviewed underwater radiation measurements in different marine environments, and calculated the depth where 10% of the surface radiation is still present in the water column (also called $Z_{10\%}$ depth). They found that in general coastal environments show relatively low UVR penetration (average $Z_{10\%}$ of 2.2 and 5.3 m in for UVB and UVA, respectively). This is mainly due to attenuation by terrestrial and autochthonous sources of CDOM [37-39] and by abundant phytoplankton in typically highly productive coastal waters [40]. On the other hand, oceanic waters showed a relatively higher UVR penetration (average $Z_{10\%}$ depths of 11.2 and 25.5 m for UVB and UVA, respectively). For example, in the Atlantic Ocean $Z_{10\%}$ was measured at 15.5 and 46.5 m (for UVB and UVA, respectively), and at 12 m into the water column the estimated effects of UVR on DNA still were 10% of those at the surface ($Z_{10\%DNA}$). Tedetti and Sempéré [36] also reported that Antarctic waters were characterized by a significant penetration of UVR, particularly during low ozone events (maximum $Z_{10\%DNA}$ of 16.4 m). These depths are relatively shallow when compared to average oceans depth, which is probably why it is traditionally assumed that solar UVR would not constitute a serious threat to plankton. However, it is in the first few meters of the photic layer that most primary production occurs (marine photosynthesis drives half of the carbon fixation at a planetary scale [41] and eggs and larvae of several marine organisms spend a critical part of their life cycle [42-44]). Therefore this layer is likely to convey the impact derived from exposure to UVR and affect other pelagic organisms. In an example from oligotrophic waters, biologically-damaging UVR doses for important picocyanobacteria (*Prochlorococcus* sp) were found between 10 and 26 m [3] which may affect the microbial community in these environments. In Antarctic waters some species can be found above the $Z_{10\%}$ in summer [45] and as mentioned above, in some oceanic regions, UVA wavelengths (which some zooplankton species respond to with negative phototaxis) can penetrate tens of meter into the water column [36] and thus may be involved in driving zooplankton diel vertical migrations [46]. For example one the deepest penetration of UVR in oceanic waters was observed in hyper-oligotrophic waters of the South Pacific Gyre, where Tedetti et al. [47] observed $Z_{10\%}$ depths of 28 and 110 m for UVB and UVA, respectively. According to this study, these values are similar to those in the clearest fresh waters.

What We Learned from Freshwater Systems

UVR research on marine zooplankton is by far less abundant in marine systems than in freshwater environments. UVR affects many aspects of biosphere, including biogeochemical cycles [48] but this is mostly studied in freshwater systems, with lakes as "*sentinels of climate change*" [49].

The studies of UVR on freshwater zooplankton are numerous and include (i) direct effects such as damage on DNA and other molecules [50], modification of feeding [51,52] and respiration rates [53] and increased mortality and reduced fecundity [52,54-56], and (ii) indirect effects such as alteration of community composition [57], behavioral changes [58-62], and impact on food quality [63-66]. Besides, UVR protection strategies have been well documented in freshwater zooplankton, and involve the production of photoprotective compounds (sunscreens and dietary antioxidants, MAAs and carotenoids) [67,68] and cellular antioxidants (e.g., Glutathione S-transferases) [69-74], DNA repair mechanisms (photoenzymatic repair) [75-79], and behavioral defenses (vertical and horizontal migration) [80,81]. Note that the interested reader can get further information regarding the effects of UVR on zooplankton in freshwater systems from a range of recent reviews; see for example [1,57,82].

Comparison of UV Exposure in Freshwater and Marine Systems

As UVR penetration varies widely between different water bodies, it is difficult to classify "marine systems" under a single category. For example, CDOM in coastal waters such as estuaries is regulated by land-ocean interactions, whereas for open oceans it is present mostly as breakdown products of marine productivity [83]. Thus coastal systems share some general characteristics with freshwater environments: shallow waters, higher productivity and more influence from terrigenous materials (which would result in higher attenuation of solar radiation). Therefore the underwater UVR exposure in coastal waters may be closer to freshwater environments than to oceanic regions. The distinction between oceanic and coastal waters is supported by underwater measurements [36]. Table 2 summarizes the main factors affecting UVR exposure in freshwater vs. marine (oceanic) systems and the known general responses of zooplankton.

At the planetary scale, marine underwater radiation profiles are still too scarce to obtain a clear picture. Some attempts have, however, been made to produce an average map of global UVR penetration in the global ocean (Figure 3) [84]. The results suggest that UVB reaches depths ($Z_{10\%}$) between 10 and 20 m for the vast majority of the oceans with maxima at oligotrophic regions[2]. This is in agreements with the deepest marine measurements of UVR, which were made at the south Pacific gyre [47]. One could then speculate that oceanic regions may be more affected by UVR than coastal sites, as in average the penetration of UVB is significant in all oceans.

Table 2. Factors affecting UVR exposure and responses in UV-protection strategies

Abiotic factors							
	Depth	Max depth (m)	DOC	(median/range) [a]	UVR penetration	Maximum 1%-UVB depth (m) [b]	Temperature
Open ocean	Deep	~11000	Low	0.45 (open ocean median)	High	35-40	Variable
Freshwater	Shallow	~1600	High	< 0.5 to > 30 (clear-bog water)	Variable	< 33	Variable

UVR defenses					
	Vertical migration	Accumulation of MAAs	Accumulation of carotenoids	PER	Internally produced anti-oxidants
Marine	No or some UV-effects	Demonstrated in few copepod species	To our knowledge not shown	Demonstrated in few copepod species	To our knowledge not demonstrated in marine copepods
Freshwater	Several cladoceran and copepod species	Demonstrated in several copepod species	Shown in several copepod species	Both cladocerans and copepods but with variable efficiency	Demonstrated for both cladocerans and copepods

a) Wetzel RG (2001) Limnology: Lake and River Ecosystems. London, UK: Academic Press (and references therein Table 23-1).

b) Morris et al.[34] and Smith C and Baker KS (1981) Optical properties of the clearest natural waters (200-800 nm). Appl Opt **20**: 177-184.

WHAT WE KNOW ABOUT MARINE SYSTEMS

Effects of UVR on Marine Copepods

Briefly, direct effects in marine copepods include the same as in freshwater copepods such as mortality [3,31,45,85-91], damage on DNA [31,87], lipid oxidation [89], reduction in feeding rates [86,89] and impaired reproduction [86-88,90,92-94] and low egg quality [95].

[2] However, since these results are obtained from satellite measurements there are still challenges to overcome for better estimations, see Vasilkov, A.P., Herman, J., Krotkov, N.A., Kahru, M., Hsu, C. & Mitchell, B.G. (2002) Problems in assessment of the ultraviolet penetration into natural waters from space-based measurements. Optical Engineering, **41**, 3019-3027.

Additionally the suppression of protein synthesis has been studied [96]. Respiration in marine pelagic copepods also seems to be significantly impaired when exposed to artificial UVR in two species (*Pseudodiaptomus marinus* and *Labidocera bipinnata*) [133].

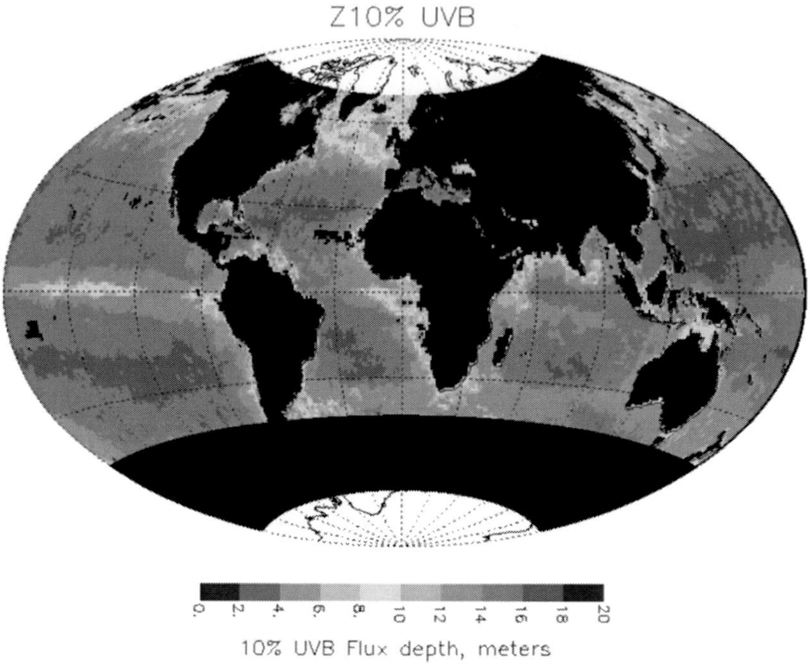

Figure 3. Modeled monthly underwater penetration of UVB in global ocean. Z10% is the depth where 10% of surface UVB reaches into the water column. From [84], with permission from John Wiley and Sons.

Important indirect effects of UVR on copepods can appear on grazing rates (i.e., via affecting the quality and quantity of food). For example UVR can induce high taxon-specific, phytoplankton mortality which may drive changes in the community structure of autotrophs and affect the dynamics of the microbial food web in clear, oligotrophic, oceanic waters [97, 98]. However there are even fewer studies of changes in community composition of marine species including copepods. For example Keller et al. [99] found that phytoplankton and nauplii abundance was reduced under UVB enhanced treatments, although the effects were not observed at higher trophic levels such as fish larvae. It is stressed that effects on feeding can be somehow contradictory. Specifically, most direct effects studies show a decrease in feeding rates of copepods exposed to UVR, while increased [100] or lower [101] grazing rates are observed when phytoplankton has been previously exposed (and not the grazer itself). These apparently contradictory results and the relative scarcity of indirect effects studies indicate that this study field is still underdeveloped and further research is critically needed to disentangle the complexity of reported responses on UVR effects on copepod grazing.

UVR Protection Strategies

Vertical Distribution/Migration

Vertical migration in response to UVR has been demonstrated for a number of freshwater zooplankton species, mostly cladocerans; see e.g., [60,62]. There are a few examples available on UVR responses in marine copepods. For example *Calanus finmarchicus* shows a downward swimming response to UVR in small (250 ml) beakers [102] or in 32 cm high cylinders [5], but this effect was not observed in deeper containers [102]. Copepods show both positive or negative phototaxis toward yellow or red light as well as low UVR levels [103], and artificial UVR was observed to induce horizontal movements in *Acartia pacifica* [104]. When comparing three groups (*Tortanus dextrilobatus, Acartiura* spp. and *Acanthacartia* spp) Speekman et al. [91] found that only *T. dextrilobatus* showed a 50 cm deeper distribution in the presence of UVB. Coincidentally *T. dextrilobatus* was the only species showing an increase in mortality when exposed to UVR, so apparently the other two would not need to avoid the high-UVR surface. In a field experiment with *Acartia hudsonica* in a marine lagoon, no differences were observed in the average vertical distribution of copepods with or without UVB [105]. Saito et al. [88] observed that migratory species (*Acartia omorii, Paracalanus* sp and *Calanus sinicus*) were less tolerant to UVR than the ones which are more permanent inhabitangs of surface waters (*Pontella rostraticuada, Pontellopsis tenuicauda, Labidocera madurae*) in terms of survival, egg production and egg hatching rates. This suggests that UVR avoidance by means of vertical migration may be a factor to consider in marine copepods. However, our knowledge is far from complete as many of these studies used artificial sources of radiation which makes it difficult to extrapolate to nature. In addition, in freshwater studies it is commonly found than vertical migration of copepods in response to UVR is less pronounced than in other zooplankter such as Daphnids, which suggests that community composition is important to evaluate overall UVR effects in zooplankton [106]. More research is needed in natural marine systems to complement the laboratory studies.

Swimming deep enough would reduce the potential threat since UVR attenuates with depth. Another benefit of leaving well-lit surface waters would be a reduced risk of being captured by visual predators such as planktivorous fish and fish larvae. However, there is a potential cost involved in leaving the surface: copepods spend energy migrating and also leave the warm (and usually food-rich) surface layers. These trade-offs have likely shaped the evolution of diel vertical migration but it is not clear yet whether fish avoidance, UVR-avoidance, food, temperature or a combination of these factors shape the adaptive significance for zooplankton. Predator avoidance has for long been the established hypothesis explaining the adaptive significance of diel vertical migration in zooplankton [107,108]. UVR avoidance has, however, also been demonstrated in several zooplankton species [57], and in some fishless lakes zooplankton also perform diel vertical migration [109] suggesting UVR may be an important factor as well. More recently, Williamson et al. [46] included UVR in a broader framework (transparency-regulator hypothesis) to study zooplankton diel vertical migration. Both UVR and visual predation were suggested as acting as "*dynamic drivers*" (i.e., factors which show great daily variations) in contrast to "*structural drivers*" such as temperature and food, which show less daily variations in the water column. Thus, zooplankton diel vertical migration would be more affected by UVR in clear waters but in less transparent waters it would be more affected by visual predation (fish) [46]. Hylander and Hansson [110] compared vertical distribution of marine Antarctic copepods and found that copepods avoided surface waters at the most transparent sites. However, there was little actual migration and copepods were distributed well below the depth where UVR had vanished to negligible levels, suggesting that they were either using deep food reserves or avoiding visual predators [110]. The transparency-regulator hypothesis [46] has mainly been developed based on data from freshwaters and rigorous comprehensive tests of the framework are still needed in both freshwater and marine systems.

Other factors may also influence the relative effects of UVR in different developmental stages. Turbulence in surface waters may prevent downward migration of nauplii which would therefore be exposed to fluctuating UVR due to vertical mixing [5]. If individuals of a UVR-sensitive species cannot avoid the exposure to high levels of solar radiation, then the strategy seems to be minimizing the damage by means of physiological traits. These fall basically into three categories: photoprotection, photorepair and induction of internal antioxidants.

Photoprotection: Protective Compounds and Trade-Offs

UV Protective Compounds

One way of minimizing the effects of UVR consist in synthetizing or acquiring UVR-protective compounds. These molecules include red carotenoids which are strong dietary antioxidants reducing the negative effects of UVR-induced harmful agents (i.e., reactive oxygen species, ROS). Freshwater studies have demonstrated that several copepod species accumulate carotenoids during exposure of UVR [67,111,112] and pigmented copepods are more tolerant to UVR compared to transparent conspecifics [113]. Few marine studies on zooplankton accumulation of carotenoids are available. So far there is little evidence of accumulation of carotenoids upon UVR exposure [95,110]. This may suggest other adaptive benefits of carotenoids in marine copepods, e.g., camouflage as suggested by Vestheim and Kaartvedt [114]. Collecting the available measurements of carotenoids in freshwater and marine copepods, Hylander et al. [95] found that carotenoid levels are an order of magnitude lower in marine compared with freshwater systems. Since these pigments make the individual more conspicuous, there is a trade-off between pigment accumulation as protection against UVR (survival) and vulnerability to visual predators (mortality). Freshwater copepods with low carotenoid contents typically come from high predation systems [68,115,116], which would suggest intense selective predation on pigmented specimens in the ocean.

On the other hand, there is a second group of substances, secondary metabolites, which are colorless and show maximum absorbance between 310 and 360 nm. These are called collectively mycosporine-like amino acids (MAAs) and have been identified in several marine species [117]. UV-absorbing compounds acting as sunscreens are another way to minimize UVR-induced damage. Similarly to carotenoids few studies exist assessing the importance of MAAs in marine copepods. Two marine copepod species have been demonstrated to accumulate MAAs when exposed to UV [95,110]. Freshwater copepods with high MAA levels do have higher tolerance against UVR [111]. Life table studies of marine copepods confirm that accumulation of MAAs do enable copepods to maintain high egg quality and fitness when exposed to UVR. There are generally few measurements available in different marine copepod species but the available ones indicate approximately the same levels as in freshwater systems [95]. For example the neustonic species *Pontella rostraticuada* and *Pontellopsis tenuicauda* may rely on MAAs to overcome UVR exposure [88]. Newman et al. [118] also found that UV-sensitive Antarctic krill (*Euphausia superba*) acquire MAAs from the microalgae *Phaeocystis antarctica*. However, copepods (and more generally animals) are not able to synthetize MAAs and carotenoids, hence they need to acquire them from diet [111,112,118] which emphasizes the importance of food quality for an optimum defense based on UV-protective compounds.

Antioxidants

Freshwater zooplankton have also been shown to induce internal antioxidant mechanisms when exposed to UVR as a means of reducing the negative effects of photoproduced molecules such as reactive oxygen species (ROS) [17,69,70,119,120]. To our knowledge there are no comprehensive studies showing whether this is an important UVR protection strategy among marine copepods.

Photorepair: Photo-Enzymatic Repair

Several organisms have evolved a way to use sunlight to activate a mechanism of enzymatic repair (called photorepair or photoreactivation) of UVB-induced damage in their genetic material. In this process, PAR and near-UVA wavelengths activate an enzyme (photolyase) which directly reverts the mutation in DNA [121-123]. In natural environments, solar UVA, UVB and PAR act simultaneously and, again, their relative contribution will be of great importance to the net biological effect. There are other DNA repair mechanisms (mainly nucleotide-excision repair, see [124,125]. They are, however, not directly related to sunlight and are rather universal; hence they are not discussed here. The ability of photorepair has been demonstrated in several freshwater species of zooplankton [77,78,126-128]. In marine copepods, evidence of photorepair is so far only available for a few *Calanus* species, i.e., *C. finmarchicus* [129], *C. pacificus* [130] and *C. sinicus* [93].

Gaps in Knowledge and Future Directions

We identified two major gaps in our current knowledge and understanding of UVR photobiology of marine copepods. First, it is known that UVR may interact with other factor associated to global change [32], therefore there is a need for more integrative studies on how these interactions affect copepods and other marine zooplankton. For example, due to climate-driven changes in vertical stratification in most of the world's open ocean, reductions in chlorophyll concentrations led to increases in water transparency [131]. This will presumably affect underwater UVR exposure of copepods. Second, our knowledge would be increased by further efforts in scaling up laboratory behavioral studies in order to evaluate how important is UVR in the field for a given community. In particular, a thorough quantitative assessment of the transparency regulator hypothesis is critically needed in marine systems. More studies are needed to test if marine copepods have the ability to behaviorally avoid solar UVR (or artificial UVR lamps with ecologically relevant output spectra) and the mechanisms involved. Note that some of the earlier marine studies used artificial UVR sources with which is difficult to adequately reproduce the actual solar spectrum [52,76,85,86,88,92-94].

A few other fields of research also need improvement, in particular when compared to what has been achieved in freshwater systems. These include assessing the effects of UVR on (i) species interactions at community level, and (ii) copepod respiration. We support the notion that in natural environments indirect effects are likely to be much more important than direct effects [31,132], yet they are poorly studied. In addition, some studies of indirect effects [100] may point in somewhat opposite direction to those from direct effects [86, 89]. It is also stressed that a better understanding of the relative importance of induction of internally-produced antioxidants, and the role of photorepair as protection mechanism is critically needed to advance our knowledge. These constitute essential questions in the context of global change where water transparency is expected to increase in the upper ocean [131], which would result in a deeper penetration of UVR and more negative effects on plankton [3].

CONCLUSION

Compared to freshwater systems, UVR research in marine copepods is still scarce. Especially studies about photorepair, internal antioxidant enzymes and vertical avoidance of natural sunlight are areas of interest for future studies. At the global scale, the indirect effects of UVR on marine copepods (e.g., through trophic interactions and environmental changes) are likely to be more important than direct effects on individuals (e.g., mutations and production of oxidative stress). Zooplankton from oceanic regions are likely to be much more influenced by UVR than species in coastal waters due to differential underwater UVR exposure of these environments. This difference may be enhanced in the future since eutrophication is expected in coastal waters while more transparency is expected in oceanic open waters [131]. Since photobleaching of CDOM largely determine the penetration of UVR, this is a factor of greater importance to consider in future studies than, for example, ozone concentration. Studies on indirect effects are scarcer than those on direct effects. However, overall indirect UVR effects are likely to be more important when considered in combination with other global change processes such as global warming and ocean acidification.

ACKNOWLEDGMENTS

R. J. G. wishes to thank Dr. V. Villafañe, Dr. Walter Helbling and Fundación Playa Unión (Argentina) for support and underwater radiation profiles. Dr. A. Vasilkov provided the figure on UVB penetration and Dr. Keith Emery provided guidance with the ASTM usage. Funding to R. J. G. was partially provided by Consejo Nacional de Investigaciones Científicas y Tecnológicas, FONCyT (PICT 2010-2438) (Argentina) and Centre for Ocean Life (Villum Kann Rasmussen Centre of excellence for the study of life in a changing ocean.) (Denmark). Funding to S. H. was provided by the H.C. Ørsted postdoc program of the Technical University of Denmark and by a grant from the Carlsberg Foundation.

REFERENCES

[1] Rautio M & Tartarotti B (2010) UV radiation and freshwater zooplankton: damage, protection and recovery. *Freshwater reviews: a journal of the Freshwater Biological Association*, 3, 105-131.

[2] Blumthaler M, Ambach W & Rehwald W (1992) Solar UV-A and UV-B radiation fluxes at two alpine stations at different altitudes. *Theoretical and Applied Climatology*, 46, 39-44.

[3] Llabrés M, Agustí S, Fernández M et al. (2013) Impact of elevated UVB radiation on marine biota: a meta-analysis. *Global Ecology and Biogeography*, 22, 131-144.

[4] Häder DP, Lebert M, Schuster M et al. (2007) ELDONET - A decade of monitoring solar radiation on five continents. *Photochemistry and Photobiology*, 83, 1384-1357.

[5] Wold A & Norrbin F (2004) Vertical migration as a response to UVR stress in *Calanus finmarchicus* females and nauplii. *Polar Research*, 23, 27-34.

[6] Buma AGJ, Boelen P & Jeffrey WH (2003) UVR-induced DNA damage in aquatic organisms. *In*: Helbling EW & Zagarese HE (eds.), UV Effects in Aquatic Organisms and Ecosystems. The Royal Society of Chemistry, Cambridge, 291-327.

[7] Malloy KD, Holman MA, Mitchell DL & Dietrich III HW (1997) Solar UVB-induced DNA damage and photoenzymatic DNA repair in antarctic zooplankton. *Proceedings of the National Academy of Sciences of the United States of America*, 94, 1258-1263.

[8] Sinha RP & Häder D-P (2002) UV-induced DNA damage and repair: a review. *Photochemical and Photobiological Sciences*, 1, 225-236.

[9] Arts MT & Rai H (1997) Effects of enhanced ultraviolet-B radiation on the production of lipid, polysaccharide and protein in three freshwater algal species. *Freshwater Bio*logy, 38, 597-610.

[10] Martínez R (2007) Effects of ultraviolet radiation on protein content, respiratory electron transport system (ETS) activity and superoxide dismutase (SOD) activity of Antarctic plankton. *Polar Biology*, 30, 1159-1172.

[11] Noaman N (2007) Ultraviolet-B irradiation alters amino acids, proteins, fatty acids contents and enzyme activities of *Synechococcus leopoliensis*. *International Journal of Botany*, 3, 109-113.

[12] Buma AGJ, Zemmelink HJ, Sjollema K & Gieskes WWC (1996) UVB radiation modifies protein and photosynthetic pigment content, volume and ultrastructure of marine diatoms. Mar Ecol Prog Ser 142: 47-54.

[13] Brodhun B & Häder D-P (1993) UV-induced damage of photoreceptor proteins in the paraflagellar body of *Euglena gracilis*. *Photochemistry and Photobiology*, 58, 270-274.

[14] Zika RG & Cooper WJ (1987) *Photochemistry of Environmental Aquatic Systems*. American Chemical Society, Washington.

[15] Sies H & Stahl W (2004) Carotenoids and UV Protection. *Photochemical and Photobiological Sciences*, 3, 749-752.

[16] Stahl W & Sies H (2003) Antioxidant activity of carotenoids. *Molecular Aspects of Medicine*, 24, 345-351.

[17] Souza MS, Balseiro E, Laspoumaderes C & Modenutti B (2010) Effect of ultraviolet radiation on acetylcholinesterase activity in freshwater copepods. *Photochemistry and Photobiology*, 86, 367-373.

[18] Souza MS, Hansson LA, Hylander S et al. (2012) Rapid enzymatic response to compensate UV radiation in copepods. *PLoS One*, 7, e32046.

[19] Farman JC, Gardiner BG & Shanklin JD (1985) Large losses of total ozone in Antarctica reveal seasonal ClOx/NOx interaction. *Nature*, 315, 207-210.

[20] McKenzie RL, Aucamp PJ, Bais AF et al. (2007) Changes in biologically-active ultraviolet radiation reaching the Earth's surface. *Photochemical and Photobiological Sciences*, 6, 218-231.

[21] McKenzie RL, Björn LO, Bais A & Ilyasd M (2003) Changes in biologically active ultraviolet radiation reaching the Earth's surface. *Photochemical and Photobiological Sciences*, 2, 5-15.

[22] Björn LO, Callaghan TV, Gehrke C et al. (1998) The problem of ozone depletion in northern Europe. *Ambio*, 27, 275-279.

[23] Feister U & Grewe R (1995) Higher UV radiation inferred from low ozone levels at northern mid-latitudes in 1992 and 1993. *Global Planet Change*, 11, 25-34.

[24] Kerr JB & McElroy CT (1993) Evidence for large upward trends of ultraviolet-B radiation linked to ozone depletion. *Science*, 262, 1032-1034.

[25] Waugh DW, Oman L, Kawa SR et al. (2009) Impacts of climate change on stratospheric ozone recovery. *Geophysical Research Letters*, 36, L036223.

[26] Bais AF, Tourpali K, Kazantzidis A et al. (2011) Projections of UV radiation changes in the 21st century: Impact of ozone recovery and cloud effects. *Atmospheric Chemistry and Physics Discussions*, 11, 10769-10797.

[27] McKenzie R, Aucamp PJ, Bais A et al. (2011) Ozone depletion and climate change: impacts on UV radiation. *Photochemical and Photobiological Sciences*, 10, 182-198.

[28] World Meteorological Organization (2011) Scientific assessment of ozone depletion: 2010. Global ozone research and monitoring project - Report No 52. Geneva.

[29] Kuhn P, Browman HI, McArthur B & St-Pierre JF (1999) Penetration of ultraviolet radiation in the waters of the estuary and Gulf of St. Lawrence. *Limnology and Oceanography*, 44, 710-716.

[30] Kuhn PS, Browman HI, Davis RF et al. (2000) Modeling the effects of ultraviolet radiation on embryos of *Calanus finmarchicus* and Atlantic cod (*Gadus morhua*) in a mixing environment. *Limnology and Oceanography*, 45, 1797-1806.

[31] Browman HI, Alonso Rodríguez C, Béland F et al. (2000) Impact of ultraviolet radiation on marine crustacean zooplankton and ichthyoplankton: a synthesis of results from the estuary and Gulf of St. Lawrence, Canada. *Marine Ecology Progress Series*, 199, 293-311.

[32] Häder DP, Helbling EW, Williamson CE & Worrest RC (2011) Effects of UV radiation on aquatic ecosystems and interactions with climate change. *Photochemical and Photobiological Sciences*, 10, 242-260.

[33] Rex M, Salawitch RJ, von der Gathen P et al. (2004) Arctic ozone loss and climate change. *Geophysical Research Letters*, 31, L04116.

[34] Morris DP, Zagarese HE, Williamson CE et al. (1995) The attenuation of solar UV radiation in lakes and the role of dissolved organic carbon. Limnology and Oceanography, 40, 1381-1391.

[35] Williamson CE, Neale PJ, Grad G et al. (2001) Beneficial and detrimental effects of UV on aquatic organisms: Implications of spectral variation. *Ecological Applications*, 11, 1843-1857.

[36] Tedetti M & Sempéré R (2006) Penetration of ultraviolet radiation in the marine environment. A review. *Photochemistry and Photobiology*, 82, 389-397.

[37] Branco AB & Kremer J (2005) The relative importance of chlorophyll and colored dissolved organic matter (CDOM) to the prediction of the diffuse attenuation coefficient in shallow estuaries. *Estuaries*, 28, 643-652.

[38] Kostoglidis A, Pattiaratchi CB & Hamilton DP (2005) CDOM and its contribution to the underwater light climate of a shallow, microtidal estuary in south-western Australia. *Estuarine, Coastal and Shelf Science*, 63, 469-477.

[39] Zhao J, Cao W, Wang G et al. (2009) The variations in optical properties of CDOM throughout an algal bloom event. *Estuarine, Coastal and Shelf Science*, 82, 225-232.

[40] Chavez FP, Messié M & Pennington JT (2011) Marine primary production in relation to climate variability and change. *Annual Review of Marine Science*, 3, 227-260.

[41] Field CB, Behrenfeld MJ, Randerson JT & Falkowski P (1998) Primary production of the Biosphere: Integrating terrestrial and oceanic components. *Science*, 281, 237-240.

[42] Shanks AL & Walters K (1997) Holoplankton, meroplankton, and meiofauna associated with marine snow. *Marine Ecology Progress Series*, 156, 75-86.

[43] Pechenik JA (1999) On the advantages and disadvantages of larval stages in benthic marine invertebrate life cycles. *Marine Ecology Progress Series*, 177, 269-297.

[44] Marcus NH & Boero F (1998) Minireview: the importance of benthic-pelagic coupling and the forgotten role of life cycles in coastal aquatic systems. *Limnology and Oceanography*, 43, 763-768.

[45] Ban S, Ohi N, Leong SCY et al. (2007) Effect of solar ultraviolet radiation on survival of krill larvae and copepods in Antarctic Ocean. *Polar Biology*, 30, 1295-1302.

[46] Williamson CE, Fischer JM, Bollens *SM* et al. (2011) Toward a more comprehensive theory of zooplankton diel vertical migration: Integrating ultraviolet radiation and water transparency into the biotic paradigm. *Limnology and Oceanography*, 56, 1603-1623.

[47] Tedetti M, Sempéré R, Vasilkov A et al. (2007) High penetration of ultraviolet radiation in the south east Pacific waters. *Geophysical Research Letters*, 34, L12610.

[48] Zepp RG, Erickson III DJ, Paul ND & Sulzberger B (2007) Interactive effects of solar UV radiation and climate change on biogeochemical cycling. *Photochemical and Photobiological Sciences*, 6, 286-300.

[49] Adrian R, O'Reilly CM, Zagarese H et al. (2009) Lakes as sentinels of climate change. Limnology and Oceanography, 54, 2283-2297.

[50] Rocco VE, Oppezzo O, Pizarro R et al. (2002) Ultraviolet damage and counteracting mechanisms in the freshwater copepod *Boeckella poppei* from the Antarctic Peninsula. Limnology and Oceanography, 47, 829-836.

[51] de Lange HJ & Lürling M (2003) Effects of UV-B irradiated algae on zooplankton grazing. *Hydrobiologia*, 491, 133-144.

[52] Lacuna DG & Uye SI (2000) Effect of UVB radiation on the survival, feeding, and egg production of the brackish-water copepod, *Sinocalanus tenellus*, with notes on photoreactivation. *Hydrobiologia*, 434, 73-79.

[53] Fischer JM, Fields PA, Pryzbylkowski PG et al. (2006) Sublethal exposure to UV radiation affects respiration rates of the freshwater cladoceran *Daphnia catawba*. *Photochemistry and Photobiology*, 82, 547-550.

[54] Williamson CE, Zagarese HE, Schulze PC et al. (1994) The impact of short-term exposure to UV-B radiation on zooplankton communities in north temperate lakes. *Journal of Plankton Research*, 16, 205-218.

[55] Zagarese HE, Cravero W, González P & Pedrozo F (1998) Copepod mortality induced by fluctuating levels of natural ultraviolet radiation simulating vertical water mixing. *Limnology and Oceanography*, 43, 169-174.

[56] Zagarese HE, Williamson CE, Mislivets M & Orr PS (1994) The vulnerability of *Daphnia* to UV-B radiation in the northeastern United States. *Archiv für Hydrobiologie*, 43, 207-216.

[57] Hansson LA & Hylander S (2009) Effects of ultraviolet radiation on pigmentation, photoenzymatic repair, behavior, and community ecology of zooplankton. *Photochemical and Photobiological Sciences*, 8, 1266-1275.

[58] Hessen DO (1994) *Daphnia* responses to UV-light. *Archiv für Hydrobiologie*, 43, 185-195.

[59] Storz UC and Paul RJ (1998) Phototaxis in water fleas (*Daphnia magna*) is differently influenced by visible and UV light. *Journal of Comparative Physiology A*, 183, 709-717.

[60] Leech DM and Williamson CE (2001) *In situ* exposure to ultraviolet radiation alters the depth distribution of *Daphnia*. Limnology and Oceanography, 46, 416-420.

[61] Alonso C, Zagarese HE, Rocco V et al. (2004) Surface avoidance by freshwater zooplankton: Field evidence on the role of ultraviolet radiation. *Limnology and Oceanography*, 49, 225-232.

[62] Rhode SC, Pawlowski M & Tollrian R (2001) The impact of ultraviolet radiation on the vertical distribution of zooplankton of the genus *Daphnia*. *Nature*, 412, 69-72.

[63] van Donk E & Hessen DO (1995) Reduced digestibility of UV-B stressed and nutrient-limited algae by *Daphnia magna*. *Hydrobiologia*, 307, 147-151.

[64] de Lange HJ & van Reeuwijk PL (2003) Negative effects of UVB-irradiated phytoplankton on life history traits and fitness of *Daphnia magna*. *Freshwater Biology*, 48, 678-686.

[65] Zellmer ID, Arts MT & Šustr V (2006) Food chain effects of sublethal ultraviolet radiation on subarctic *Daphnia pulex* - a field and laboratory study. *Archiv für Hydrobiologie*, 167, 515-531.

[66] Hessen D, De Lange H & Van Donk E (1997) UV-induced changes in phytoplankton cells and its effects on grazers. *Freshwater Biology*, 38, 513-524.

[67] Hansson LA (2000) Induced pigmentation in zooplankton: a trade-off between threats from predation and ultraviolet radiation. *Proceedings of the Royal Society of London B: Biological Sciences*, 267, 2327-2331.

[68] Hansson LA (2004) Plasticity in pigmentation induced by conflicting threats from predation and UV radiation. *Ecology*, 85, 1005-1016.

[69] Borgeraas J & Hessen DO (2000) UV-B induced mortality and antioxidant enzyme activities in *Daphnia magna* at different oxygen concentrations and temperatures. *Journal of Plankton Research*, 22, 1167-1183.

[70] Borgeraas J & Hessen DO (2002) Variations of antioxidant enzymes in *Daphnia* species and populations as related to ambient UV exposure. *Hydrobiologia*, 477, 15-30.

[71] Hessen D, Borgeraas J & Ørbaek JB (2002) Responses in pigmentation and antioxidant expression in Arctic *Daphnia* along gradients of DOC and UV exposure. *Journal of Plankton Research*, 24, 1009-1017.

[72] Hessen DO, Borgeraas J & Ørbæk JB (2002) Responses in pigmentation and antioxidant expression in Arctic *Daphnia* along gradients of DOC and UV exposure. *Journal of Plankton Research*, 24, 1009-1018.

[73] Souza MS, Hansson LA, Hylander S et al. (2012) Rapid enzymatic response to compensate UV radiation in copepods. *PLoS One*, 7, 1-6.

[74] Souza MS, Modenutti BE & Balseiro EG (2007) Antioxidant defences in planktonic crustaceans exposed to different underwater light irradiances in Andean lakes. *Water, Air and Soil Pollution*, 183, 49-57.

[75] MacFadyen EJ, Williamson CE, Grad G et al. (2004) Molecular response to climate change: temperature dependence of UV-induced DNA damage and repair in the freshwater crustacean *Daphnia pulicaria*. *Global Change Biology*, 10, 408-416.

[76] Siebeck O & Bohm U (1994) Challenges for an appraisal of UV-B effects upon planktonic crustaceans under natural radiation conditions with a non-migrating (*Daphnia pulex* obtusa) and a migrating cladoceran (*Daphnia galeata*). *Archiv für Hydrobiologie*, 43, 197-206.

[77] Grad G, Burnett BJ & Williamson CE (2003) UV damage and photoreactivation: Timing and age are everything. *Photochemistry and Photobiology*, 78, 225-227.

[78] Grad G, Williamson CE & Karapelou DM (2001) Zooplankton survival and reproduction responses to damaging UV radiation: A test of reciprocity and photoenzymatic repair. Limnology and Oceanography, 46, 584-591.

[79] Huebner JD, Loadman NL, Wiegand MD et al. (2009) The effect of chronic exposure to artificial UVB radiation on the survival and reproduction of *Daphnia magna* across two generations. *Photochemistry and Photobiology*, 85, 374-378.

[80] Fischer JM, Nicolai JL, Williamson CE et al. (2006) Effects of ultraviolet radiation on diel vertical migration of crustacean zooplankton: An *in situ* mesocosm experiment. *Hydrobiologia*, 563, 217-224.

[81] Hansson LA & Hylander S (2009) Size-structured risk assessments govern *Daphnia* migration. *Proceedings of the Royal Society B: Biological Sciences*, 276, 331-336.

[82] Gonçalves RJ, Souza MS, Aigo J et al. (2010) Responses of plankton and fish from temperate zones to UVR and temperature in a context of gobal change. *Ecología Austral*, 20, 129-153.

[83] Siegel DA, Maritorena S, Nelson NB et al. (2002) Global distribution and dynamics of colored dissolved and detrital organic materials. *Journal of Geophysical Research*, 107, 3228.

[84] Vasilkov AP, Krotkov N, Herman J et al. (2001) Global mapping of underwater UV irradiances and DNA-weighted exposures using Total Ozone Mapping Spectrometer and Sea-viewing Wide Field-of-view Sensor data products. *Journal of Geophysical Research: Oceans*, 106, 27205-27219.

[85] Karanas JJ, Van Dyke H & Worrest RC (1979) Midultraviolet (UV-B) sensitivity of *Acartia clausii* Giesbrecht (Copepoda). *Limnology and Oceanography*, 24, 1104-1116.

[86] Lacuna DG & Uye SI (2001) Influence of mid-ultraviolet (UVB) radiation on the physiology of the marine planktonic copepod *Acartia omorii* and the potential role of photoreactivation. *Journal of Plankton Research*, 23, 143-155.

[87] Browman HI, Vetter RD, Alonso Rodríguez C et al. (2003) Ultraviolet (280 - 400 nm) induced DNA damage in the eggs and larvae of *Calanus finmarchicus* G. (Copepoda) and atlantic cod (*Gadus morhua*). *Photochemistry and Photobiology*, 77, 2-29.

[88] Saito H, Uye SI & Taguchi S (1998) Effects of ultraviolet radiation (UVB) on marine zooplankton. *Global Environmental Research*, 2, 203-210.

[89] Ma Z, Li W & Gao K (2012) Impacts of solar UV radiation on grazing, lipids oxidation and survival of *Acartia pacifica* Steuer (Copepod). *Acta Oceanologica Sinica*, 31, 126-134.

[90] Kouwenberg JHM, Browman HI, Runge HA *et al.* (1999) Biological weighting of ultraviolet (280-400 nm) induced mortality in marine zooplankton and fish. II. *Calanus finmarchicus* (Copepoda) eggs. *Marine Biology*, 134, 285-293.

[91] Speekmann CL, Bollens SM & Avent SR (2000) The effect of ultraviolet radiation on the vertical distribution and mortality of estuarine zooplankton. *Journal of Plankton Research*, 22, 2325-2350.

[92] Karanas JJ, Worrest RC & Van Dyke H (1981) Impact of UV-B radiation on the fecundity of the copepod *Acartia clausii*. Mar Biol 65: 125-133.

[93] Naganuma T, Inoue T and Uye S (1997) Photoreactivation of UV-induced damage to embryos of a planktonic copepod. *Journal of Plankton Research*, 19, 783-787.

[94] Saito H & Taguchi S (2003) Influence of UVB radiation on hatching success of marine copepod *Paracalanus parvus* s.l. *Journal of Experimental Marine Biology and Ecology*, 282, 135-147.

[95] Hylander S, Grenvald JC and Kiørboe T (2013) Fitness costs and benefits of ultraviolet radiation exposure in marine pelagic copepods. *Functional Ecology*, 28, 149-158.

[96] Tartarotti B & Torres JJ (2009) Sublethal stress: Impact of solar UV radiation on protein synthesis in the copepod *Acartia tonsa*. *Journal of Experimental Marine Biology and Ecology*, 375, 106-113.

[97] Llabrés M & Agustí S (2006) Picophytoplankton cell death induced by UV radiation: Evidence for oceanic Atlantic communities. *Limnology and Oceanography*, 51, 21-29.

[98] Llabrés M, Agustí S, Alonso-Laita P & Herndl GJ (2010) *Synechococcus* and *Prochlorococcus* cell death induced by UV radiation and the penetration of lethal UVR in the Mediterranean Sea. *Marine Ecology Progress Series*, 399, 27-37.

[99] Keller AA, Hargraves P, Jeon H et al. (1997) Effects of ultraviolet-B enhancement on marine trophic levels in a stratified coastal system. *Marine Biology*, 130, 277-287.

[100] Fields DM, Durif CM, Bjelland RM et al. (2011) Grazing rates of *Calanus finmarchicus* on *Thalassiosira weissflogii* cultured under different levels of ultraviolet radiation. *PLoS One*, 6, e26333.

[101] Kouwenberg JHM & Lantoine F (2007) Effects of ultraviolet-B stressed diatom food on the reproductive output in Mediterranean *Calanus helgolandicus* (Crustacea; Copepoda). *Journal of Experimental Marine Biology and Ecology*, 341, 239-253.

[102] Aarseth KA & Schram TA (1999) Wavelength-specific behaviour in *Lepeophtheirus salmonis* and *Calanus finmarchicus* to ultraviolet and visible light in laboratory experiments (Crustacea: Copepoda). *Marine Ecology Progress Series*, 186, 211-217.

[103] Martynova DM & Gordeeva AV (2010) Light-dependent behavior of abundant zooplankton species in the White Sea. *Journal of Plankton Research*, 32, 441-456.

[104] Ma Z, Li W & Gao K (2010) Horizontal migration of *Acartia pacifica* Steuer (copepoda) in response to UV-radiation. *Journal of Photochemistry and Photobiology B: Biology*, 101, 233-237.
[105] Bollens SM & Frost BW (1990) UV light and vertical distribution of the marine planktonic copepod *Acartia hudsonica* Pinhey. *Journal of Experimental Marine Biology and Ecology*, 137, 89-93.
[106] Hylander S & Hansson LA (2010) Vertical migration mitigates UV effects on zooplankton community composition. *Journal of Plankton Research*, 32, 971-980.
[107] Lampert W (1989) The adaptive significance of diel vertical migration of zooplankton. Functional Ecology, 3, 21-27.
[108] Hays GC (2003) A review of the adaptive significance and ecosystem consequences of zooplankton diel vertical migrations. *Hydrobiologia*, 503, 163-170.
[109] Williamson CE, Olson OG, Lott SE, Walker ND, Engstrom DR, et al. (2001) Ultraviolet radiation and zooplankton community structure following deglaciation in Glacier Bay, Alaska. *Ecology*, 82, 1748-1760.
[110] Hylander S & Hansson LA (2013) Vertical distribution and pigmentation of Antarctic zooplankton determined by a blend of UV radiation, predation and food availability. Aquatic Ecology, 47, 467-480.
[111] Moeller RE, Gilroy S, Williamson CE et al. (2005) Dietary acquisition of photoprotective compounds (mycosporine-like amino acids, carotenoids) and acclimation to ultraviolet radiation in a freshwater copepod. *Limnology and Oceanography*, 50, 427-439.
[112] Hylander S, Boeing WJ, Graneli W et al. (2009) Complementary UV protective compounds in zooplankton. *Limnology and Oceanography*, 54, 1883-1893.
[113] Ringelberg J, Keyser AL & Flik JG (1984) The mortality effect of ultraviolet radiation in a translucent and in a red morph of *Acanthodiaptomus denticornis* (Crustacea, Copepoda) and its possible ecological relevance. *Hydrobiologia*, 112, 217-222.
[114] Vestheim H & Kaartvedt S (2006) Plasticity in coloration as an antipredator strategy among zooplankton. *Limnology and Oceanography*, 51, 1931–1934.
[115] Hylander S, Larsson N & Hansson LA (2009) Zooplankton vertical migration and plasticity of pigmentation arising from simultaneous UV and predation threats. *Limnology and Oceanography*, 54, 483-491.
[116] Hylander S, Souza MS, Balseiro E et al. (2012) Fish-mediated trait compensation in zooplankton. *Functional Ecology*, 26, 608-615.
[117] Karentz D, McEuen FS, Land MC & Dunlap WC (1991) Survey of mycosporine-like amino acid compounds in Antarctic marine organisms: potential protection from ultraviolet exposure. Marine Biology, 108, 157-166.
[118] Newman SJ, Dunlap WC, Nicol S & Ritz D (2000) Antarctic krill (*Euphausia superba*) acquire a UV-absorbing mycosporine-like amino acid from dietary algae. *Journal of Experimental Marine Biology and Ecology*, 255, 93-110.
[119] Balseiro E, Souza MS, Modenutti B & Reissig M (2008) Living in transparent lakes: Low food P :C ratio decreases antioxidant response to ultraviolet radiation in *Daphnia*. *Limnology and Oceanography*, 53, 2383–2390.
[120] Souza MS, Modenutti BE, Carrillo P et al. (2010) Stoichiometric dietary constraints influence the response of copepods to ultraviolet radiation-induced oxidative stress. *Limnology and Oceanography*, 55, 1024–1032.

[121] Ahmed FE, Setlow RB, Grist E & Setlow N (1993) DNA damage, photorepair, and survival in fish and human cells exposed to UV radiation. *Environmental and Molecular Mutagenesis*, 22, 18-25.

[122] Yasuhira S, Mitani H & Shima A (1991) Enhancement of photorepair of ultraviolet-damage by preillumination with fluorescent light in cultured fish cells. *Photochemistry and Photobiology*, 53, 211-215.

[123] Sancar GB & Sancar A (1987) Structure and function of DNA photolyases. *Trends in Biochemical Sciences*, 12, 259-261.

[124] Friedberg EC, Walker GC & Siede W (1995) *DNA Repair and Mutagenesis*. American Society of Microbiology Press, Washington.

[125] Sancar A and Sancar GB (1988) DNA repair enzimes. *Annual Review of Biochemistry*, 57, 29-67.

[126] Persaud AD & Williamson CE (2005) Ultraviolet and temperature effects on planktonic rotifers and crustaceans in northern temperate lakes. *Freshwater Biology*, 50, 467-476.

[127] Connelly SJ, Moeller RE, Sanchez G & Mitchell DL (2009) Temperature effects on survival and DNA repair in four freshwater cladoceran *Daphnia* species exposed to UV radiation. *Photochemistry and Photobiology*, 85, 144-152.

[128] Zagarese HE, Feldman M & Williamson CE (1997) UV-B induced damage and photoreactivation in three species of *Boeckella* (Copepoda, Calanoida). *Journal of Plankton Research*, 19, 357-367.

[129] Aarseth KA & Schram TA (2002) Susceptibility to ultraviolet radiation in *Calanus finmarchicus* and *Lepeophtheirus salmonis* and the adaptive value of external filtering (Crustacea: Copepoda). *Journal of Plankton Research*, 24: 661-679.

[130] Dey D, Damkaer D & Heron G (1988) UV-B dose/dose-rate responses of seasonally abundant copepods of Puget Sound. *Oecologia*, 76, 321-329.

[131] Boyce DG, Lewis MR & Worm B (2010) Global phytoplankton decline over the past century. Nature, 466, 591-596.

[132] Browman HI (2003) Assessing the impacts of solar ultraviolet radiation on the early life stages of crustacean zooplankton and ichthyoplankton in marine coastal systems. *Estuaries*, 26, 30-39.

[133] Ma Z, Li W and Gao K (2013) Impacts of UV radiation on respiration, ammonia excretion, and survival of copepods with different feeding habits. Hydrobiologia, 701: 209-218.

In: Copepods: Diversity, Habitat and Behavior
Editor: Laurent Seuront
ISBN: 978-1-63117-846-7

Chapter 6

SWIMMING, FEEDING AND MATING BEHAVIOR IN COPEPODS: A BRIEF REVIEW

***Luciana Sabia*[1,2,*]*, Marco Uttieri*[2]*, Giacomo Zagami*[1] *and Enrico Zambianchi*[2]**

[1]Department of Animal Biology and Marine Ecology, University of Messina, Salita Sperone, Messina, Italy

[2]Department of Sciences and Technologies, University of Naples "Parthenope", Centro Direzionale di Napoli, Napoli, Italy

ABSTRACT

Animal behavior is an important aspect for the knowledge of the ecological role of a species: it permits to understand the interactions between individuals, and on a larger scale on population dynamics. Copepod behavior occurs at very different scales, ranging from millimeters to tens of meters and over time spans ranging from milliseconds to hours. In the first case, events such as food encounter, creation of feeding currents and escape responses develop over very short times (milliseconds to seconds) and occur over a few millimeters or less. On the opposite extreme large scale processes like vertical migrations take place. At intermediate scales (minutes and centimeters), it is possible to obtain information on how an individual copepod moves during longer periods while feeding or how its behavior is modified in relation to unfavorable environmental conditions or to chemical or mechanical stimuli. In this contribution, we review the most critical aspects of copepod behavior, with a peculiar emphasis on the swimming, feeding and mating behavioral adaptations of these organisms, which, despite the small dimensions, display a great repertoire of modalities of interaction with the surrounding environment.

[*] E-mail address: luciana.sabia@uniparthenope.it.

INTRODUCTION

Life for a copepod is quite different from what is familiar to us as "they are small, at least in comparison to us, they inhabit a fluid environment, and while they may not be 'thinking' creatures, they are endowed by natural selection with behavioural algorithms that are open to rational interpretation. Small, wet and rational" [1].

In this chapter several aspects of copepod behavior are reviewed, where behavior is intended, following Mauchline [2], as all the actions of the individual and of the population that are not directly connected with the physical and chemical environment. These actions enable the individuals and/or the population to successfully reproduce and to protect from predation and extinction.

Behavior is important to comprehend the ecological role of a species: the main task for a copepod is surviving and survival probability of individual, hence their species, depends on their perceptual abilities, probability of being predated and feeding requirements [1]. Survival probability thus largely depends on the strategies adopted for searching food and mates and hiding or escaping predators. These factors have consequences on the spatial pattern of distribution and on the interactions with other species. From the study of these factors it is possible to have an in depth comprehension of population dynamics [1]. Behavioral studies are indeed a "bottom-up" approach to the subject: starting from the individual they may reach ecosystem level conclusions.

The first definition of plankton considered organisms as passively transported by the fluid in which they are immerged, with no ability to perform any decision about the direction to be followed in it. Indeed the word plankton derives from a Greek term and means "wandering, drifter", hence zooplankton are "wandering animal". Passive drifting is a very important factor in the displacement into the water column, since zooplankters are not able to swim against currents. For this reason it is quite difficult to observe groups of organism behaving like a single one like fishes can do (schools) [2]. In contrast with this definition, a great number of studies show how, despite water currents being a very strong forcing contributing to spread these organisms in different areas, zooplankters are able to perform "a wide diversity of movements and behaviour" [3]. These behaviors allow them to move independently from the flow, and not only be passively transported in it.

Motions are modulated by the presence of food, predators, conspecific, mates, and by the information obtained from the surrounding environment. Zooplankters are able to obtain information they need through the interpretation of hydromechanical and chemical signals transmitted through the fluid in which they live. All these actions are made possible through motion, so a relevant part of the study of the copepod behavior is represented by the study of the swimming behavior.

This chapter focuses on the individual behavior of copepods and how this behavior can influence phenomena at the ecosystems level. Focusing on copepods, both Schminke [4] and Kiørboe [5] stressed that the capability of moving in the fluid is an explanation to the fact that copepods are the most abundant animals on earth. Following Scminke's [4] opinion, secondary factors in the ability of copepods to colonize every aquatic habitat are copepods' dimensions, typically ranging from 0.5 to 5.0 mm [6], and their ability of having different transitional stages through metamorphosis. Kiørboe [5], instead, focused on the torpedo-shaped body of copepods, and their ability to have a "very detailed three-dimensional view of the signals coming from the fluid" [7] through the sensory antennulae and on the great swimming speed [8], ranging between 1 and 1000 mm s^{-1} [9].

In recent years, considerable advances have been made in the understanding of the behavior of zooplankton in general and of copepods in particular, but much remains to be done. This contribution is intended as a comprehensive review of copepod behavior (swimming, feeding and mating), though not exhaustive owing to the great number of references available in the literature.

SWIMMING BEHAVIOR

Copepods exhibit a wide diversity of movement, accordingly to different strategies and external stimuli. A three-dimensional trajectory describes at best their motion, because they live in a three-dimensional environment. However, in some cases (e.g., a nearly completely isodirectional kind of motion, or a unidirectional one) a two-dimensional trajectory may be enough to describe their motion. Their swimming speeds and motion patterns are, however, highly variable for a given species considered under different biotic and abiotic conditions, but also between species.

Swimming Mechanisms

Copepods live at the edge between the turbulent and the viscous worlds [1]: as their dimensions are typically of a few millimeters, depending on their swimming speed they perceive the fluid in which they live as a viscous medium (when moving slowly, in the order of a few body lengths per second). In other circumstances they perceive it as inertial (for example, during fast escape responses). Most of the time the medium is perceived as viscous and fast propulsion is more difficult in this case.

Copepods move beating different appendages depending on the kind of motion they perform: the vibration of their cephalic appendages near the mouthparts creating feeding currents, leads to slow displacement as the effect of the motion of water directly near the mouth [11-13]. In this case their energetic expenditure is low and their motion is characterized by low Reynolds number [14,15]. When swimming, instead, the energetic expenditure is higher because they have to face the inertia due to the viscous nature of their environment. In this case the swimming legs are used [2]. Antennulae are also used to allow stronger propulsion in jumping and escaping [3]. In these cases copepods can reach very high speeds, up to 500 body lengths per second, and the acceleration can reach a value of 1200 cm s^{-2} [3,16]. When performing such fast motions, copepods perceive the medium as turbulent, characterized by high Reynolds numbers [16]. This kind of motion obviously requires much more energy than the previous ones [17] and is only rarely used, sometimes by males to catch the female after following her signals and often as an escape reaction to predators. Escape reactions are very common in copepods and are often the answer to hydormechanical cues. For instance, Green et al. [18] studied the escape reactions of various species of copepod nauplii to the flow field generated by blue mussels and stressed the differences in escape reaction among species. The main part of the capture of the mussels occurred when copepods were in the sector where the deformation rate of the fluid was weakest.

Swimming Patterns

Copepod swimming behavior lead to a great variety of trajectories: they can be a smooth linear path or a very convoluted track. The trajectories can allow the exploration of large volumes of water or allow remaining in a restricted area for a long time [3]. Calanoid copepods in particular usually swim at speed ranging from 1 to 20 mm s^{-1} [2]. Jumps are of the order of 10 to 20 mm [2], although instantaneous swimming speed can reach much higher values, often exceeding 500 m s^{-1} in escape reactions [19].

Several modes of motion are distinguishable in copepod swimming and divided into four classes [2]:

- cruise: it consists of swimming at moderate velocities and the path can be almost rectilinear or more convoluted, depending on the environmental factors. In some cases copepods adopt a looping behavior, cruising in a helicoidal trajectory, and often following a chemical stimulus perceived in the fluid. The path and the average speed obviously depend on the species-specific motion patterns. This swimming mode is considered optimal when foraging at low turbulence levels [20].
- cruise and sink: it is characterized by an alternation between the slow swimming described above and periods of non activity, in which the animal sinks vertically into the water column. Usually copepods are slightly negative buoyant, although their lipid content may affect their buoyancy [21] and in some cases they use antennulae to slow down their sinking speed [22] which usually may vary with the species and with the stage. Sinking velocities usually range between 0.3 and 14.8 mm s^{-1}, depending on the different density of the copepod [2].
- hop and sink: when performing a hop, absolute swimming speed usually becomes really high, of the order of hundreds of millimeters per second (e.g., [19] jump speed of two species of *Acartia* ranging between 121 and 573 mm s^{-1}) and consequently highly energetically expensive. Consequently, this motion needs to be alternated with an energy saving kind of motion. Copepods perform rapid jumps into the water with different directions, alternated with periods of sinking. In this case the motion is quite jerky.
- jump and rest: fast jumps, already described in this section, are interposed by periods of rest, but also as a rare motion made as a reaction to a stimulus coming from the outside, at the end of which the copepods go back swimming in the previous swimming mode.

In Figure 1 are represented the simulated trajectories of two different copepod motion patterns: the black track represent the swimming behavior of a continuously cruising copepod (e.g., *Temora longicornis*), while the grey one represent the swimming behavior of a hop and sink species (e.g., *Oithona plumifera*). Both trajectories have the same number of data points. Cruise and sink is a mixture between these two extremes, while often in jump and rest behavior, most of the time the copepod is not moving.

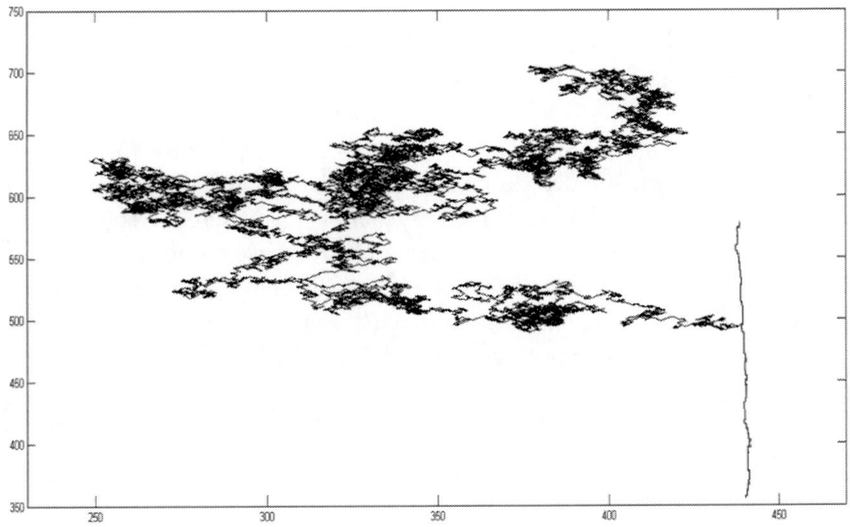

Figure 1. Examples of simulated tracks of two typical copepod behavior, a continuously cruising species (black), and a species exhibiting a hop and sink behavior (grey). The *x* and *y* axes are expressed in virtual spatial units.

Table 1. A short review of swimming modes frequently used by some copepod species

Species	Swimming mode	Reference
Acartia clausi	Hop and sink	[66]
Acartia tonsa	Hop and sink	[52]
Centropages hamatus	Slow cruise and break	[131]
Centropages typicus	Fast cruise and sink	[66]
Centropages velificatus	Hop and sink	[131]
Clausocalanus furcatus	Cruising	[116]
Eurytemora affinis	Cruise / cruise and sink	[28]
Oithona davisae	Jump-swimming	[132]
Oithona nana	Break , occasional motion	[133]
Oithona plumifera	Sink and occasional hop	[134]
Oithona similis	Break, occasional motion	[133]
Oncaea venusta	Hop and break	[133]
Paracalnaus parvus	Slow cruise and break	[66]
Pseudocalanus elongatus	Slow cruise and break	[66]
Pseudodiaptomus annandalei	Cruise and break (with occasional jumps)	[53]
Pseudodiaptomus marinus	Cruise and break	[54]
Temora longicornis	Slow cruise and break	[66]
Temora stylifera	Cruise	[131]
Temora turbinala	Cruise	[133]
Undinula vulgaris	Upward cruise, break	[133]

As mentioned above, each copepod species can adopt just one or more of these swimming modes. Some species cruise continuously, while other adopt a hop and sink behavior. A non-exhaustive review of the swimming modes described in the literature for a range of copepod species is provided in Table 1.

Swimming Mode and Alterations as an Adaptation to Biotic and Abiotic Conditions

Each of the swimming modes described above is adopted depending on the external conditions and forcing received by each species, each with a different frequency.

During the resting phase, copepods get rid of those food particles and detritus accumulated near the mouthparts and the appendices (grooming phase) which could reduce their efficiency in searching and capturing food [17]. Furthermore, stopping their appendages they are in the best condition to perceive chemical and hydromechanical cues coming from the environment and react consequently.

A classification first made by Fraenkel & Gunn [23] divides copepods motion into two kinds of movements:

- random motions are classified as kinesis and comprise all the small scale motion made in any direction. These movements may allow a rapid dispersion of the individuals, as long as may lead to accumulation of organisms [24,25].
- some directed movements are instead classified as taxis. Examples can be vertical migrations, more common in species that live in tidal and/or hyperbenthic environments [2,26], made daily to avoid visual predators during daytime [27] and to feed on fresh food along the water column at night time or, as an example, to follow the optimal salinity concentration [28]. These migrations can be driven by diverse factors, the most important of which seems to be the quantity of light [29] but also temperature [30], dissolved oxygen, food availability [31].

Although swimming can be very important in copepods' life, passive drifting is a very important factor in the formation and maintenance of aggregation of copepods [25,32]. Copepod behavior can be affected by turbulence in many different ways: they can change their speed, the geometry of their trajectories and the relative residence time in swimming or in other states [17,33]. The effect on the animals varies depending on the speed, the spatial and temporal scales of the turbulent flow and on the size of the animals [20,34]. The distribution of some copepodites could be independent of turbulence until a certain threshold, above which their swimming abilities are not strong enough to swim against the turbulent forces of water [35].

Small-scale turbulence may act in different ways on copepod behavior. It may enhance escape responses, while persistent condition of turbulence may cause habituation and a decrease in the frequency of elicited escape responses [36,37] or predator sensing ability [38], while at the same time reduced turbulence levels may increase prey-predator encounter rate [39]. For spatial scales lower than 10 m, turbulence and swimming behavior are the main factors acting on copepod distribution and patch formation [3]. At larger scales, structures such as fronts, Langmuir cells and internal waves are much more effective in the formation of patches [25], creating regions of aggregation ranging from a few meter to many kilometers where the copepod densities can be 3-5 times higher than the background density [2].

Among the external factors influencing copepod behavior, temperature is one of the most obvious. From a strictly behavioral point of view, all the crustaceans tend to avoid extreme temperatures (behavioral thermoregulation) for a klinokinetic response to the stimulus instead of a tactic response [30]. A sudden reduction in temperature may act as a stimulus and the copepod becomes more active, but this effect lasts for a limited amount of time, then the activity decreases rapidly [31]. From a metabolic point of view, temperature directly affects the activity of the copepod.

Another factor strongly influencing swimming activity is salinity: some studies conducted on the estuarine copepod *Eurytemora affinis* [28,40] outlined how the activity of the copepods change with a change in salt concentration. It influences the time spent swimming with respect to the time of inactivity, the swimming speed, the main direction and the complexity of the trajectory. These works also stress changes in the overall swimming path and directions when the salinity is away from the optimal range.

Non-lethal concentrations of pollutants may alter the swimming behavior of copepods, including naupliar stages [2]. These pollutants can modify the capability of avoiding predators, mating, and reduce their fitness. This happens by altering the typical turning angles of the animals, their swimming speeds and swimming/breaking rhythms or altering the sensing and locating capability of mates or prey [41-43]. In numerous ecotoxicological studies, changes in behavior have recently been used as biomarkers [44]. The first effect on the swimming behavior of some species is the tendency to avoid areas of higher concentration of pollutants like oil [44]. A higher concentration of hydrocarbons could reduce the complexity of the tracks and in some cases alter the swimming speeds [45,46]. It has also been shown [47] that a high presence of APEs (Alkylphenol-polyethoxylate) in the water may cause an increase in the swimming speed and an alteration of the relative residence time and frequency in the swimming of the *E. affinis.*

In addition animal concentration and that of predators may also be factors influencing swimming activity [48,49]. In conditions of high predator abundance, copepods tend to increase the inter-individual distances to avoid being predated altogether at the same time. In this case they change their swimming behavior leading to a higher dispersion of the individuals [50].

Swimming Changes Through Stages and Sex

Among the factors influencing the swimming patterns and speeds of copepods, the developmental stage is one of the most evident: different stages swim and behave differently. The naupliar stages can have a motion quite different from that of adults, with differences in the type of the trail, frequency, duration and velocities [51]. Titelman & Kiørboe [51] grouped them into two swimming types: the type A moves with a jump-sink, and a more active type B swims more or less continuously with occasional jump and sink events. Swimming varies both with the species and with the stage, usually ranging from 0.5 to 2 mm s^{-1}. By contrast, copepodids behave in a similar manner to adults, but often with reduced speeds, which increase as the copepodids grow through the stages [52].

Among the adult stages, swimming speeds are often higher in males than in females, possibly due to the mate-seeking activity of the male. In other cases (e.g., [53,54]) males are slower than females when no external stimuli, such as the presence of females or predators, are present, due probably to the reduced dimensions of males. In the same way, the time spent swimming by the males is much higher with respect to the time spent swimming by females. In both cases males afford a greater risk of predation.

FEEDING BEHAVIOR

Most copepod species, in particular the smaller ones, have historically been defined as herbivorous. However, a given copepod species is usually not exclusively herbivorous [55] although they are the major consumers of phytoplankton at sea. There is a small group of obligate carnivores [56], but most of copepods are omnivorous. They feed mainly on phytoplankton, but also on heterotrophic microplankton (ciliates, dinoflagellates) or often on detritus [55]. All of them are able to search and select the food they ingest, choosing the size of the particles or the plankton species they ingest[57]. In general copepods prefer eating motile preys with respect to non-motile ones [57], and each species is characterized by a different feeding strategy. In this way, copepods enhance the probability of obtaining a nutritionally complete ration and thus enhance the fitness of the population [55]. For example, monospecific diets may result in sterility [58]. Feeding can be an energetically expensive activity due to the very diluted environment in which copepods live [59].

Feeding activity is mainly influenced by temperature [55] while ingestion rate varies linearly with particle concentration and depends on particle size [55]. Moderate levels of turbulence may enhance the encounter rate, but when turbulence becomes higher, it may be detrimental for the pursuit of the prey [55]. The filtration rate depends on the sensory and motor capabilities of the copepod species [56].

Feeding Modes

Copepods feed either by filtering water and capturing little particles, or with raptorial actions to catch prey. Some copepods create feeding currents moving rhythmically and at high speed using the cephalic appendages: exo- and endopodites of maxilipeds, 2nd antennae and maxillae [17]. The flow generated by the cephalic appendages carries algae into the capture area, a well defined area around the mouthparts in which algae can be caught [12,60]. The movements of these appendages are species-specific and from these movements it is possible to retrieve the sensory feeding behavior (mechanoreception, chemoreception or both) [61]. Through high speed micro-cinematography and often using tethered animals it is possible to observe in detail the mechanism of creation of a feeding current which consists of three steps [60]: (i) searching for food, (ii) encounter and food recognition, and (iii) food capture, handling and ingestion or rejection of particle. The handling consists only of two or three movements. If the copepod is not able to catch the alga with these movements, it will be rejected [60].

Feeding currents usually move water from a distance of the order of millimeters and the capture area is usually smaller than 1 mm^2, whit a maximum speed of 1 cm s^{-1} near the capture area [60]. The flow field of the feeding current is viscous and thus the motion around the mouthparts is laminar, and not all the algae that are in the feeding current can be captured. The copepod has the ability to sense (feeding currents are usually associated with chemical cues to recognize particles) the algae coming and redirect the current to capture a chosen particle [11] by beating maxillipeds and endopodites of the second antennae [60]. In some case, particles are captured, tasted through chemical detectors directly on the mouthparts and then rejected if not palatable [13]. Particles can be grabbed both passively and actively. Usually small sized particles are captured passively, while the others can be both passively and actively captured [62]. Some evidences, indeed, show that copepods are able to capture particles smaller than the intersetal distance [55] and that they are able to choose their food, and the capture is made possible through the motion of different appendages [63,64]. In this case mechanoreception is an important component [61]. Some experiments also demonstrated that rejection of food decreases when the copepods are exposed for a long time to the same food species [55].

The filter feeding strategy is commonly used by copepods for little particles suspended in the fluid, like phytoplankton cells. For larger moving food, like dinoflagellates and microzoooplankton, which are energetically more conspicuous but are also able to escape from the feeding current, copepods can actively move, search and catch their prey. By moving they are able to increase their probability of encountering food [65]. Also in this case, different strategies have been observed to maximize the intake.

A strategy commonly used by cyclopoids and some calanoid copepods consists in passively slow sinking into the water column and perceive a possible prey through mechanoreceptors [66,67] and then actively capture them [59]. This strategy is called ambush feeding and is used to capture prey like ciliates [67]. In this case, the hydromechanical signal created by the deformation of the fluid owing to the copepod is not high, helping the copepod not being perceived both in the capture of prey and in the avoidance of predators. Another feeding strategy consists of actively swimming in search for prey, trying to reduce as much as possible the fluid deformation, and to approach the prey in a quieter way than while filtering.

Foraging Strategies

The overall feeding strategy of a copepod species can be a combination of feeding currents and one of the two other motion strategies (ambush feeding and cruising).

There is evidence that naupliar stages, usually feeding on smaller particles, are also able to choose between motile and non-motile prey, showing changes in their behavior in relation to the prey [68]. Other studies on species that at the adult stage are ambush feeders, showed how the earlier copepodites stages are more dependent on the suspension feeding due to their low capability of fast sinking and to their not yet completely developed sensing apparatus. As long as they grow they enhance their capability of sinking faster and eat more efficiently while suspension feeding [69].

Foraging strategies are among the most important forcing in the formation and maintenance of patches and aggregations: on a larger scale, ranging from centimeters to meters (the typical scale of the phytoplankton patches in the ocean [70]). The behavior displayed when searching for preys is very different compared to the behavior when inside a high density food patch (the food density can be three times higher than in the surrounding fluid [71]). Copepods tend to remain inside a patch as long as possible in order to maximize their food intake; this is achieved by changing their swimming behavior. They decrease their swimming speed and adopt a more looping and convoluted pathway, with horizontal circles and in general a more horizontal trajectory [72,73]. The efficiency in capturing prey, thus the fitness, is maximal when the amplitude of meanders is comparable to the copepod detection radius [74]. Swimming speed is also important when searching for food: the higher the food concentration, the lower the optimal swimming speed. This swimming mode is defined *area restricted search* (ARS) [53] and is adopted by animals belonging to different taxa all with the aim of maximizing the exploitation of food resources when non homogeneously distributed; see e.g., [75,76]. Leising & Franks [77] showed that an area-restricted search behavior increases the foraging efficiency of zooplankton feeding on a theoretical distribution of phytoplankton cells

On the contrary, when the animal is not inside a high density food area, it tends to reach it in the minimum time required and with the minimum energetic expenditure, but at the same time it needs to explore a large volume of water with the lowest risk of being predated [78]. As mentioned before, copepods live in a very diluted environment thus lacking an effective search strategy might be very risky for the survival of the species. The same swimming behavior can be very efficient at exploiting food resources in the patch, but not equally efficient at finding food patches. The searching success depends crucially on the characteristics of the motion: the move lengths, the turning angle distribution and the prey distribution, always taking into account the costs of an ineffective search [79] and the risk of predation while searching. A strategy to avoid being predated is slowing down the maximal velocity to decrease the probability of being predated while searching [10]. In general, more linear trajectories result to be the best strategy in an uniform landscape, while in clumped landscapes a relatively more tortuous path could be a better strategy, although the optimal search strategy may be strongly affected by the patch density as much as by the total number of patches [80].

On a large scale, many copepods exhibit a rhythmic behavior also in feeding, increasing their activity at night. This behavior is quite difficult to distinguish from the environment-induced endogenous rhythms, and seems to be controlled by light intensity: during the day copepods remain near the bottom, feeding on detritus to avoid visual predators, while at night, when the risk of being predated is lower, they can feed living phytoplankton cells in the water column [81]. Furthermore, some species making vertical migrations avoid the surface high chlorophyll region also at night to avoid being predated, or avoid consuming toxic algae [82].

MATING BEHAVIOR

Another crucial problem for copepod survival is mating; it is fundamental for the success of a species and the proliferation of new generations. At the same time it can be dangerous because during the time spent by a mating couple to swim together, they are more visible to visual predators. Furthermore, before actual mating can take place, similarly to prey-predator interactions [65] three different actions need to be achieved [83]: (i) mate search, (ii) mate encounter, and (iii) mate recognition. Trivially, encounter rates depend on the concentration of males and females and the ability in searching for mate [84].

Mate Searching

Usually males search for females following cues that will be described below; in the absence of cues, they adopt more similar swimming strategies than the ones described for copepods searching for food, trying to scan as much volume possible and to avoid rescanning the same area. Thus encounter depends on the main part on the swimming behavior of searching males and on the modalities of communication between the two genders. As a general rule males and females exhibit different swimming behavior [40,85]. In most cases, males swim faster than females [86,87]. In some species, especially the demersal ones, males often swim slower than females, but their capability of exploring the water volume in which they are immersed is considerably higher than the females one. In this case, indeed, males adopt a much more explorative behavior, being more active than females, swimming almost continuously and with more linear tracks [42]. In both cases, however, when encountering a female signal, males are able to swim much faster than non-ovigerous females, in order to reach them for mating [88]. Such fast swimming behavior is adopted only when females are present, while when they are absent males reduce the swimming speed as a saving energy strategy. There are some copepod species in which males do not feed when mature for mating. This phenomenon occurs when mate searching strategies are in conflict with food searching strategies, like in ambush feeders [89]. These males swim at high speed when searching for mate and their swimming behavior is much different from the females and the non-mature males swimming behavior. In suspension feeding species, instead, males can feed and search for mate at the same time, although feeding is more efficient when hovering and searching for mate is more efficient when cruising. For this reason, in species like *Pseudocalanus elongatus*, *Temora longicornis* or *Temora stylifera* although the swimming behavior of males and females is generally similar, males spend more time cruising while females are hovering [89].

In several cases searching behavior on its own may not be efficient enough to allow the survival of a species. For this reason, according to some theories and experimental evidences, some copepod species tend to aggregate at a certain depth in the water column, forming aggregations or swarms to facilitate mating [84]. An example is given by the dense swarms (tens of individuals per milliliter) formed by the copepod *Dioithona oculata* [83]. Kiørboe & Bagøien [84] related the ability of the male searching behavior to the area where a copepod species can live. They found that if the mate searching activity is efficient enough, copepods can live in a large area, even when population density is low. In periods when the population density is too low or the searching is not efficient, the population is constrained to a smaller area to allow the maintenance of the population.

Recognition

The second step before mating can have place is recognition: it is necessary that males recognize a mature conspecific female and it is necessary that the females recognize their conspecific males to avoid any escape mechanism when the male approaches. Males usually remotely detect chemical pheromones produced by females, but mechanoreception is also involved in some cases.

The presence of pheromones involved in mating was described in a large number of species; see e.g., [90-93]. In a series of three papers published in 2005 [92,94,95], titled '*Blind dating*', choosing the behavior of three different copepod species, the authors described different modalities for finding mates, analyzing by means of three-dimensional video recordings the behavior of three species and concluded that mate finding in copepods can be divided into three main categories.

In the first one they analyzed the swimming behavior of the calanoid copepod *Centropages typicus* [92], and showed that males were able to sense a pheromone trail left by females up to ten centimeter long and 31 s old. When they found such a trail, they left their path and exhibit a 3-fold increase in their swimming speeds (from 10 mm s^{-1} to 30 mm s^{-1}) to reach the female following exactly the same track. In some cases, due to turbulence or the age of the track, males could lose it. In these instances, they further increased their swimming speed and made little circles around the point where they lost the track to find it again. When the male was a few body lengths away from the female and could detect her hydromechanically, he stopped following the chemical trail and jumped directly at her. The pursuit, however, was not always successful and in some cases the females escaped and the male lost its interest. A similar behavior is common in many other copepods species such as *Calanus marshallae* or *Temora longicornis* [92] with some differences in the modalities of leaving pheromones to allow a quicker recognition of the signal also in case of turbulence. The distance at which a male can detect the pheromone trail of the female varies with turbulence and with the species ability.

In the second study the behavior of another calanoid copepod, *Pseudocalanus elongatus*, was described [95]. Their behavior is typical for species in which females spend a lot of time hovering; females were less motile than in the previous case and there was a remarkable difference in the shape of the signal released. Females moved slowly and often reoriented their feeding current in the water, with the result that the pheromones released remained all around the body of the females, with a cloud shape instead of a trail. When the male perceived such a cloud, he swam at high speed around and inside it with looping and zig-zag behavior, sometimes slowing down and swimming directly to the female and touching her. During this period, the female reoriented often her feeding current and made a sort of slow dance allowing her to spread the pheromones in every direction, helping the male to precisely locate her.

These findings were also reported by Jacoby & Youngbluth [96], who described the differences in the mating behavior of three species of *Pseudodiaptomus.* In the three species the male searching movements when a female signal was present were similar and characterized by tight turns and loops, while in other cases males had a slower and calm swimming behavior. Females did not need to move or create feeding currents to elicit the response from males. The same behavior described by males searching for females was observed when males were placed in female preconditioned water.

In the third case, in *Acartia tonsa* [94], mate detections occur through hydromechanical cues. Below a critical distance, males were able to recognize the hydromechanical disturbance generated by a swimming female and replied with a sort of dance to recognize each other and decrease the distance. The dance lasted a few seconds and the copepod were able to recognize each other from a distance of few millimeters, the typical detection distance in hydromecanical sensing [97,98]. Mates of different species differ in frequency of jumping and in jump length, and this may be useful for the female to understand that the follower is a mate and not a predator. However when jumping the female cannot perceive signals from the surroundings because of the use of antennulae for locomotion, and this could be the reason why grasping of the female is executed at high speed [99]. Strickler [99] observed the same behavior in the freshwater copepod *Cyclops scutifer*. A few number of species use hydromechanical signals to recognize the mate,. This may be due to the possible error induced by noise in mechanical detection [100]. The species using hydromechanical signals are usually small, because the sensitivity to hydromechanical stimuli generally decreases with size [94].

Copulation

After the preliminary steps described above, mating can take place. In contrast to the other steps, the sequences of mating behavior are in general similar in all the species and consist of four successive phase [2]: (i) the attraction of the male to the female, (ii) the initial capture of the female by the male, (iii) the adoption of the mating position, and (iv) the transfer and attachment of spermatophore.

At the end the mates separate. Some males have a right geniculate antenna to grasp the female's urosome. In general the capture of the females takes place at high speeds and then the couple makes several rapid jumps [88]. During actual mating the female continues swimming and receives the spermatophore, while the male is not able to make any own propulsive movement [100]. Sometimes both the mates slowly sink to the bottom or stay motionless while in other cases a feeding current is generated and the couple makes distinctive spirals [88].

In some laboratory observations, males have been observed grasping and transferring spermatophores to other males, probably an error in recognizing the right mate in water overcrowded by females releasing pheromones, although this phenomenon needs further study [100].

Eggs produced can be freely released in the water and in this case females produce a consistently great number of eggs [102,103], while other species carry their eggs until they are ready to hatch; e.g., [104-107]. Although the number of eggs produced by egg carrying species is much lower [108], in general the survival rate of nauplii is much higher than the number for free spawning species. The rate of egg production in both cases varies with temperature, season, and with the availability and quality of food [104]. In herbivorous species the maximum egg production rate is associated with phytoplankton blooms, so that nauplii can take advantage of phytoplankton high abundance, but this timing is not clearly demonstrated [104]. Species living at low latitudes reproduce during all the year, while copepod species living at higher and middle latitudes usually show seasonality in breeding or, in some cases of high population densities, these species are also able to produce resting eggs, which do not develop immediately but sink to sea bed and remain viable for a long time (up to 40 years) [2]. The mortality of eggs and larval stages is usually high due to predation or unfavorable environmental conditions.

Techniques to Observe Copepod Behavior

Field Observations

Being able to observe copepod behavior *in situ* would be the best way to understand it, in a non-invasive way, but it is not an easy task, though some techniques have been developed to this aim.

Some techniques try to observe zooplankton movements directly in the water column using an underwater video camera. One of the first to adopt this solution was Froese [109] who used a submerged camera integrated in the cod end of a plankton net showing on board, together with CTD data, the images of the organisms being sampled to test the hypotheses on small-scale distribution of zooplankton. Six years later, McGehee & Jaffe [32] used a three-dimensional acoustic imaging system with the aim of non-invasively examining zooplankton swimming behavior. Soon after Tiselius [110] was using an *in situ* video camera to record zooplankton aggregation, with a system able to reach 100 m depth.

The Video Plankton Recorder [111] has also been extensively used in the field. The system uses four cameras and different volume sizes simultaneously with forward scattered light to image the zooplankters and provide information on the microscale distribution of such organisms (the field of view ranges between 0.5 and 10 cm). Underwater holography was also used to obtain a three-dimensional image of the field of view [112]. A more detailed summary about the progresses made in the field of underwater technologies for biological oceanography is given in Dahms & Hwang [113].

The technical difficulties in observing *in situ* aggregation phenomena and more in general copepods swimming behavior led to the beginning of laboratory observations of copepod behavior in small microcosms, trying to deduce the behavior in the water from these observations.

Laboratory Observations

It is very difficult to notice the details of swimming behavior only by direct observation, due to their reduced dimensions and high speed at which events occur. Particular techniques are needed to observe such processes [11,13,16,19], or alternatively appropriate models can be implemented to study it. One possibility is using cinematographic techniques, although various technologies can be used for diverse observation purposes. One approach consists of using high sampling rate camera to record the swimming behavior, in the order of hundreds of frames per second [14,19,64,86,114,115]. This technique allows to study all movements made at very high speed, such as the motion of the feeding appendages in a feeding current, or the fast reaction to a stimulus or a capture event. When using these techniques, copepods are often tethered to keep them in focus for all the time of observation.

A different cinematographic technique, with specific video apparatuses, observes copepods freely swimming in an aquarium. This technique, in general, has a much lower sampling rate sufficient to observe the behavior and the path followed by the animal in a microcosm. Several studies made in the last twenty-five years used this method to study copepod behavior in a range of conditions: searching for mate, searching for food or in different physical and chemical conditions [37,42,68,72,116-120]. Some critics were raised by Carpenter [121] about the dimensions of the aquarium that usually are not larger than a few liters, a constriction which does not adequately represent the ecosystem and community conditions. There are, anyway, some devices that could be made to minimize the effect of such a little environment and avoid any disturbance: light sources, streams, presence of food, too many individuals in the same environment could be only some examples of the factors that could influence copepod motion. The aquarium size needs to be large enough to consider negligible the wall effects at the center of it, but not too big to allow to maintain focus on each single organism and in relation with the organism's dimension [49]. In addition, a great number of replicates need to be observed to avoid errors due to the individual variability [33] but at the same time an overcrowded environment could alter the behavior.

Both the above-mentioned techniques can be used to record two-dimensional images, but three-dimensional observations of the swimming behavior are also possible. This method allows a more accurate estimation of looping behavior, swimming speeds and sinking. It is a more complicated method because it requires two different cameras and the synchronization of their signal to have an accurate definition of copepod track, but copepods usually swim in a 3D environment and thus, when possible this technique is preferable because it allows more information on isotropic conditions of the swimming. Anyway when a high number of trajectories are recorded, with a 2D observation it is possible to have statistically consistent information on velocity and thus it is good enough for investigations regarding the effect of given factors, although some swimming patterns can be misinterpreted [49].

From the analysis of the trajectories it is possible to obtain several kind of information: as an example it is possible to deduce swimming velocities, the kind of swimming adopted by the investigated organism, their capability of filling the space and the influence of biotic and abiotic factors on their swimming.

Theoretical Approaches

A system is composed by a relevant number of components often interacting in ways difficult to be studied simultaneously and to be predicted. Numerical modeling is often a good solution to study such processes and in particular to study behavioral ones. Numerical modeling, indeed, allows an in-depth comprehension of the factors involved in the phenomena and their reconstruction. In such cases, numerical modeling starts from experimental data and analyses processes that would be difficult to observe in the field or even in laboratory, such as the effect of turbulence on swimming behavior or on the encounter rate [122].

In particular individual based models (IBMs), taking into account the single individual involved in the studied phenomenon, are particularly useful in behavioral studies. Zooplankton motion, and copepod behavior in particular, is often described through a random walk, following the idea that motion is largely determined by random events. From this random walk it is then possible to understand the causes of a particular distribution along the water column. Random walks are able to simulate diffusive-like processes and turbulent dispersion and to link it to macro scale dynamics [1].

Some theories do not consider animal movement as a purely random walk but, taking into account the episodic long excursions in a precise direction, compare their movement to a

Lévy distribution, a probability distribution in which both Gaussian and the so-called "long-tailed" distributions are taken into account [1].

Theoretical models can be a useful tool to investigate features influencing zooplankton ecology related to swimming behavior. An example is given by the model of Visser & Thyghesen [25], who analyzed the spatial distribution of plankton and their tendency to aggregate. Several other studies used both theoretical models to describe copepod motion (e.g., [85,123,124]) and random walk applied to the characteristics of a real copepod species; see e.g., [125-129]. Theoretical models are designed to study processes and are thus less flexible than random walk models. Random walk models, instead are better suited to describe specific dynamics and allow a much higher flexibility: the same model can be adapted, by changing some key variables, to different species and conditions.

Conclusion

Behavior can help to understand the autoecology of a species, its ecological role and its differences with the other species coexisting in the same environment. In this way it is possible to understand possible niche separation. It allows an in-depth vision of the possible strategies that make a species successful in an ecosystem. Copepods display a great variety of adaptive behavior, which may have direct consequences on the strategies for escaping predators and foraging. Swimming behavior can be considered as the result of the adaptive processes to the surrounding environment and is consequently an important source of information on the biological responses of the animal to different conditions.

Thus starting from the individual scale (how each animal moves, finds resources, is able to detect and capture prey and mates, and to escape from predators) and from the interaction with the environment, such as the distribution of patches, the chemical, physical and biological conditions experienced by each individuals, it is possible to observe patterns that characterize the population

This bottom-up trend is particularly important when dealing with copepods, which occupy a key position in the functioning of the pelagic food web. Thus through the adequate comprehension of the ecology of this group and of the mechanisms that determine the answers of copepods to environmental forcing, it is possible to go up in the food web and understand and predict the possible effects on the whole ecosystem of a particular event at the macroscopic level.

Although in the last 30 years our comprehension of copepod behavior and its consequences at the population and ecosystem levels has made giant leaps, much remains to be done. For example, in recent years copepod behavior has been used as a sub-lethal biomarker in ecotoxicology studies, while at the same time this peculiar aspect may be helpful in understanding the invasive dynamics of alien copepod species. These examples clarify the importance of pursuing this line of research, which will provide extremely useful details on the adaptations of these small crustaceans to a variable environment.

ACKNOWLEDGMENTS

The research of M. Uttieri was supported by the MOKA project (Modelling and Observation of zooplanKtonic orgAnisms; ID: RBFR10VF6M) financed by the Italian Ministry of Education, University and Research.

REFERENCES

[1] Visser AW (2011) *Small, Wet and Rational*. Technical University of Denmark, Copenhagen.

[2] Mauchline J (1998) *The Biology of Calanoid Copepods*. Academic Press, San Diego.

[3] Strickler JR (1977) Observation of swimming performances of planktonic copepods. *Limnology and Oceanography*, *22*, 165-170.

[4] Schminke HK (2007) Entomology for the copepodologists. *Journal of Plankton Research*, *29*, 1149-1162.

[5] Kiorbøe T (2011) What makes pelagic copepods so succesful? *Journal of Plankton Research*, *33*, 677-685.

[6] Huys R & Boxshall GA (1991) *Copepod Evolution*. The Ray Society, London.

[7] Fields DM, Shaeffer DS & Weissburg MJ (2002) Mechanical and neural responses from the mechanosensory hairs on the antennule of *Gaussia princeps*. *Marine Ecology Progress Series*, *227*, 173-186.

[8] Lenz J, Hower AE & Hartline DK (2004) Force production during pereiopod power strokes in *Calanus finmarchicus*. *Journal of Marine Systems*, *49*, 133-144.

[9] Yen J (2000) Life in transition: balancing inertial and viscous forces by planktonic copepods. *Biological Bulletin*, *198*, 213-224.

[10] Zollner PA & Lima SL (2005) Behavioural tradeoffs when dispersing across a patchy landscape. *Oikos*, *108*, 219-230.

[11] Alcaraz M, Paffenhofer GA & Strickler JR (1980) Catching the algae: a first account of visual observations on filter feeding calanoids. *American Society of Limnology and Oceanography Special Symposium*, *3*, 241-248.

[12] Strickler JR (1982) Calanoid copepods, feeding currents, and the role of gravity. *Science*, *218*, 158-160.

[13] Paffenhöfer G.A, Strickler JR & Alcaraz M (1982) Suspension-feeding by herbivorous calanoid copepods: a cinematographic study. *Marine Biology*, *67*, 193-199.

[14] Alcaraz M & Strickler JR (1988) Locomotion in copepods: pattern of movements and energetics of *Cyclops*. *Hydrobiologia*, *167/168*, 409-414.

[15] Vlymen WJ (1970) Energy expenditure of swimming copepods. *Limnology and Oceanography*, *15*, 348-356.

[16] Strickler JR (1975) Intra- and interspecific information flow among planktonic copepods: receptors. *Verhandlungen Internationalen Verein Limnologie*, *19*, 2951-2958.

[17] Alcazar M, Saiz E & Calbet A (2007) *Centropages* behaviour: swimming and vertical migration. *Progress in oceanography*, *72*, 121-136.

[18] Green S, Visser AW, Titelman J & Kiørboe T (2003) Escape responses of copepod nauplii in the flow field of the blue mussel *Mytilus edulis*. *Marine Biology*, *142*, 727-733.

[19] Buskey EJ, Lenz PH & Hartline DK (2002) Escape behaviour of planktonic copepods in response to hydrodinamic disturbances: high speed video analysis. *Marine Ecology Progress Series*, *235*, 135-146.

[20] Visser AW, Mariani P & Pigolotti S (2009) Swimming in turbulence: zooplankton fitness in terms of foraging efficiency and predation risk. *Journal of Plankton Research*, *31*, 121-133.

[21] Sargent JR & Falk-Petersen S (1998) The lipid biochemistry of calanoid copepods. *Hydrobiologia*, *167/168*, 101-114.

[22] Jiang H & Strickler JR (2005) Mass density contrast in relation to the feeding currents in calanoid copepods. *Journal of Plankton Research*, *27*, 1003-1012.

[23] Fraenkel GS & Gunn DS (1961) *The Orientation of Animals*. Dover Publications, London.

[24] Flierl GR, Grunbaum D, Levin SA & Olson D (1999) From individuals to aggregations: the interplay between behaviour and physics. *Journal of Theoretical biology*, *196*, 397-454.

[25] Visser AW & Thygesen UH (2003) Random motility of plankton: diffusive and aggregative contributions. *Journal of Plankton Research*, *25*, 1157-1168.

[26] Zagami G (2010) Copepoda calanoidi iperbentonici. *Biologia Marina Mediterranea*, *17*, 432-434.

[27] Bollens SM & Frost BW (1991) Diel vertical migration in zooplankton: rapid individual response to predators. *Journal of Plankton Research*, *13*, 1359-1365.

[28] Seuront L (2006) Effect of salinity on the swimming behaviour of the estuarine calanoid copepod *Eurytemora affinis*. *Journal of Plankton Research*, *28*, 805-813.

[29] Lampert W (1989) The adaptative significance of diel vertical migration of zooplankton. *Functional Ecology*, *3*, 21-27.

[30] Lagerspetz KYH & Vainio LA (2006) Thermal behaviour of crustaceans. *Biological Reviews*, *81*, 237-258.

[31] Bezerra-Neto JF & Pinto-Coelho RM (2007) Diel vertical migration of the copepod *Thermocyclops inversus* (Kiefer, 1936) in a tropical reservoir: the role of oxygen and the spatial overlap with *Chaborus*. *Aquatic Ecology*, *41*, 535-545.

[32] Mc Gehee D & Jaffe JS (1996) Three dimensional swimming behaviour of individual zooplankters: observations using the acoustical imaging system FishTV. *ICES Journal of Marine Science*, *53*, 363-369.

[33] Seuront L, Yamazaki H & Souissi S (2004) Hydrodynamic disturbance and zooplankton swimming behaviour. *Zoological Studies*, *43*, 376-387

[34] Moison M, Schmitt FG, Souissi S et al. (2009) Symbolic dynamics and entropies of copepod behaviour under non turbulent and turbulent conditions. *Journal of Marine Systems*, *77*, 388-396.

[35] Maar M & Nielsen Tisser AW (2003) Distribution of zooplankton in relation to turbulent diffusion. *Limnology and Oceanography*, *48*, 1312-1325.

[36] Saiz E, Alcaraz M & Paffenhöfer GA (1992) Effects of small-scale turbulence on feeding rate and gross-growth of three *Acartia* species (Copepoda: Calanoida). *Journal of Plankton Research*, *14*, 1085-1097.

[37] Fields DM & Yen J (1997) The escape behaviour of marine copepods in response to a quantifiable fluid mechanical disturbance. *Journal of Plankton Research*, *19*, 1289-1304.

[38] Gilbert OM & Buskey EJ (2005) Turbulence decreases the hydodynamic predator sensing ability of the calanoid copepod *Acartia tonsa*. *Journal of Plankton Research*, *27*, 1067-1071.

[39] MacKenzie BR, Miller TJ, Cyr S & Leggett WC (1994) Evidence for a dome shaped relationship between turbulence and larval fish ingestion rate. *Limnology and Oceanography*, *39*, 1790-1799.

[40] Michalec FG, Souissi S, Dur G et al. (2010) Differences in behavioural responses of *Eurytemora affinis* (Copepoda, Calanoida) reproductive stages to salinity variations. *Journal of Plankton Research*, *32*, 805-813.

[41] Sullivan BK, Buskey EJ, Miller DC & Ritacco PJ (1983) Effects of copper and cadmium on growth, swimming and predator avoidance in *Eurytemora affinis* (Copepoda). *Marine Biology*, *77*, 299-306.

[42] Weis JS, Samson J, Zhou T et al. (2001) Effects of contaminants on behaviour: biochemical mechanisms and ecological consequences. *Bioscience*, *51*, 209-217.

[43] Clement B & Zaid S (2004) A new protocol to measure the effects of toxicants on daphnid–algae interactions. *Chemosphere*, *55*, 1429-1438.

[44] Seuront L (2010) Zooplankton avoidance behaviour as a response to point sources of hydrocarbon-contaminated water. *Marine and Freshwater Research*, *61*, 263-270.

[45] Seuront L & Leterme SC (2007) Increased zooplankton behavioural stress in response to short-term exposure to hydrocarbon contamination. *The open Oceanography Journal*, *1*, 1-7.

[46] Seuront L (2011) Hydrocarbon contamination decreases mating success in a marine planktonic copepod. *PLoS ONE*, *6*, 26283.

[47] Cailleaud K, Michalec FG, Forget-Leray J et al. (2011) Changes in the swimming behaviour of *Eurytemora affinis* (Copepoda, Calanoida) in response to a sub-lethal exposure to nonylphenols. *Aquatic Toxicology*, *102*, 228-231.

[48] Woodson CB, Webster DR, Weissburg MJ & Yen J (2007) Cue hierarchy and foraging in calanoid copepods: ecological implications of oceanographic structure. *Marine Ecology Progress Series*, *330*, 163-177.

[49] Dur G, Souissi S, Schmitt FG et al. (2011) Effect of animal density, volume and the use of 2D/3D recording on behavioural studies of copepods. *Hydrobiologia*, *666*, 197-214.

[50] Tinbergen N, Impekoven M & Franck D (1967) An experiment on spacing-out as a defence against predation. *Behaviour*, *28*, 307-321.

[51] Titelman J & Kiørboe T (2003) Motility copepod nauplii and implications for food encounter. *Marine Ecology Progress Series*, *247*, 123-135.

[52] Buskey EJ (1994) Factors affecting feeding selectivity of visual predators on the copepod *Acartia tonsa*: locomotion visibility and escape responses. *Hydrobiologia*, *292/293*, 447-453.

[53] Dur G, Souissi S, Cheng SH & Hwang JS (2010) The different aspects in motion of the three reproductive stages of *Pseudodiaptomus annandalei* (Copepoda, Calanoida). *Journal of Plankton Research*, *32*, 423-440.

[54] Sabia L, Uttieri M, Pansera M et al. (2012) Swimming behaviour of the calanoid copepod *Pseudodiaptomus marinus* from Lake Faro. *Biologia Marina Mediterranea, 19*, 240-241
[55] Kleppel GS (1993) On the diet of calanoid copepods. *Marine Ecology Progress Series, 99*, 183-195.
[56] Kiorboe T & Saiz E (1995) Planktivorous feeding in calm and turbulent environments, with emphasis on copepods. *Marine Ecology Progress Series*, *122*, 133-145
[57] Schultz M & Kiorboe T (2009) Active prey selection in two pelagic copepods feeding on potentially toxic and non toxic dinoflagellates. *Journal of Plankton Research*, *31*, 553-561.
[58] Ianora A, Poulet SA & Miralto A (1995) A comparative study of the inhibitory effect of diatoms on the reproductive biology of the copepod *Temora stylifera*. *Marine Biology*, *121*, 533-539.
[59] Kiørboe T (2010) How zooplankton feed: mechanisms, traits and trade offs. *Biological Reviews*, *86*, 311-339.
[60] Strickler JR (1985) Feeding currents in copepods: two new hypotheses. *In*: Laverack MS (ed.), *Physiological Adaptations of Marine Animals*. Society for Experimental Biology Symposium 39, Research Books, 460-485.
[61] Poulet SA & Gill CW (1988) Spectral analyses of movements made by the cephalic appendages of copepods. *Marine Ecology Progress Series*, *43*, 259-267.
[62] Vanderploeg HA & Paffenhofer GA (1985) Modes of algal capture by the freshwater copepod *Diaptomus sicilis* and their relation to food-size selection. *Limnology and Oceanography*, *30*, 871-885.
[63] Price HJ & Paffenhöfer GA (1986) Effects of concentration on the feeding of a marine copepod in algal monocultures and mixtures. *Journal of Plankton Research*, *8*, 119-128.
[64] Price HJ, Paffenhöfer GA & Strickler JR (1983) Modes of cell capture in calanoid copepods. *Limnology and Oceanography*, *28*, 116-123.
[65] Gerritsen J, Strickler JR (1977) Encounter probabilities and community structure in zooplankton: a mathematical model. *Journal of fish Research Board Canada*, *34*, 73-82.
[66] Tiselius P & Jonsson P (1990) Foraging behaviour of six calanoids copepods: observations and hydrodinamical analysis. *Marine Ecology Progress Series*, *66*, 23-33.
[67] Kiørboe T, Saiz E & Viitasalo M (1996) Prey switching behaviour in the planktonic copepod *Acartia tonsa*. *Marine Ecology Progress Series*, *143*, 65-75.
[68] Henriksen CI, Saiz E, Calbet A & Hansen B (2007) Feeding activity and swimming patterns of *Acartia grani* and *Oithona davisae* nauplii in the presence of motile and non motile prey. *Marine Ecology Progress Series*, *331*, 119-129.
[69] Takahashi T & Tiselius P (2005) Ontogenic change of foraging behaviour during copepodite development of *Acartia clausi*. *Marine Ecology Progress Series*, *303*, 213-233.
[70] Cowles TJ (2004) Planktonic layers: physical and biological interactions on the small scale. *In*: Seuront L & Strutton PG (eds.), *Handbook of Scaling Methods in Aquatic Ecology - Measurements, Analysis, Simulation*. CRC Press, Boca Raton, 31-49.
[71] Dekshenieks MM, Donaghay PL, Sullivan JM et al. (2001) Temporal and spatial occurrence of thin phytoplankton layers in relation to physical processes. *Marine Ecology Progress Series*, *223*, 61-71.

[72] Tiselius P (1992) Behaviour of *Acartia tonsa* in patchy food environments. *Limnology and Oceanography*, *37*, 1640-1651.

[73] Leising AW (2001) Copepod foraging in patchy habitats and thin layers using a 2-D individual-based model. *Marine Ecology Progress Series*, *216*, 167-179.

[74] Visser AW (2007) Motility of zooplankton: fitness, foraging and predation. *Journal of Plankton Research*, *29*, 447-461.

[75] Fauchald P & Tveraa T (2003) Using first-passage time in the analysis of area-restricted search and habitat selection. *Ecology*, *84*, 282-288.

[76] Bailleul F, Pinaud D, Hindell M et al. (2008) Assessment of scale-dependent foraging behaviour in southern elephant seals incorporating the vertical dimension: a development of the First Passage Time method. *Journal of Animal Ecology*, *77*, 948-957.

[77] Leising AW & Franks PJS (2000) Copepod vertical distribution within a spatially variable food source: a simple foraging-strategy model. *Journal of Plankton Research*, *22*, 999-1024.

[78] Zollner PA & Lima SL (1999) Search strategies for landscape-level interpatch movements. *Ecology*, *80*, 1019-1030.

[79] Cain ML (1985) Random search by herbivorous insects: a simulation model. *Ecology*, *66*, 876-888.

[80] Heinz SK & Strand E (2006) Adaptative patch searching strategies in fragmented landscapes. *Evolutionary Ecology*, *20*, 113-130.

[81] Stearns DE (1986) Copepod grazing behaviour in simulated natural light and its relation to nocturnal feeding. *Marine Ecology Progress Series*, *30*, 65-76.

[82] Leising AW, Pierson JJ, Cary S & Frost BW (2005) Copepod foraging and predation risk within the surface layer during night-time feeding forays. *Journal of Plankton Research*, *27*, 987-1001.

[83] Buskey EJ (1998) Components of mating behaviour in planktonic copepods. *Journal of Marine Systems*, *15*, 13-21.

[84] Kiørboe T & Bagøien E (2005) Motility patterns and mate encounters rates in planktonic copepods. *Limnology and Oceanography*, 50, 1999-2007.

[85] Sabia L (2012) *An Integrated Approach to the Study of the Swimming Behaviour of Pseudodiaptomus marinus*. PhD thesis, University of Messina.

[86] Doall MH, Colin, S P, Strickler JR & Yen J (1998) Locating a mate in 3D: the case of *Temora longicornis*. *Philosophical Transactions of the Royal Society B*, *353*, 681-689.

[87] Uttieri M, Cianelli D, Strickler JR & Zambianchi E (2007) On the relationship between fractal dimension and encounters in three dimensional trajectories. *Journal of Theoretical Biology*, *247*, 480-491.

[88] Lee CH, Dahms HU, Cheng SH et al. (2011) Mating behaviour of *Pseudodiaptomus annandalei* (Copepoda, Calanoida) at calm and hydrodinamically disturbed waters. *Marine Biology*, *158*, 1085-1094.

[89] Kiørboe T (2008) Optimal swimming strategies in mate-searching pelagic copepods. *Oecologia*, *155*, 179-192.

[90] Katona S (1973) Evidence for sex pheromones in planktonic copepods. *Limnology and Oceanography*, *18*, 574-583.

[91] Griffiths AM & Frost BW (1976) Chemical communication in the marine planktonic copepods *Calanus pacificus* and *Pseudocalanus* sp. *Crustaceana*, *30*, 1-8.

[92] Bagøien E & Kiørboe T (2005) Blind dating-mate finding in planktonic copepods. I tracking the pheromone trail of *Centropages typicus. Marine Ecology Progress Series*, *300*, 105-115.

[93] Kiørboe T (2007) Mate finding, mating, and population dynamics in a planktonic copepod *Oithona davisae*: there are too few males. *Limnology and Oceanography*, *52*, 1511-1522.

[94] Bagøien E, Kiørboe T (2005) Blind dating-mate finding in planktonic copepods. III Hydromechanycal communication in *Acartia tonsa. Marine Ecology Progress Series*, *300*, 129-133.

[95] Kiørboe T, Bagøien E a Thygesen UH (2005) Blind dating-mate finding in planktonic copepods. II The pheromone cloud of *Pseudocalanus elongatus. Marine Ecology Progress Series*, *300*, 117-128.

[96] Jacoby CA & Youngbluth MJ (1983) Mating behaviour in three species of *Pseudodiaptomus* (Copepoda: Calanoida). *Marine Biology*, *76*, 77-86.

[97] Kiørboe T, Saiz E & Visser AW (1999) Hydrodynamic signal perception in the copepod *Acartia tonsa. Marine Ecology Progress Series*, *179*, 97-111.

[98] Visser AW (2001) Hydromechanical signals in the plankton. *Marine Ecology Progress Series*, *222*, 1-24.

[99] Strickler JR (1998) Observing free-swimming copepods mating. *Philosophical Transactions of the Royal Society B*, *353*, 671-680.

[100] Uchima M & Murano M (1988) Mating behaviour of the copepod *Oithona davisae. Marine Biology*, *99*, 39-45.

[101] Cheng SH, Lee, CH, Dahms HU & Hwang JS (2008) Homosexual mating in the planktonic copepod *Pseudodiaptomus annandalei* (Copepoda: Calanoida). *Journal of Crustacean Biology*, *28*, 580-582.

[102] Webb DG & Weaver A (1988) Predation and the evolution of free spawning in marine calanoid copepods. *Oikos*, *51*, 189-192.

[103] Saiz E, Tiselius P, Jonsson PR et al. (1993) Experimental records of the effects of food patchiness and predation on egg production of *Acartia tonsa. Limnology and Oceanography*, *38*, 280-289.

[104] Kiørboe T (2006) Sex, sex-ratios, and the dynamics of pelagic copepod populations. *Oecologia*, *148*, 40-50.

[105] Liang D & Uye S (1997) Population dynamics and production of the planktonic copepods in a eutrophic inlet of the Inland Sea of Japan. IV *Pseudodiaptomus marinus* the egg-carrying calanoid. *Marine Biology*, *128*, 415-421.

[106] Dur G, Souissi S, Devreker D et al. (2009) An individual-based model to study the reproduction of egg bearing copepods: application to *Eurytemora affinis* (Copepoda Calanoida) from the Seine estuary, France. *Ecological Modelling*, *220*, 1073-1089.

[107] Kiørboe T & Sabatini M (1994) Reproductive and life cycle strategie in egg carrying cyclopoid and free spawning calanoid copepods. *Journal of Plankton Research*, *16*, 1353-1366.

[108] Sabatini M & Kiørboe T (1994) Egg production, growth and development of the cyclopoid copepod *Oithona similis*. *Journal of Plankton Research*, *16*, 1329-1351.

[109] Froese R, Barthel KG, Welsch W et al. (1990) Development of an underwater video camera for recording ichthyoplankton and zooplankton. *ICES Journal of Marine science.*

[110] Tiselius P (1998) An *in situ* video camera for plankton studies: design and preliminary observations. *Marine Ecology Progress Series*, *164*, 293-299.
[111] Davis CS, Gallager SM, Berman MS et al. (1992) The Video Plankton Recorder (VPR): design and initial results. *Archiv fur Hydrobiologie Beiheft Ergebnisse Limnologie*, *36*, 67-81.
[112] Jaffe JS, Moore KD, McLean J & Strand MP (2001) Underwater optical imaging: status and prospects. *Oceanography*, *14*, 64-75.
[113] Dahms HU & Hwang JS (2010) Perspectives of underwater optics in biological oceanography and plankton ecology studies. *Journal of Marine Science and Technology*, *18*, 112-121.
[114] Cowles TJ & Strickler JR (1983) Characterization of feeding activity patterns in the planktonic copepod *Centropages typicus* Kroyer under various food conditions. *Limnology and Oceanography*, *28*, 106-115.
[115] Hwang JS, Turner JT, Costello JH et al. (1993) A cinematographic comparison of behaviour by the calanoid copepod *Centropages hamatus* Lilljeborg: tethered versus free-swimming animals. *Journal of Experimental Marine Biology and Ecology*, *167*, 277-288.
[116] Mazzocchi MG & Paffenhofer GA (1999) Swimming and feeding behaviour of the planktonic copepod *Clausocalanus furcatus*. *Journal of Plankton Research*, *121*, 1501-1518.
[117] Svensen C & Kiørboe T (2000) Remote prey detection in *Oithona similis*: hydromechanycal versus chemical cues. *Journal of Plankton Research*, *22*, 1155-1166.
[118] Kazutaka T & Tiselius P (2005) Ontogenic change of foraging behaviour during copepodite development of *Acartia clausi*. *Marine Ecology Progress Series*, *303*, 213-223.
[119] Jiang H & Paffenhofer GA (2008) Hydrodynamic signal perception by the copepod *Oithona plumifera*. *Marine Ecology Progress Series*, *373*, 37-52.
[120] Fields DM (2010) Orientation affects the sensitivity of *Acartia tonsa* to fluid mechanical signals. *Marine Biology*, *157*, 505-514.
[121] Carpenter S (1996) Microcosm experiments have limited relevance for community and ecosystem ecology. *Ecology*, *77*, 677-680.
[122] Lewis DM & Pedley TJ (2000) Planktonic contact rates in homogeneous isotropic turbulence: theoretical predictions and kinematic simulations. *Journal of Theoretical Biology*, *205*, 377-408.
[123] Mariani P, Botte V & Ribera d'Alcalà M (2005) An object-oriented model for the prediction of turbulence effects on plankton. *Deep Sea Research II*, *52*, 1287-1307.
[124] Cianelli D, Uttieri M, Strickler JR & Zambianchi E (2009) Zooplankton encounters in patchy particle distribution. *Ecological Modelling*, *220*, 596-604.
[125] Mariani, P, MacKenzie BR, Visser AW & Botte V (2007) Individual-based simulations of larval fish feeding in turbulent environments. *Marine Ecology Progress Series*, *347*, 155-169.
[126] Mariani P, Botte V & Ribera d'Alcalà M (2008) A numerical investigation of the impact of turbulence on the feeding rates of *Oithona davisae*. *Journal of Marine Systems*, *70*, 273-286.

[127] Wiggert JD, Haskell AGE, Paffenhöfer GA et al. (2005) The role of feeding behaviour in sustaining copepod populations in the tropical ocean. *Journal of Plankton Research*, *27*, 1013-1031.

[128] Wiggert JD, Hofmann EE & Paffenhöfer GA (2008) A modelling study of developmental stage and environmental variability effects on copepod foraging. *ICES Journal of Marine Science*, *65*, 379-398.

[129] Uttieri M, Sabia L, Cianelli D et al. (2010) Lagrangian modelling of swimming behaviour and encounter success in co-occurring copepods: *Clausocalanus furcatus* vs. *Oithona plumifera*. *Journal of Marine Systems*, *81*, 112-121.

In: Copepods: Diversity, Habitat and Behavior
Editor: Laurent Seuront
ISBN: 978-1-63117-846-7

Chapter 7

HIGH-SPEED IMAGING IN COPEPOD BEHAVIOR

Yuji Tanaka*

Plankton Laboratory, Graduate School of Tokyo University of Marine Science and Technology, Minato, Tokyo, Japan

ABSTRACT

Copepods, dominating the world of plankton, are capable of vibrating their appendages at high frequency that are well beyond the temporal resolution of both human eyes and brain (*ca.* 10-20 Hz) and normal video cameras (*ca.* 30 Hz). The detail in the motion of a copepod driven by such frequent vibration has thus been too difficult to observe by naked eyes or even video equipment with an ordinary performance. Historically, only a relatively small number of scientists have been disclosing the copepod motion by using high-speed imaging. Adding to their work, I here show some new observations on the motion of copepods recorded by use of microscopic high-speed video apparatuses that are capable of seeing how well the appendages are moving, and, how powerfully a copepod can swim propelled by the motion of appendages. Observations of *Calanus pacificus* CV demonstrate that the appendages of the mouthparts moves by collaborating to each other, and that the swimming legs beat powerfully in an organized manner to attain an abrupt acceleration and fast speed of swimming. Watching the free-swimming of a *Paraeuchaeta elongata* adult revealed that this animal is an amazingly powerful swimmer, even being capable of making aerial jump. These behavior could never been seen without the aid of high-speed apparatuses. Future studies of aquatic sciences should include such direct observations that may produce firsthand and quantitative information on the behavior of copepods, which are extremely abundant and important in the aquatic ecosystems.

* E-mail address: ytanaka@kaiyodai.ac.jp.

INTRODUCTION

Hearing a hundred times is not as good as seeing once.
- Qian Han Shu

This now colloquial expression is especially true in the field of copepod behavior as their actual motion is impossible to visualize without special tools. Their appendages can vibrate at such a high-frequency that human brain, eyes and nervous system cannot precisely recognize; ordinary video equipment that records *ca.* 30 frames per second are not resolving what is going on in such a high-frequency domain [1]. Driven by such a high-frequency motion of the appendages, copepods are capable of swimming so fast that nobody can easily determine the exact velocity (i.e., direction and speed) of the motion. Moreover, there exist some intrinsic difficulties in seeing copepods: they are tiny, and live in a three-dimensional fluid, which hinders optical instrumentation. Due to all these problems, making quantitative observations of copepod behavior has always been a challenge.

In spite of these difficulties, scientists have been tackling the problems in seeing copepod behavior, considering the ubiquity, abundance and dominance of copepods in the aquatic environment. In general, these animals play a key role in aquatic ecosystems: they grow by consuming smaller organisms and detritus (which are the base of aquatic primary production), and get eaten by larger animals such as zooplankton, micro-nekton and fish, hence linking to the food source of us human beings. Quantitative estimates on how copepods eat food, and how they are eaten by predators, and how they grow and how much energy they consume in the course of their life, are fundamental keystones to our understanding of marine ecosystems and to reliable evaluation of the flux of matter and energy through aquatic ecosystems.

Historically, scientists have been making significant effort to assess the behavior of copepods, that is, how copepods move. Two German pioneering papers [2, 3] provided the first microscopic high-speed strobe cinematographs of the motion of a copepod of the genus *Diaptomus*. In the 1970s, J.R. Strickler invented and established a method to film copepod motion very effectively and efficiently; one of the most important part of his work was the establishment of a way to tether the animal on the end of a dog hair so it stays in the focus range of a microscopic lens [4]. Strickler's early invention made a breakthrough in aquatic sciences and seminally inspired the generation of scientists to come. High-speed cinematography initially cost a lot of money [5], hence intrinsically limited its application to zooplankton behavior. Now, thanks to the recent advancement in technology, digital high-speed video apparatuses, which were initially developed for military and industry use and have been far too expensive to be widely used, are now reasonably affordable.

In this chapter, to stress the importance of knowing how copepods actually live (e.g., eat, escape, grow, mate) to a quantitative evaluation of the aquatic ecosystems, and further, in the future of human beings, I show some recent microscopic high-speed observations on the motion of copepod appendages, and a 3-D observations on the actual speed of free swimming or jumping of copepods, who dominate the world of plankton.

HIGH-SPEED CINEMATOGRAPHY OF A TETHERED COPEPOD

In this section, a few examples of high-speed observations on the motion of mouthpart appendages and swimming legs of tethered copepods are shown. This tethering method, invented by J.R. Strickler [4], allows us to keep the copepod within the focal depth of microscopic-lens.

Motion of Mouthpart Appendages to Hover and to Produce Feeding Currents

Firstly, I describe observations of the appendage motion of a *Calanus pacificus* copepodite stage V. I used a cubic chamber (80 mm × 80 mm × 80 mm) filled with filtered natural seawater at 20°C. In this chamber was immersed an adult *Calanus pacificus* (prosome length *ca.* 2.2 mm) tethered to the end of a piece of human hair, connected *via* platinum wire to a rod of solid glass held by a micro-manipulator (Figure 1). A Photron high-speed camera (Fastcam PCI model 100k) equipped with a Sonic zoom lens (BSL-Z50HD), allowing a working distance of 80 mm, was used. The recording was made through the side of the chamber, while keeping the animal's head oriented upward. To obtain an appropriate illumination, a Nippon P-I metal-halide cold-spot light (PCS-UMX250) was used, so the recording with frame rate and exposure time of 2000 fps and 1/103,000 s (*ca.* 9.7 μs), respectively, can be made. Images were subsequently analyzed frame by frame.

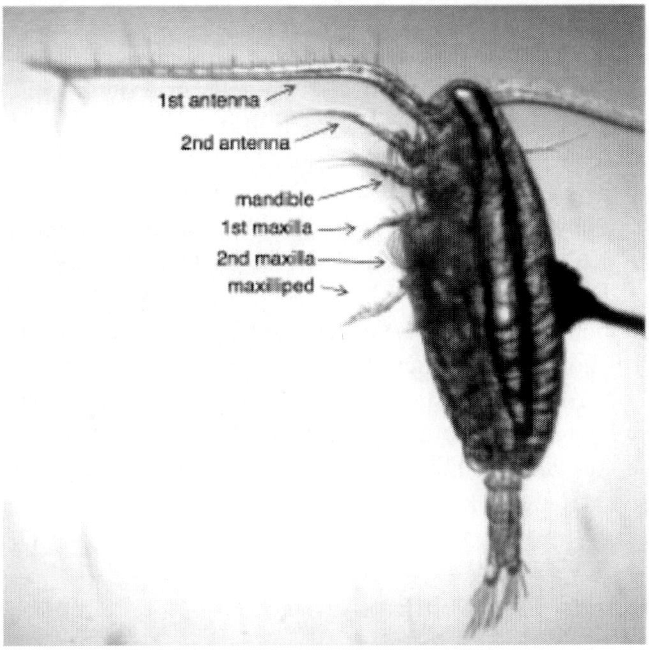

Figure 1. Close-up of a *Calanus pacificus* copepodite stage V (prosome length *ca.* 2.2 mm) tethered to the end of a human hair. This image is a view from 45° left from the ventral front.

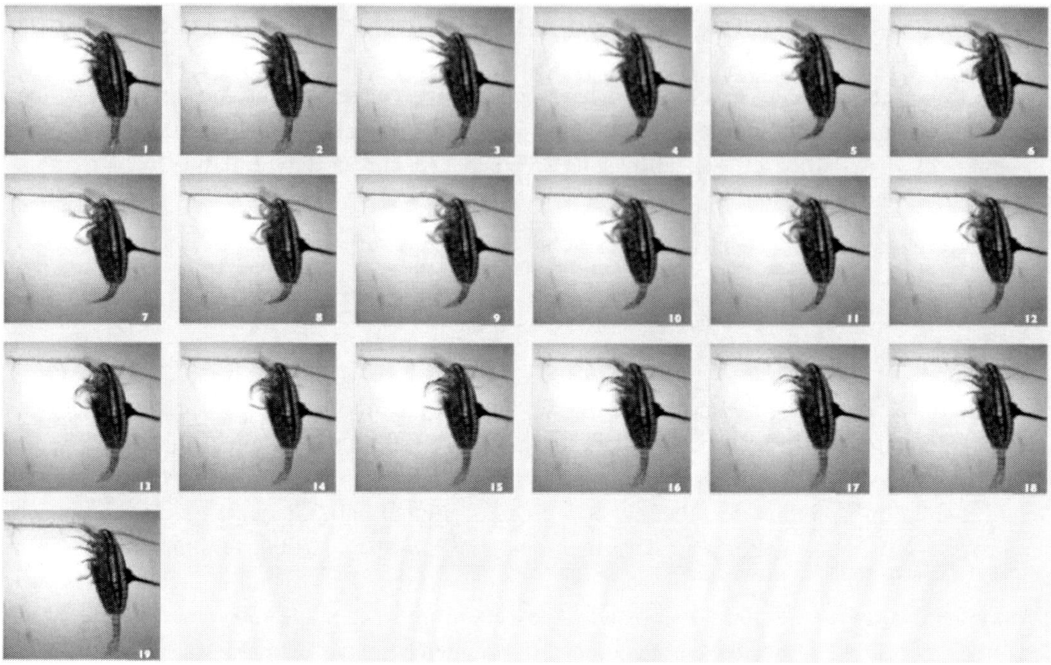

Figure 2. Illustration of a complete cycle of the motion of mouthpart appendages. Numerals in the bottom-right of each panel show the sequence. From the original images taken every 0.5 ms, those every 2 ms are picked out and shown in order. This one cycle of the appendage motion took *ca.* 34 ms to complete, giving a frequency of the motion *ca.* 30 Hz.

This animal, to hover or swim relatively slowly (compared with the quick motion involved in jumping), was flapping each appendage of the mouthpart like a scull of Temma boats (Figure 2). These appendages oscillate in an anterior-posterior direction, fluttering the each end. The motion appears to be well organized and constantly cyclic. Every pair of the appendages (2nd antennae, mandibles, first maxillae, second maxillae, and maxillipeds) oscillate at exactly the same frequency, but different phase in the flapping angle, apparently generating a constant upward thrust [6]. In this experiment, the cyclic motion had a frequency of *ca.* 30 Hz. Note that the 1st antennae and swimming legs remained motionless during the whole cycle.

This organized motion of mouthpart appendages has been reported as an effective way to stay at a certain position in the vertical (i.e., hovering; copepods are denser than seawater, hence sink in the absence of thrusting force) [6], producing feeding current, which occur while hovering, and detecting, capturing and ingesting (or rejecting) food particles [7]. On detecting food, the constancy in the motion may be changed to regulate the feeding current and manipulate food particles [7].

Motion of Swimming Legs While Jumping

Secondly, I show how a copepod maneuvers its swimming legs when making a rapid swim or jump. I used the chamber and tethering technique described in the previous section to examine the position and shape of the swimming legs, but used images from two directions (one from the left-hand side, and another from the top) were simultaneously recorded. To do so, two high-speed cameras (Photron Fastcam PCI R2) were synchronously used (Figure 3). The frame rate and exposure time were 500 fps and 1/70,000 s (*ca.* 14 μs), respectively. Lenses and illumination are as described in the previous section.

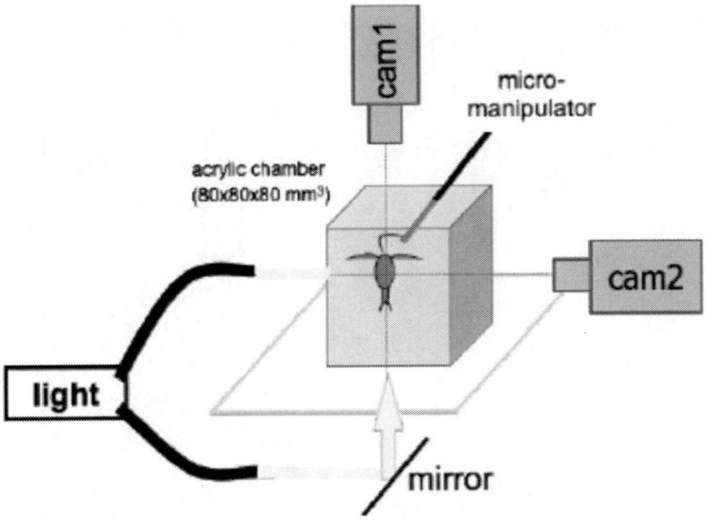

Figure 3. Experimental setup to observe a copepod simultaneously from two directions. Two synchronized high-speed cameras working are viewing the copepod from the top and from the side, respectively.

After a certain period of no motion, or of motion limited to the mouthpart appendages, the animal suddenly starts cyclic beat of the five pairs of swimming legs as shown in Figure 4. One cycle of this beat took approximately 30 ms, giving a beat frequency of *ca.* 30-35 Hz. During the fore-stroke (or power-stroke), the pair of swimming legs at the most posterior part of this animal start beating first, followed in order by the pair anterior to it until the most anterior pair. When all swimming legs made the fore-stroke, a recovery stroke starts immediately, leading to move all the legs backwards almost simultaneously. The images taken from the top show that each pair of swimming legs is widely spread during the fore-stroke, like a leaf of ginkgo. Conversely during the recovery stroke, the legs are tightly folded like a bundle of pine needles (Figure 5). The area of the legs during the recovery-stroke is only about a quarter of the area observed during the fore-stroke. This leads to maximize the propulsion during the forward strokes, while reducing the drag force generated by the recovery strokes. The motion of the swimming legs shown here has been detailed by Petra Lenz and her colleagues [8]

Note the warp in the whole body of the copepod during and after finishing the fore-stroke (Figure 6), suggesting that the body segments, each having a pair of appendages, are jointed flexibly. I presume that, this flexibility may of use in the powerful swimming; this may be a way to increase the efficiency of transferring the power generated by the muscle inside the copepod body [9] to the legs.

Figure 4. Illustration of a complete cycle of the power-stroke of a *Calanus pacificus* tethered to a human hair. The position and shape of the appendages were simultaneously recorded from the left-hand side and from the top using two synchronized high-speed cameras (Photron Fastcam PCI R2). The frame rate and exposure time were respectively 500 fps and 1/70,000 s (*ca.* 14 μs). Numerals in the bottom-left corner of each side-view panel show elapsed time in milliseconds. The two left-hand side columns shown the forestroke, while the two right-hand columns show the recovery stroke; both strokes last for *ca.* 14 ms.

The 30-35 Hz beat of swimming legs described above can be repeated 20 times or more, i.e., for a duration exceeding 0.6 s. Considering a 2 mm copepod can make a burst swimming speed of 1 m s^{-1}, a *Calanus pacificus* could make a sudden motion of half a meter or more in half of a second, which must be effective to escape from predators in natural environment [10].

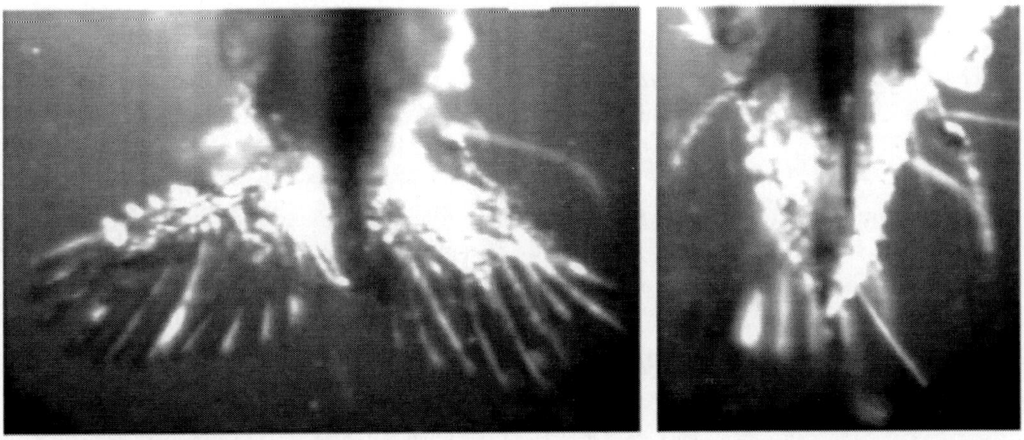

Figure 5. Shape of swimming legs of *Calanus pacificus* viewed from the top. Left: during a fore-stroke, the shape of the legs looks like a leaf of ginkgo or a web of waterfowls. Right: during a recovery-stroke, all the legs are fold together like a bundle of pine needles. The area of the legs during the recovery-stroke is only about a quarter of that during the fore-stroke. This stroke resembles the way taken by a paddling duck.

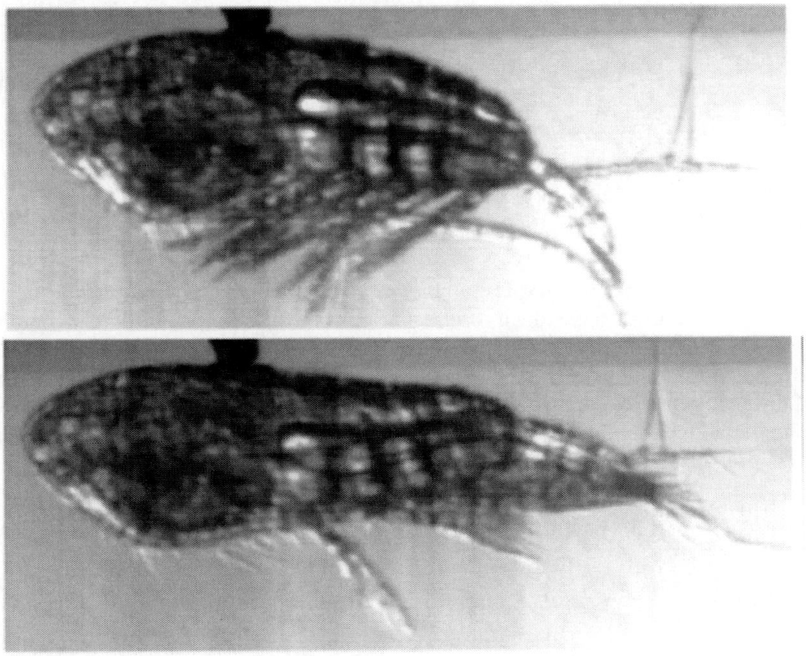

Figure 6. Side view of a *Calanus pacificus* before and after a fore-stroke. The top and bottom panels correspond to the images taken at times 0 and 14 ms in Figure 3. This crustacean animal has a hard shell, but apparently there is a warp in the whole body, suggesting that the body segments, each having a pair of appendages, are jointed to each other flexibly and are suggested as a way to enhance the transmission of the power generated by the internal muscle contraction to the legs.

Motion of the 1st Antennae

The 1st antennae of *Calanus pacificus* are fairly long. They are longer than the head to tail length of its body. When this animal is hovering or swimming slowly, the antennae are rigidly held perpendicular to the body axis. In contrast, while jumping the antennae are folded and oriented backwards along the body. The 2000 fps observation and close look at the movies suggest that the folding of the 1st antennae is an active process and may be generating some propulsion on commencement of a jump. This observation is consistent with previous studies showing that the 1st antennae move actively and generate acceleration [11, 12]. It is finally stressed that during a jump, the 1st antennae are folded against the ventral body surface and envelop all the mouthpart appendages, leading to an overall bullet-shaped body (Figure 7).

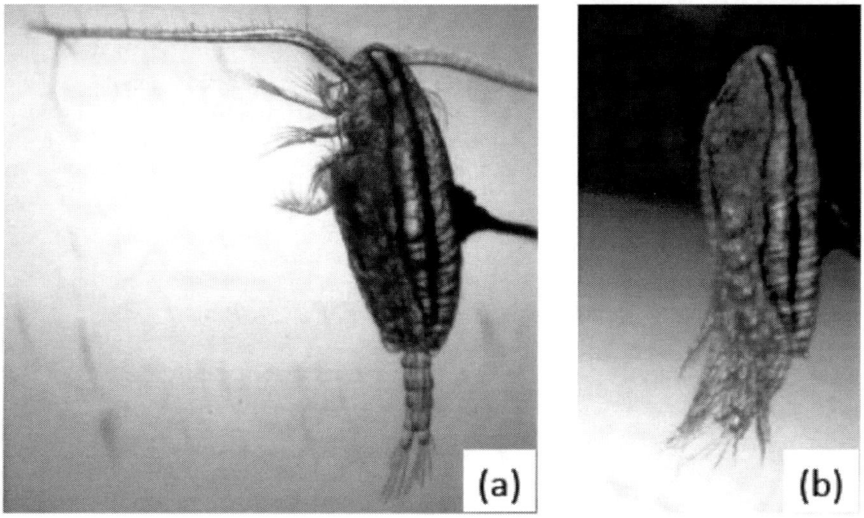

Figure 7. Two different postures of *Calanus pacificus* depending on the phase of swimming. (a): While hovering. Mouthpart appendages are in action, while all five pairs of the swimming legs are folded tightly orienting anterior on the abdominal surface. The 1st antennae are rigidly stretched side-ward perpendicularly to the body axis. (b): While jumping, reaching the final phase of a fore-stroke. All of the swimming legs are oriented posterior. The 1st antennae are folded on the ventral body surface, enveloping all the mouthpart appendages together, leading to an overall bullet-shaped body.

High-Speed Cinematography of a Free-Swimming Copepod

Here, I will show a few observations on the free-swimming behavior of a copepod. The examples shown here are of *Paraeuchaeta elongata*, a large-sized predatory copepod with strong ability in swimming. Using a system of two high-speed video cameras capable of synchronized recording, leg motion and swimming ability were recorded and analyzed three-dimensionally. Animal shown here is of *ca.* 4 mm prosome length (body length *ca.* 6-7 mm). The experimental set-up was similar to the one described in the previous section. In this experiment, the lenses were replaced with those of Fujinon-TVI: 0.85/25. Observations were recorded with a frame rate of 250 fps, and exposure time of 1/5,000 s (0.2 ms) to 1/1,000 s (1 ms).

Swimming Beahvior of *Paraeuchaeta elongata*

P. elongata, observed while swimming freely in a the acrylic chamber of 80 mm × 80 mm × 80 mm (Figure 8), hovers or cruises at 0-50 mm s^{-1}, being propelled by constant motion (*ca.* 70 Hz) of mouthpart appendages. While hovering or cruising, the animal intermittently makes a rapid swimming (or jump), with instantaneous speed exceeding 1.2 m s^{-1} (*ca.* 200 body length per second), by beating the swimming legs with a frequency between 30 and 90 Hz. The top speed is reached only after the first cycle of the leg beat, and the copepod then keep moving forward for 200-300 ms by beating multiple times, despite some deceleration occurring during recovery strokes. This rapid swimming is associated with Reynolds numbers in the 3,000-6,000 range.

Aerial Jumping Behavior of *P. elongata*

While free-swimming, a sudden stimulation of *P. elongata* induced escape reactions that can ultimately lead to aerial jumps through the water surface [13]. The sequence leading to such an extreme behavior is detailed below.

P. elongata was hovering in the water (see the circled area in Figure 9) immediately before it was stimulated by a solid glass rod reflecting strong illumination to make a sudden jump by sequentially beating the swimming legs. After the first stroke, half of the anterior part emerged from the water surface. After the second stroke, the whole body was in the air, and made an aerial jump. In the air, it made a somersault with double twist. On touching down, it made little splash.

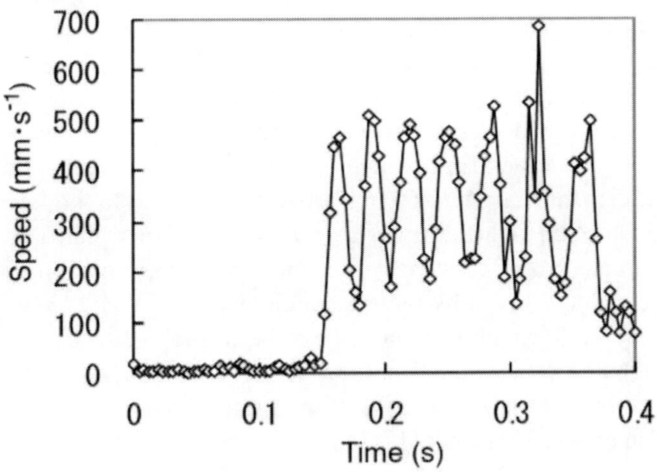

Figure 8. An example of measurements on the swimming speed of *Calanus pacificus* tracked three-dimensionally. Before 0.15 s, it was hovering (or cruising) with a speed of 0-25 mm·s^{-1}. After 0.15 s, it started to make a jump (or burst swimming) lasting for *ca.* 0.25 s. The maximum instantaneous speed reached almost 700 mm·s^{-1} in this measurement. The highest speed *ca.* 1,200 mm/s was observed in other series of measurements. Note the jump-swimming speed was fluctuating due to the fore- and recovery-stroke of the swimming legs.

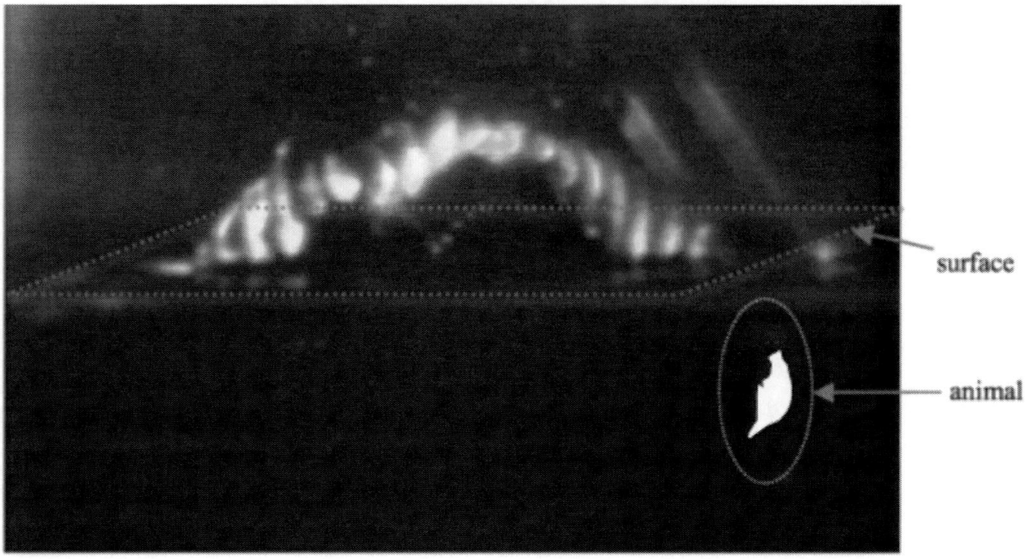

Figure 9. Aerial jump of *Paraeuchaeta elongata.* This picture is a composite image of the motion captured by a high-speed video at a frame rate of 250 fps, and corresponds to an exposure time of 0.5 ms. The dotted parallelogram indicates the surface of the seawater in an acrylic chamber of 80 mm × 80 mm × 80 mm. This copepod, first hovering in the water (in the oval circle at the right), was stimulated by a solid glass rod reflecting strong illumination and made a sudden jump. After the first stroke, half of the anterior part emerged above the water surface. After the second stroke, the whole body was in the air, leading to an aerial jump. In the air, it made a somersault with double twist. On touching down, it made a little splash.

This movie was actually recorded with two synchronized high-speed cameras, so the 3D position of the animal can be tracked to allow a precise analysis of the motion. As the result, the jump had a 45° elevation angle and an initial speed of 600 mm s^{-1} immediately after breaking the water surface; the maximum height of the trajectory was *ca.* 8 mm above the water surface. The horizontal distance traveled with this flight was *ca.* 40 mm. Since one cycle of swimming leg beat took approximately 10 to 20 ms, the acceleration for this jump by this leg beat can be estimated as 30 to 60 m s^{-2}, being equivalent to the force produced during a space rocket launch. Although the flight duration here was as short as 110 ms, this aerial jump may be an extreme, but effective way to avoid visual feeder near the surface [14].

Such escape jumps are driven by the systematic power strokes of swimming legs, which can beat with a very high speed. Copepods have been evolved to be capable of producing forces strong enough to swim fast against the viscous environment [3, 11, 15, 16]. The issue of understanding, from an adaptive and evolutionary stand-point, how and why this has been achieved is a growing area of research [17, 18].

CONCLUSION

High-speed cinematography is a device that conveniently enhances the temporal resolution of human eyes and brain. In other words, only by use of this device, we can recognize what is actually going on within such a time that is too short for a human to detect.

This is analogous to the role of microscopes in terms of enhancing the spatial resolution of human eyes and brain, which were invented more than five centuries ago. In this context, high-speed cameras can be considered '*temporal microscopes.*' The invention of microscopes was one of the greatest achievements in the history of science and technology that led us to discover the existence of tiny organisms that could have never been seen without the instruments. High-speed cameras as modern technology may be as important as microscopes. Originally being devised for military and industrial use, and having been too expensive for aquatic biologists to easily utilize, high-speed cameras are now becoming increasingly popular, thanks to the advancement in modern technology. Coupling modern high-speed imaging and microscopy together, we can further reveal how plankton are *actually* moving, hence opening wide perspectives in our understanding of the processes at play in the pelagic realm.

ACKNOWLEDGMENTS

I sincerely thank Professor J. Rudi Strickler of University of Wisconsin at Milwaukee, who inspired and encouraged me to plunge into the field of high-speed imagery of plankton. I also appreciate Emeritus Professor Takashi Ishimaru of Tokyo University of Marine Science and Technology, who spiritually and financially supported me to keep doing high-speed experiments. Professor Peter J. S. Franks of the Scripps Institution of Oceanography (SIO) provided me the opportunity to concentrate on doing such fundamental observations, facilitating the use of his lab space, equipment, and budget as well as stimulating discussions. Dr. Erdem Karaköylü of SIO greatly helped me sailing offshore to collect live copepods, culture algae for their food and working sometimes overnight with me in the laboratory. The present work was partly supported by JSPS KAKENHI Grant Number 16580148.

REFERENCES

[1] Strickler JR (1977) Observation of swimming performances of planktonic copepods. *Limnology and Oceanography*, *22*, 165-169.

[2] Storch O & Pfisterer O (1925) Der Fangapparat von Diaptomus. *Journal of Comparative Physiology A: Neuroethology, Sensory, Neural, and Behavioral Physiology*, *3*, 330-376.

[3] Storch O (1929) Die Schwimmbewegung der Copepoden, auf Grund von Mikro-Zeitlupenaufnahmen analysiert. *Verhandlungen der Deutschen Zoologischen Geselleschaft; Zoologische Anz. Suppl.*, *4*, 118-129.

[4] Barber RT & Hilting AK (2000) *Achievements in Biological Oceanography*, in *50 Years of Ocean Discovery: National Science Foundation 1950-2000.* National Academy Press, Washington, 11-21.

[5] Alcaraz M, Paffenhofer GA & Strickler JR (1980) Catching the algae: A first account of visual observations on filter-feeding calanoids. *In*: Kerfoot WC (ed.), *Evolution and Ecology of Zooplankton Communities. Special Symposium III American Society of Limnology and Oceanography*, University Press of New England, 241-248.

[6] Strickler JR (1982) Calanoid copepods, feeding currents, and the role of gravity. *Science*, *218*, 158-160.

[7] Koehl MAR & Strickler JR (1981) Copepod feeding currents: food capture at low Reynolds number. *Limnology and Oceanography*, *26*, 1062-1073.

[8] Lenz PH, Hower AE & Hartline DK (2004) Force production during pereiopod power strokes in *Calanus finmarchicus*. *Journal of Marine Systems*, *49*, 133-144.

[9] Boxshall GA (1985) The comparative anatomy of two copepods, a predatory Calanoid and a particle-feeding Mormonilloid. *Philosophical Transactions of the Royal Society of London Series B-Biological Sciences*, *311*, 303-377.

[10] Kiørboe T & Jiang HS (2013) To eat and not be eaten: optimal foraging behaviour in suspension feeding copepods. *Journal of the Royal Society Interface*, *10*, 20120693.

[11] Strickler JR (1975) Swimming of planktonic *Cyclops* species (Copepoda, Crustacea): pattern, movements and their control. *In*: Wu TY, Brokaw CJ & Brennan C (eds.), *Swimming and Flying in Nature*. Plenum Press, New York, 599-613.

[12] Borazjani I, Sotiropoulos F, Malkiel E & Katz J. (2010) On the role of copepod antennae in the production of hydrodynamic force during hopping. *Journal of Experimental Biology*, *213*, 3019-3035.

[13] Tanaka Y (2009) Planktons are wonderful athletes. *Iden: Seibutsu no Kagaku*, *63*, 17-22.

[14] Gemmell BJ, Jian H, Strickler JR & Buskey EJ (2012) Plankton reach new heights in effort to avoid predators. *Proceedings of the Royal Society B-Biological Sciences*, *279*, 2786-2792.

[15] Jiang HS & Kiørboe T (2011) Propulsion efficiency and imposed flow fields of a copepod jump. *Journal of Experimental Biology*, *214*, 476-486.

[16] Kiørboe T (2011) What makes pelagic copepods so successful? *Journal of Plankton Research*, *33*, 677-685.

[17] Murphy DW, Webster DR & Yen J (2012) A high-speed tomographic PIV system for measuring zooplanktonic flow. *Limnology and Oceanography: Methods*, *10*, 1096-1112.

[18] Bradley CJ, Strickler JR, Buskey EJ & Lenz PH (2013) Swimming and escape behavior in two species of calanoid copepods from nauplius to adult. *Journal of Plankton Research*, *35*, 49-65.

In: Copepods: Diversity, Habitat and Behavior
Editor: Laurent Seuront
ISBN: 978-1-63117-846-7

Chapter 8

THE SENSORY HORIZON OF MARINE COPEPODS

David M. Fields*
Bigelow Laboratory for Ocean Sciences, East Boothbay, ME, US

ABSTRACT

The behaviors of copepods are not random. They attack prey, find and recognize conspecifics with whom to mate and they escape from potential predators. An inappropriate response can lead to a missed opportunity to feed or mate or ultimately result in death from a predator. To behave in an ecologically appropriate manner copepods must detect environmental and biological cues. Organisms derive environmental information from activity of their sensory neurons, with each individual receptor acting as a broadband filter for biologically relevant signal properties. The suite of signals an animal can detect is determined by characteristics of individual sensors and their distribution throughout the body. Because sensory neurons frequently are tuned to preferentially detect biologically significant signals, the characteristics of sensory neurons offer valuable insight into the information that governs the organism's behavior, and determine the underlying mechanisms controlling the extent and dynamics of populations. Copepods rely primarily on chemoreception [1, 2] and mechanoreception to locate prey and detect potential predators. Distance mate detection for some copepods species involves chemical trail following until the final leap when mechanoreception plays an important role. With the variety of setal morphologies on the antennae of copepods we can identify particular morphological features that narrow the range in fluid motion to which specific seta respond. While setal length and orientation affect how the setae encode basic properties such as velocity, frequency and direction, the arrangement of setae along the antennules mediate perception of more complicated properties, such as shear. These structure–function relationships provide potential insight into trophic status, predator detection abilities or distributions, and perhaps can explain the fantastic degree of variation in setal and antennule morphology. This information is also helpful for interpreting how other mechanoreceptive organisms function. For example, how the harbor seals can locate distant objects by hydrodynamic trail following [3] or how catfish track the wakes of prey [4]. However, structure–function predictions remain largely

* E-mail address: dfields@bigelow.org

unverified, because we generally lack complimentary data on both the structure and ecological roles of the mechanosensory system for most aquatic organism. The purpose of this contribution is to apply current knowledge about the mechanosensory structures of marine copepods as a case study for understanding sensory ecology from a structure-function vantage point. We ask how sensor design (i.e., morphology) shapes sensor response properties and thus the behavioral and/or ecological function of particular sensor types. The focus is on planktonic copepods as model organisms for studying the role of mechanoreception because (i) they are a critical component of aquatic environments, (ii) fluid signals often act as the final proximate cue for copepod behavioral responses, and (iii) the basic properties of the copepod mechanosensory system are relatively easy to identify, and may be broadly general across a diverse range of species.

Introduction

The implicit assumptions of studying animal behavior are that organisms can discern different signals and that the uniqueness of the signals is the basis for their response. Interpreting behavior of copepods requires an understanding of the characteristics of all the potential signals, the physical and physiological response of the sensor to the fluid signal, and the relationship between the physiology and animal behavior. All information about an organism's perceivable world comes from the real-time activity of its sensory neurons. Sensory input provides organisms with a running commentary on the environmental conditions as they occur. Analyzing the string of data produced by an organism's sensors is a critical step towards understanding how organisms collect and process sensory information in order to respond in ecologically appropriate ways. Of the six common sensory modalities, touch, smell, vision, taste, hearing and magnetic orientation, only the first four appear to be important to the behavior of copepods. Numerous studies have shown that copepods respond remotely to potential mates [2,5-7], predators [8], and prey [9-12]. Although the stimuli used to identify and locate other individuals may involve a cocktail of chemical, mechanical and visual signals, previous studies show that in most cases mechanical signals are sufficient for eliciting feeding responses [11] and escape reactions [11,13-15], and are likely to be important for the final steps of mate capture [5]. These observations suggest that fluid mechanical signals convey information on the identity, the 3D location [16] as well as the relative motion of the signal source [12]. Further, recent literature suggests that for most planktonic copepods light [17] and chemical signals modulate the mechano-sensitivity of copepods but do not evoke a prey capture response or an escape reaction.

Initiating a behavioral response to a fluid mechanical signal requires a cascade of events (Figure 1). By definition for mechanoreception to occur there must be fluid motion. As an organism (prey, predator or mate) moves through the water (or water moves past the organism) it creates a fluid disturbance that imparts structure on the surrounding fluid. If the intensity of the fluid motion surpasses a mechanical threshold it causes the individual mechanoreceptor to move relative to the body. Not all motion in the mechanoreceptor will cause a neural response. Sensory perception occurs only if the displacement and the motion of the setae surpass the physiological threshold. This causes the associated neuron(s) to depolarize sending an action potential along the sensory axon. If the firing patterns surpass the behavioral threshold it causes a behavioral response. Different behaviors are initiated in response to different patterns of sensory input. In this chapter we discuss recent advances in our understanding of the mechanosensory ability of copepods, how signals interact with individual sensors, the transduction of signals into neurophysiological signals and ultimately how this cascade of events lead to the behavior of copepods in ecologically appropriate ways.

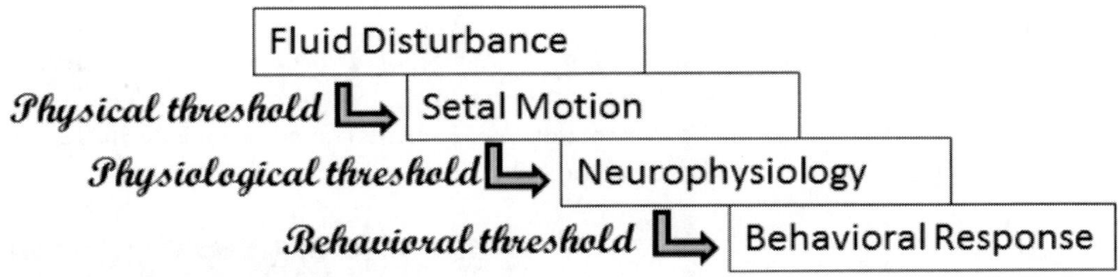

Figure 1. Sequence of events causing a behavioral response in a copepod.

STIMULUS PROPERTIES

All organisms live in a fluid environment. As they move relative to the fluid (or the fluid moves by them), they create a disturbance in the flow. Depending on the magnitude of the disturbance, the density and viscosity of the fluid and the sensitivity of the mechanoreceptor the disturbance can persist for 10s to 100s of seconds and travel centimeters to meters. Two important characteristics of fluid signals created at low to intermediate Reynolds numbers (i.e., Re < 3000) are (i) they have structure and (ii) once produced they behave in a predictable fashion that is dictated by the physical properties of the fluid (i.e., viscosity, temperature, and density). Within the past decade new visualization techniques such as PIV (particle imaging velocimetry) and highspeed video have become sophisticated and economical enough to map the complex flow fields produced by mm-sized moving animals (Figure 2). What was immediately evident is that the fluid disturbances are often orders of magnitude larger than the organisms themselves [18]. In addition the fluid motion was asymmetrical in magnitude and direction, which provided information about the location, size and speed of the source [19]. In general, fluid mechanical signals can be described by their frequency, amplitude and the initial size of the disturbance, with different organisms producing signals with different characteristics. For example, locomoting fish and large benthic crustaceans produce fluid disturbances with power spectra characterized by strong frequencies below 10 Hz with some higher frequency components (25 - 200 Hz) present in vortices shed from the caudal fins [20]. There is a growing body of literature focused on describing the fluid signals created by copepods [11,12,18,21-24]. Fluid disturbances created by swimming copepods are typically within the range of 10-50 Hz with frequencies of 30-300 Hz within the 1-2 cm in diameter small vortices release during rapid escape reactions [19]. Pulses from the independent movement of the pereiopods can achieve frequencies in excess of 500 Hz in the near-field [25]. The maximum amplitude of the disturbance is dependent on the size and behavior of the organisms. For swimming fish in the size range of 1-6 cm (the size of a typical copepod predator), the volume of water displaced due to the caudal fin motion is on the order of 0.02 ml to 3.3 ml and travels at speeds of 5 to 50 mm s^{-1}. These disturbances can be detected by copepods at a distance of over 11.8 mm. The much smaller ciliate (*ca.* 40 um), a typical copepod prey, produce maximum fluid speeds in the range of 2-8 mm s-1 when propelling themselves through the water [12] and are detected at distances of *ca.* 0.8 mm by predatory copepods. This distance is similar to the detection of individual passive particles (50 μm polystyrene beads are detected at a distance of *ca.* 1 mm) such as phytoplankton [26]. As the fluid disturbance travels from the source, its amplitude varies inversely with the distance cubed [11,27] creating a temporal pattern to the amplitude and structure with typical deformation rates of 5 to 10 s^{-1} [18,24]. This structure is persistent for several body lengths and provides a fading footprint of the organisms that created it [28]. Mimicking these signals is critical for laboratory experiments aimed at understanding how complex fluid signals are perceived and interpreted.

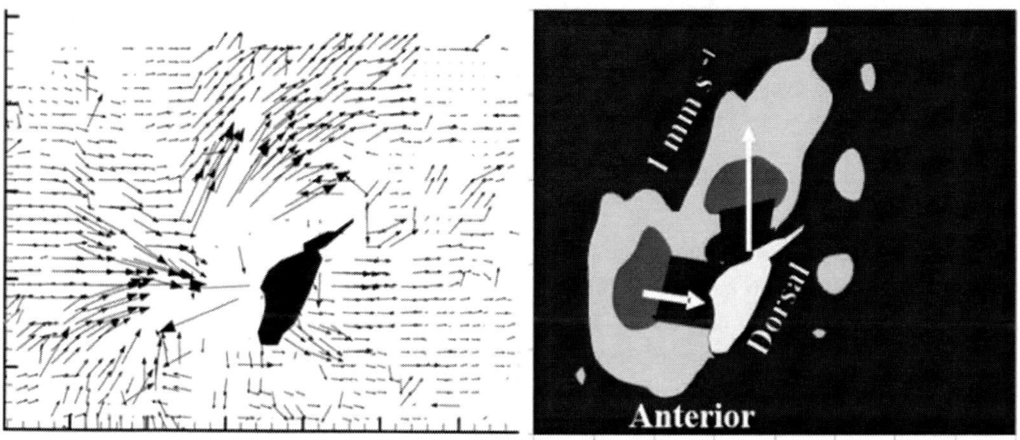

Figure 2. Left panel: particle imaging velocimetry (PIV) of the fluid disturbance created by the feeding current of an adult female *Acartia tonsa* (0.98 mm). Note how water is drawn from the region anterior and ventral to the animal and expelled ventral and posterior to the animal. *Right panel*: contours of the fluid speed of *A. tonsa*. Outer contour line marks 1 mm s^{-1}, inner contour is at 5 mm s^{-1}.

INDIVIDUAL SENSOR PROPERTIES

The literature provides several sources describing the ultra-structure of individual sensors on the antennules of copepods. Readers are directed to early work [29] for morphological characteristics of the mechanosensors, and [30] for morphological characteristics of chemoreceptors. More recent work [31] provide an excellent study on the external (SEM, light microscopy) and internal (TEM, serial section) of the setae of a deep-ocean calanoid copepod. Briefly, most mechanoreceptive setae are conical or spine shaped, with a large diameter at the base and narrowing at the tip (Figure 3). The chitinous sheath covering the antennae is continuous along the seta. The bases of individual mechanoreceptive setae are housed in an asymmetrical cuticular socket that may limit angular displacement of the sensor in a particular direction and provide protection for the setae during rapid behavioral responses when the setae are retracted along the antennule. Setal lengths within an individual animal can vary by over two orders of magnitude from short (*ca.* 10 μm) to very long (greater than 1mm) [32-35].

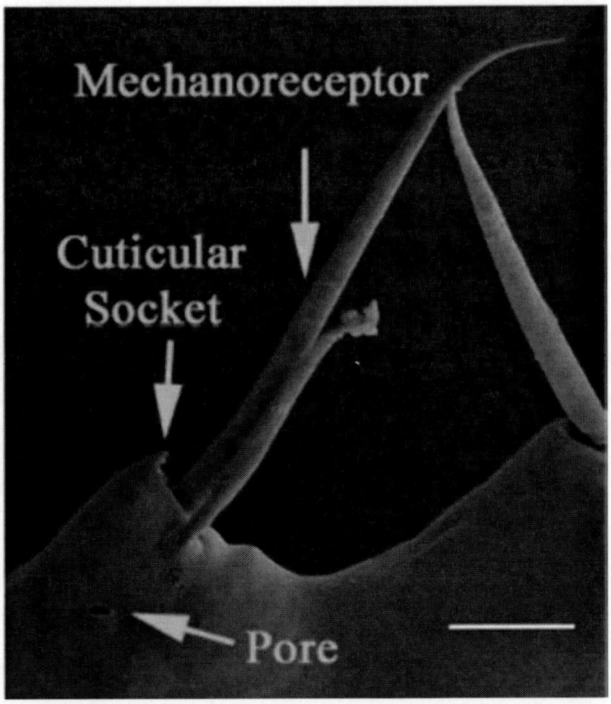

Figure 3. Setae located on the 3rd and 4th segments of *Gaussia princeps*. Note the cuticular socket housing the mechanoreceptive seta. Scale bar represents 100 μm.

The ultrastructure of the individual mechanosensory setae of the copepod is similar to the cuticular mechanosensilla of other crustaceans and arthropods. In arthropods, including the copepods, the dendrites from the associated mechanosensory neurons attach to the cuticle of the sensory hair that they innervate; for review see [36,37]. This attachment site is presumably the location where mechanotransduction occurs [33,38-40]. Fundamentally, the transduction of a mechanical signal into electrical/chemical (neurophysiological) signals requires the mechanical opening of membrane bound ion channels. The short latency of the response argues against activation of a second-messenger cascade and favors direct activation of a mechanically sensitive channel [41]. Although the exact mechanism is still unknown [31], it is hypothesized that the membrane bound channels are attached to rigid structures inside (microtubules) and outside (inner sheath cells) of the sensory cell *via* linking proteins (Figure 4). As the mechanoreceptive seta bend relative to the internal microtubules, the individual channels are pulled open *via* the linking proteins to allow the influx of positively charged ions into the membrane. This influx generates the local depolarization, which leads to the propagation of an action potential (*via* voltage gated Na ion channels; [42] in the associated axon.

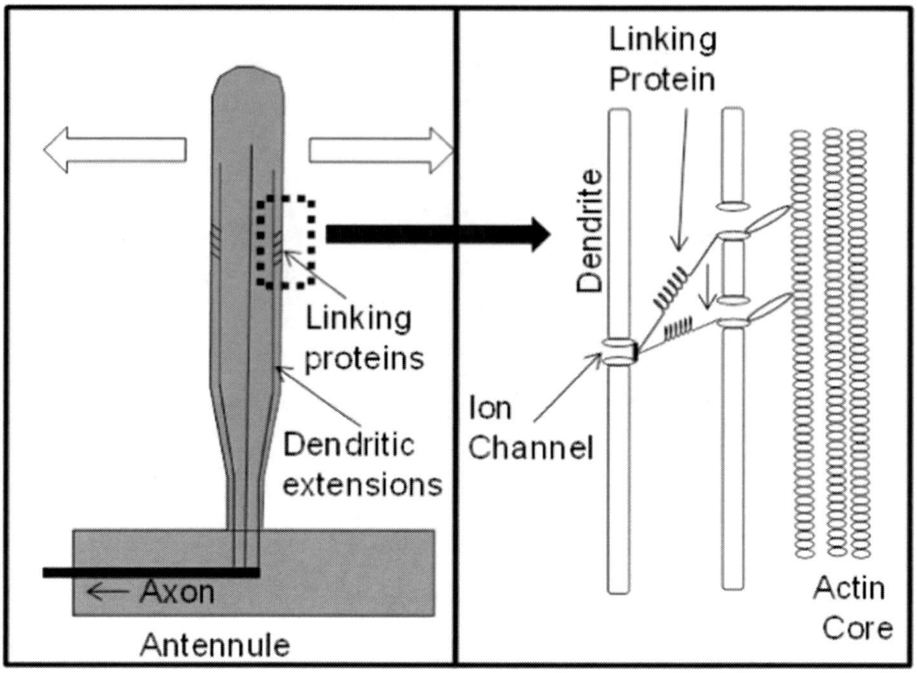

Figure 4. *Left panel*: copepod mechanoreceptive setae. When the seta is deflected to the left (white arrows), linking proteins pull open membrane bound ion channels causing local depolarization. *Right panel*: if the displacement is slow, the origin of the linking protein can move within the membrane decreasing the tension on the ion channel. If the angular displacement is rapid, the origin of the linking protein cannot move quickly enough and the channels are pulled open. Modified from [84].

There is ample evidence that the sensitivity and particular characteristics of the fluid signals detected are, in part, governed by the auxiliary structures associated with the individual neurons [35,43-45]. Across different phylum, these associated structures vary enormously in their complexity ranging from simple naked neurons to extremely elaborate structures such as those found in the mammalian ear. Slight modifications in the morphology of associated structures [46-48] or in the subtleties of their attachments [31,49] [4] [5] can give rise to a large spectrum of different stimuli that can be perceived. It is likely that the structure and function of copepod mechanosensors have been selected to maximize their effectiveness given the significant ecological consequences of perceiving (or misperceiving) mechanosensory information. As a group, copepods present a spectacular diversity of setal morphologies, orientations and degree of ornamentation [50]. The causes and consequences of this diversity remain unexplored, but the staggering degree of morphological variation suggests structure–function relationships between mechanosensor properties and their sensory roles.

Physical Responses of Seta to Fluid Motion

Different copepod species, genders and age classes show variations in the shape of their individual setae and in the overall configuration of their sensory array that reflect the dominant ecological pressure (e.g., prey capture *vs.* escape) they face in their environment. Different setae respond to a given fluid stimulus with a characteristic angular displacement [35]. To fully appreciate what the sensor morphology reveals about the environment and ecological pressures faced by these animals it is imperative to characterize, in a mechanistic way, how sensors respond to quantified, biologically relevant fluid disturbances

By definition mechanoreceptors require displacement sufficient to open the associated mechanically gated ion channels to initiate a neurophysiological response [51]. Once the seta surpasses the minimum displacement threshold, the firing rate of the associated neuron provides information about the magnitude of the motion.

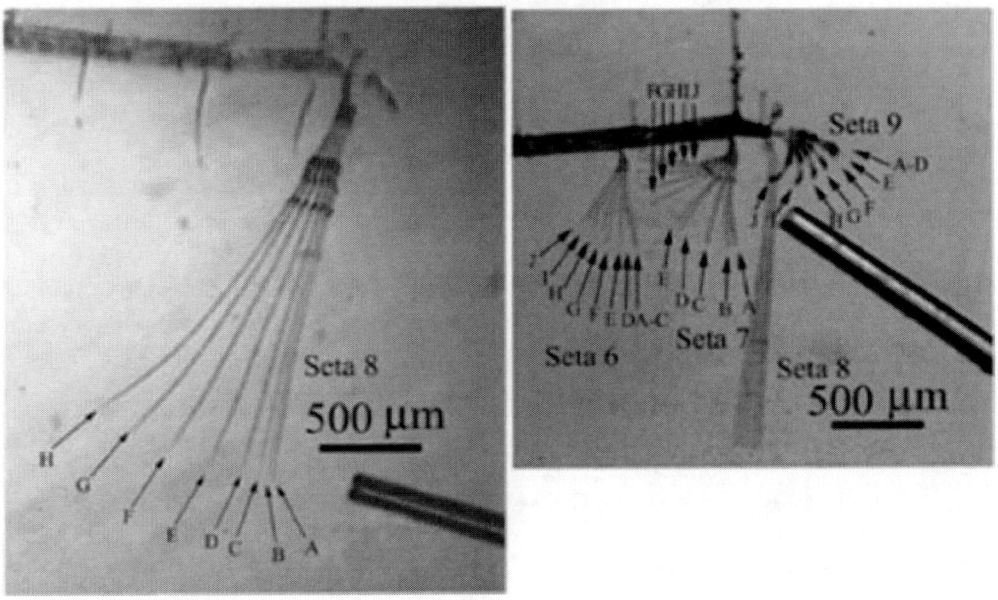

Figure 5. *Left panel*: Stimulation of the large distal setae of *Gaussia princeps* using a small water jet. Letters mark the maximum displacement of the seta at different fluid speeds. The letter “A “ in both panels represents the seta orientation with no stimulation. *Right panel*: water jet is aimed at the smaller setae #7 (note that the other setae showed a delayed response).

Although only a few measurements exist, the data suggests that copepods require a minimum of 0.1 to 2.3° of angular displacement [35,42,52] to initiate an action potential in the associated nerve cell. Achieving threshold displacement depends on the velocity of the fluid (amplitude) and the duration of the pulse (frequency). However, not all setal morphologies respond similarly to the same fluid signal (Figure 5). The response a function of physical features such as the diameter and length of the seta, its mechanical properties (e.g., torsional spring and damping constant; [53]), and the viscous drag associated with the medium in which it moves [54]. For example, analytical solutions [50] (supported by experimental data [35]) show that the stimulus frequency that gives rise to the maximum angular displacement varies with the inverse of length cubed. Long setae reach the greatest maximum displacement at low frequencies while short setae respond best to high frequency stimuli. Further, since the optimal frequency is proportional to L^{-3}, small decreases in setal length give rise to large changes in the optimal frequency. One of the ramifications of this size dependent displacement is that for a neural response to occur, the tip of the longer setae must travel further and at slower speeds. This combination leads to a longer response time for longer seta. Since rapid responses are critical to the survival of copepods [6] the use of long setae may be constrained to gathering information for behavioral responses that are relatively less time sensitive.

The physical constraints on setal bending permit a number of predictions regarding associations between setal morphology and the role of mechanosensory information. Predatory copepods that use high frequency signals from escaping prey (> 200 Hz; [11,50]) will have short setae. In contrast, copepods that detect slow moving, low frequency disturbances will have longer setae. While the optimal length might reflect the dominant signal the animal needs to detect, in reality, the complex tasks performed by most copepods involve a range of signals, and so require that they possess a complement of setal types. For instance, predatory copepods depend on the ability to detect their prey at a distance typically greater than their capture area; the larger the perceptive field, the greater the opportunity for successful capture.

Extensive perceptive fields require relatively longer, more sensitive setae, since fluid disturbances from distant sources become attenuated as they travel towards the receiver. The predator also is required to respond with directional accuracy within milliseconds when fast moving preys are within the capture region. The signals produced in close proximity to the copepod are strong, with substantial high-frequency components. Thus, predatory strikes likely depend on short setae. The relative frequency of these setal morphs may reflect the relative importance of particular signal features.

Physiological Responses of Seta to Fluid Motion

The transduction of a mechanical signal into electrical/chemical (neurophysiological) signals requires the mechanical opening of membrane bound ion channels. The short latency of the response argues against activation of a second-messenger cascade and favors direct activation of a mechanically sensitive channel [41]. Although the exact mechanism is still unknown, it is hypothesized [31] that the membrane bound channels are attached to rigid structures inside (microtubules) and outside (inner sheath cells) of the sensory cell *via* linking proteins (Figure 4). As the mechanoreceptive seta bend relative to the internal microtubules, the individual channels are pulled or stretched open allowing the influx of positively charged ions into the membrane. This influx creates the local depolarization which can lead to the generation of an action potential in the associated axon (*via* voltage gated ion channels; [42]). The frequency of the action potentials and the relative timing of firing between cells inform the broader nervous system of the direction and intensity of the external mechanical perturbation.

There are only a few studies aimed at quantifying the characteristics of setal motion that give rise to the neurophysiological response in copepods. Previous studies [56], which have used a vibrating bead as their stimulus, report that *G. princeps, Euchaeta rimana* and *Pleuromamma xiphias* can detect fluid displacement as low as 10-20 nm at corresponding fluid velocity of 20 to 40 $\mu m\ s^{-1}$.

However these values were based on calculated movements of the seta with the assumption that the seta accurately followed the movement of the surrounding fluid. Yen et al., [52] stimulated individual seta and found similar sensitivities for *Labidocera madurae*, with the sensitivity to displacement decreasing with decreasing stimulation frequency. To compare sensitivities of different seta it is helpful to normalize the data to individual setal length and report the displacement characteristics in terms of angular displacement and speed. Recalculating the data of Yen et al., [52], shows that at frequencies above 1000 Hz, the distal seta of *Labidocera madurae* have a displacement and velocity threshold of 10 nm and *ca.* 20 μm s^{-1}, respectively. For a setal length of 200 μm this is equivalent to an angular displacement of 0.003 degrees, and an angular velocity of 6 degrees s^{-1} (assuming maximum velocity is reached at half the oscillation cycle; [52]). This is extremely sensitive given that the smooth hairs of a crayfish are 3 orders of magnitude less sensitive [49].

At frequencies between 250 Hz and 1000 Hz the threshold displacement for *L. madurae* remained fairly constant at *ca.* 50 nm. This tip displacement corresponds to an angular displacement of 0.01 degrees and an angular velocity of 5-20 degrees s^{-1}. At frequencies more likely to occur in nature (< 200 Hz [57]; 10-50 Hz [19]) the threshold displacement required to elicit a neurological response varied inversely with frequency, with the seta becoming relatively insensitive, requiring larger displacements and velocities before a physiological response is initiated [52].

The data of Yen et al., [52] was extrapolated to more "natural" frequencies to predict that at 85 Hz and 10 Hz the 200 μm seta of *L. madurae* would require a displacement of 0.1 and 2.3 degrees, respectively, and corresponding angular velocities of 17.0 degrees s^{-1} and 41.4 degrees s^{-1}. At these lower frequencies the threshold values are up to 2 orders of magnitude greater for displacement and require a 3-fold increase in the angular velocity. This suggests that copepods may be significantly less sensitive to displacement and velocity at lower frequencies than previously assumed. However, these values are likely to be the realm of biologically relevant stimuli. Furthermore, if the displacement and velocity thresholds vary with frequency, they are not likely to be velocity or displacement sensors as much of the literature has concluded. Additional work in this area is certainly warranted.

NEUROLOGICAL CODING OF FLUID MOTION

When the motion of the setae surpasses the physiological threshold it generates a series of action potentials (Figure 6). Since the spike amplitude of an individual neurons is constant (see [58] for exception), information from a single sensory cell is provided by the timing of the first action potential, the length of time that the cell fires and the frequency at which it fires. Comparing the movement of the setae with the initiation of a spike train from a single neuron, we can determine the minimum displacement required to initiate the action potential in the associated neuron.

The length of time that the cell fires, provides information on the duration of the stimulus while the instantaneous frequency of the spike train, provides information about the intensity of the signal.

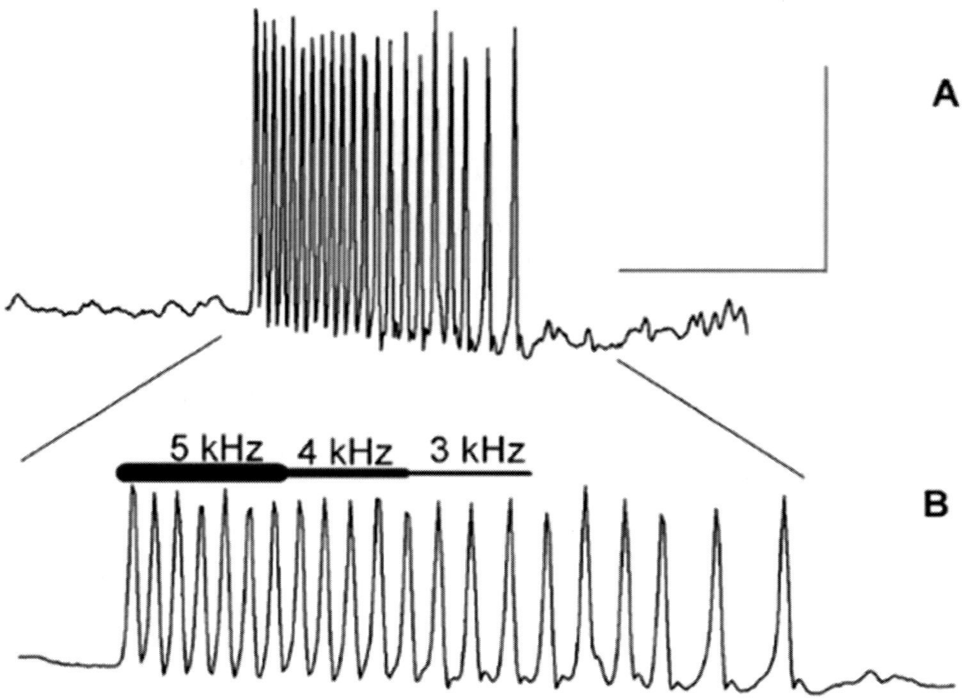

Figure 6. Neural recording from a seta of *G. princeps* (A) Series of the spikes recorded in response to a small water jet. Scale bar represents 500 μV on the ordinate and 5 ms on the abscissa. (B) Expanded view of the spike train. Note the high frequency response (3 kHz - 5 kHz) was sustained for over 3 ms. Modified from [42].

Rapid Behavioral Response

Detecting prey or avoiding predators is fundamental to the survival of individual copepods, and many animals derive a selective advantage by increasing their ability to rapidly evade predators or attack quickly moving prey. The pressure for rapid responses has led to special mechanical and neurophysiological solutions [59] to increase speed and diminish response time. Many of these solutions have evolved independently across different taxa [60]. The biological and physical environment of pelagic copepods requires highly discriminate and yet rapid behavioral responses. Living at low Reynolds numbers, chemical stimuli are transported to animal sensors largely through the slow process of laminar fluid displacement and Fickian diffusion [61]. Similarly, fluid mechanical stimuli are attenuated quickly by viscous dampening causing fluid velocity to decrease with distance cubed [11]. As such, copepods often do not detect other individuals until they are within a few body lengths of each other providing little time and information on which to base behavioral responses.

Responding in an ecologically appropriate manner with a 1 to 5 ms behavioral latency presents a unique challenge. Rapid and accurate behavioral responses of copepods require that three fundamental conditions be met. First, the neural impulses must have time to travel from the receptor site to the target motor region. Second, the neural impulses must contain adequate information to differentiate stimuli from each other and, third, they must provide information to determine the 3-D location of the stimulus source. How does the neural system in copepods generate enough information in such a short time interval? Calanoid copepods possess numerous antennal sensors [22,29,30,35]. Depolarization of the nerve cell results from the deflection of the associated mechanosensory hair. Disturbance size is encoded by the number of setae simultaneously stimulated, whereas fluid velocity, at the level of the individual sensor, is encoded by the firing rate [62]. The rate and angle of deflection are likely to provide the animal with information regarding the velocity and potentially the acceleration of the fluid signal. Compared to other crustaceans [63], copepod mechanoreceptors are at least an order of magnitude more sensitive, which provides them with early detection of a potential predator allowing maximum time to collect information and initiate a behavioral response. To achieve both stimulus differentiation and behavioral latencies of 1 to 5 ms, individual neurons must fire at a minimum frequency of 250 to 1,000 Hz to generate the requisite 2-spike minimum to differentiate a signal from the spontaneous firing of a single action potential. Graded responses to fluid velocity, required for finer discrimination of signal strength, require higher frequencies. For example, coding a second or third magnitude of fluid velocity within the same 1-3 ms constraint would demand minimum frequencies of 2,000 and 3,000 Hz, respectively.

One mechanism to decrease the response time is to increase the rate that information travels along sensory neurons. For example, large diameter nerve fibers have evolved in numerous invertebrate species, such as the giant squid [64], crayfish [65], cockroaches [66] and older lineages of copepods [31]. Similarly, many vertebrates, and some more recent evolutionary lines of invertebrates such as copepods [67], have independently evolved myelin or myelin-like structures surrounding nerve cells to increase the speed of sensory or motor information. Doubling the speed, doubles the amount of information transferred during the same time period. Increasing the transfer of sensory information is analogous to an increase in "bus" speed in computer architecture. A second solution for moving sensory information quickly is by densely packaging information. Continuing with the computer analogy, this solution is analogous to increasing the size of the information packet (bit-size). By doubling the spike rate the transfer time decreases in half. Again it appears that copepods have evolved mechanisms to produce high frequency action potentials. Action potential frequencies in copepods can reach in excess of 5 kHz [42]. Little is known about the relationship between axonal conduction speeds and the action potential frequency and even less is known about the underlying mechanisms at the level of the ion channels within the nerves that permit such rapid firing rates. Further work in this area is hence certainly warranted.

DIRECTIONAL SENSITIVITY

Accurate attack and escape responses require not only detecting the magnitude of the stimulus but also its direction. In principle, direction can be discerned from a single seta housing multiple neurons that respond specifically to bending in certain directions. Anatomical observations of [33] for *Pleuromamma xiphias* suggest multiple innervation of some setae, but other studies found no physiological evidence of multiple innervation per seta [11]. However, significant asymmetry of the physiological response to proximal *vs.* distal deflections, in combination with directionally specific displacement-velocity curves can produce stimulus-response functions that incorporated directional specificity. In some cases, morphological features on the antennule are key in producing these effects. For example, a cuticular socket that restricts motion of the seta in a particular direction [52,63,68] can provide a directional bias to the response [69]. Similarly morphological features of the setae can increase rigidity in one direction over another. Setae have been shown to have sharp angles located along the shaft, which offers increased rigidity when bent in one direction over another. Feathered setae are more responsive to flows contacting the brad face of the sensor rather than flows coming from the side. Similar results were found for insect hairs where the direction of best mobility accurately predicted the directional specificity of the neuronal response [70].

For copepods that have setae with single innervation it is unlikely that reliable directional signals can be garnered from the response of a single seta even with a mechanical directional bias. Since the activity of a single neuron depends on both the magnitude and the direction of the stimulus, different combinations of these two signal characteristics may lead to a similar neural response [71,72]. The signals from multiple setae, however, provide an excellent mechanism for spatial location. The temporal delay in firing between adjacent seta [35] can be used to both evaluate the speed of the disturbance and the direction. Increased fluid speed would decrease the lag time between the depolarization events and the order that the specific seta fire could distinguish the direction.

Mathematical Modeling of the Interaction between Medium Flow and Mechanosensory

Mathematical modeling has been employed to study the rather complex interaction between media flow (mostly air flow) and sensory hairs [43,71,73,74]; reviewed by Humphrey et al., [44,53]. The mechanics of stimulus coupling and transduction have been studied for numerous organisms including insects [73] and arachnids [44]. Tautz [73] was the first to suggest modeling the filiform hairs as forced damped harmonic oscillators to establish the dynamic characteristics of thoracal hairs in caterpillars. Typically the models have assumed the hairs to be smooth cylinders that respond as a stiff rod. Such simplified models seem to fit empirical data for many of the terrestrial organisms tested (although see Dechant et al., [45]. Humphrey et al., [48] provided an analytical mathematical-mechanical model describing the steady-state motion of a sensory hair protruding from a solid substrate in oscillating media (air or water) flows. Essentially, the hair is modeled as a rigid shaft with boundary conditions at the base. Using this model, Humphrey et al., [44] considered hair deflection for two biologically significant cases: (i) air flow oscillating parallel to the longitudinal axis of the spider leg (the substrate), and (ii) air flow oscillating perpendicular to the longitudinal axis of the spider leg. They found that the relative orientation between the respective directions of the fluid flow and the substrate axis has a significant effect on the magnitudes of hair deflection, angular velocity and acceleration but not on the resonance frequency of the hair. One potential problem of applying this model to copepods is that it calculated the displacement characteristics of the mechanosensory at resonant frequency. Copepods, because of their rapid behavioral responses, do not have time to respond to multiple cycles of an oscillating stimulus. Copepods can respond to fluid signals within 2 ms of being generated. During this short time frame their seta would be exposed to only 1 complete cycle of a 500 Hz signal and only ½ a cycle for signals at 250 Hz [42]. This is insufficient to reach steady-state motion and the seta will certainly not achieve the same maximum displacements predicted at their resonant frequency. Rather than allowing the model seta to reach steady-state oscillations, using a modified Humphrey & Barth [48] model we consider only the initial displacement in a single direction. If the sensors are not allowed to reach steady-state (resonant frequency) the displacements are much smaller than analytical solutions would predict. Since copepods respond to fluid signals within milliseconds, the numerical model more accurately describes conditions experienced by the copepods and suggests that they must be much more sensitive to fluid motion (require much smaller displacement) than the analytical solutions would predict.

Spatial Detection of Fluid Signals

Positioned at the front of the copepod are two highly innervated antennae. For most pelagic copepods the antennae are a linear array of approximately 100 mechanoreceptive setae [50,52,75] that vary in length (Figures 7 and 8), thickness and orientation [76]. The individual setae on the antennule of the copepod are similar to the hairs cells of the human inner ear and the sensory cells within the lateral lines organ of fish. In all cases the mechanosensors serve as extremely sensitive mechano-electric transducers converting mechanical force into an electro-chemical signal. The signals are subsequently transmitted to the brain or peripheral ganglia where information from several sites may be integrated. One primary difference, however, is how complex signals that contain many superimposed frequencies and magnitudes are decomposed into their constituent properties. The human ear changes far field vibrations in the outer ear into a near field fluid disturbance in the inner ear. Within the inner ear, hair cells are arranged along the spiraling cochlea, which by virtue of its shape spatially separates the different frequencies such that they stimulate different sets of hair cells arranged along the length of the organ. The different clusters of hair cells are mapped in the brain to provide information on the narrow range in frequency that the spiral structure of the cochlea permits them to experience. In contrast, copepods do not have an external mechanism to spatial decompose fluid signals. The setae of copepods project directly into the fluid and are exposed to the entire symphony of frequencies. Thus for the animal to detect a wide range of signals with sufficient detail requires that individual hairs are be sensitive within a narrow range yet are adjacent to sensors that have complimentary tuning.

The antennules of copepods with a series of closely spaced flow sensors, provides an ideal structure for detecting fine scale heterogeneity in a local flow field (Figures 8 and 9). Setal morphology along the antennules of copepods and between species of copepods is highly diverse [30,33,77]. As discussed above, these differences in morphology allow copepods to detect a wide range of fluid characteristics at discrete locations along the antennules [35]. Encoding of spatial properties of a fluid disturbance, in contrast, occurs by comparing stimuli from numerous mechanoreceptive sites. Individual mechanosensors are arranged along the antennule much like an oceanographic deployment of a series of specifically tuned flow meters connected linearly along a single tether (Figure 9).

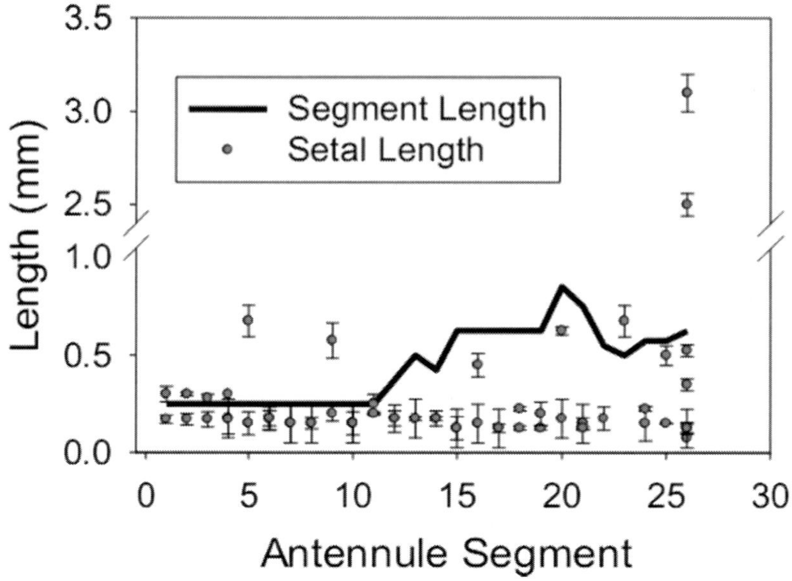

Figure 7. Antennule morphology of *Gaussia princeps*. Open circles represent the length of individual seta (±SE, $n = 6$). Solid line represents segment length. Note the very long mechanoreceptive setae on the final 2 segments of the antennule and the increase in segment length at segment 11-26.

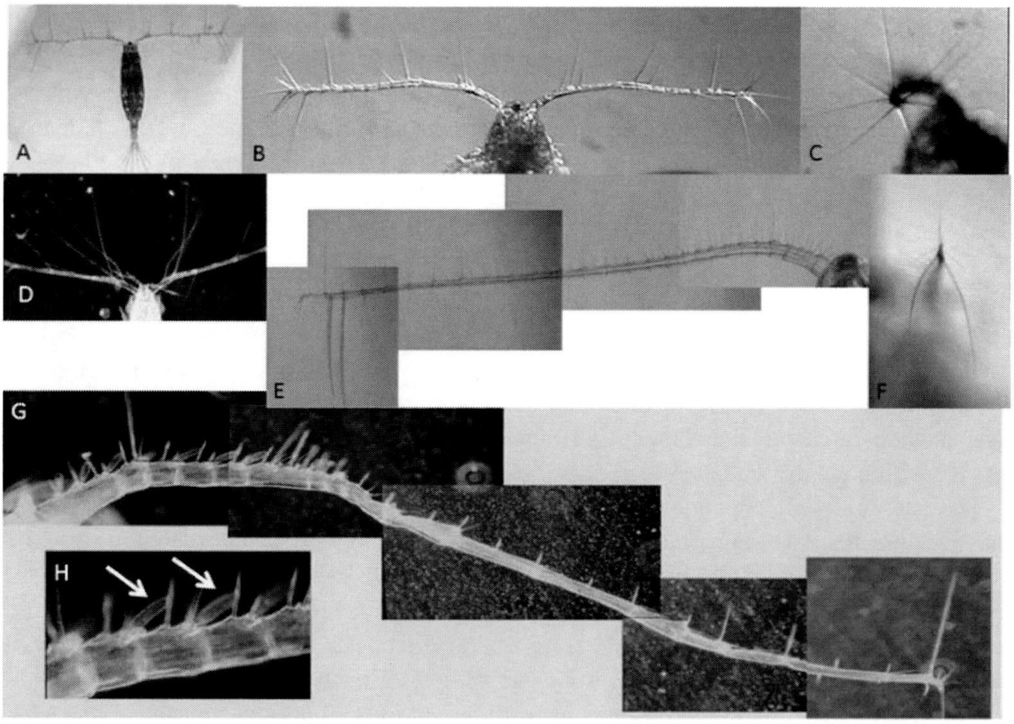

Figure 8. Antenules of different copepod species. *Acartia tonsa* (940 µm) dorsal (A-B) and lateral view (C). Oithona sp. (D). Calanus *finmarchicus dorsal* (E) and lateral view (F). *Pleuromamma* xiphias (4.2 mm) male right antennule (G-H). Arrows mark putative chemoreceptors.

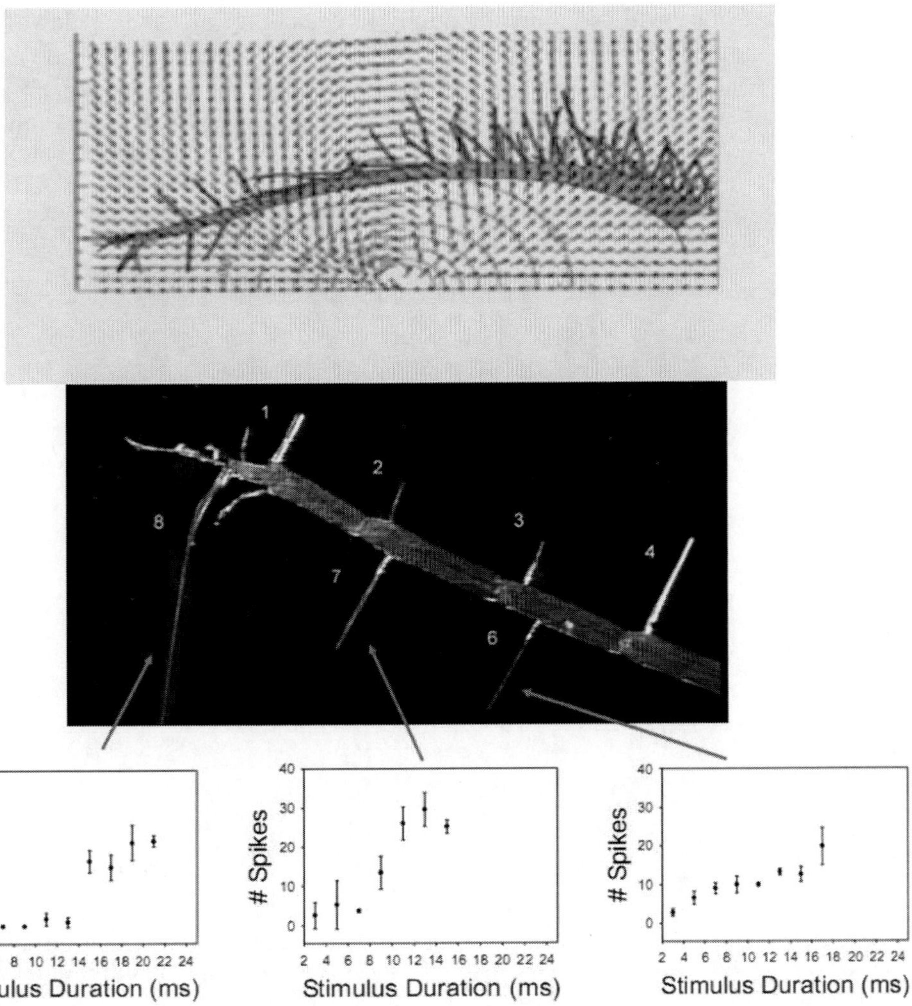

Figure 9. Upper panel: a schematic (based on real data) of the antennule of *Eurytemora* sp. made from light and SEM images (Fields, unpublished data) superimposed on a CFD solution for a copepod jumping [24]. Arrows represent magnitude of fluid motion. Note that each seta is exposed to a different (time dependent) fluid intensity. Lower panel: neurophysiology as a function of signal strength for 3 different setae on the antennae of *G. princeps*. Each seta has different physical characteristics, response profile and directional sensitivities (not shown).

Sensors are densely packed in regions where measurements of fine scale structure are needed, and distributed over a greater distance when their task is the detection of larger spatial features. A long antennule span provides greater spatial coverage, but decreases resolution of finer scale features, assuming a constant number of sensors. Behavioral responses of copepods to fluid motion suggest that they detect spatial gradients in fluid velocity [14]. Copepods often are smaller than the smallest turbulent eddies in their environment [78,79] and frequently experience fluid disturbances as 3-D laminar shear. Directional sensitivity of individual setae and their linear distribution along the antennule suggest that detection of velocity gradients is maximized within the plane defined by the body and the antennules [80]. Thus the relative orientation of the copepod within a larger hydrodynamic feature will determine its ability to detect the surrounding flow. If individual setae all have the same sensitivity, the length of the antennule would determine the maximum distance over which a given velocity gradient is detected (Figure 10). Small increases in antennule length give rise to a large difference in the velocities at either end of the antennule, since fluid velocity decays exponentially with distance.

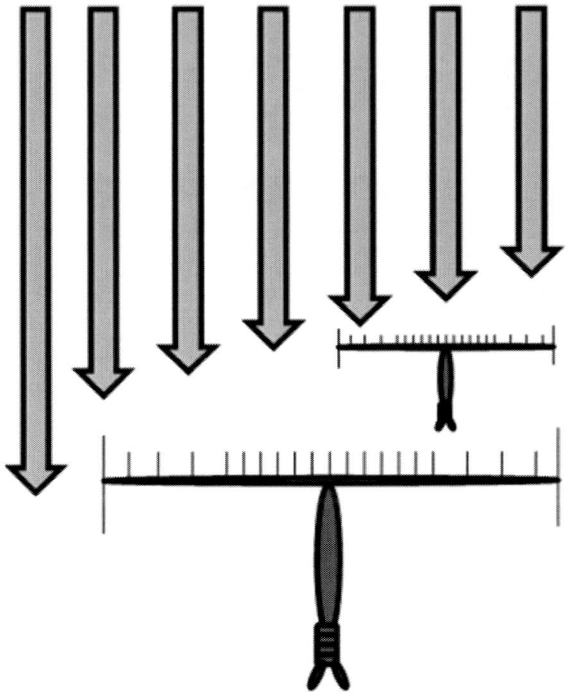

Figure 10. Two copepods with different antennule lengths exposed to the same fluid disturbance. The copepod with the short antennules cannot detect the small gradient in fluid velocity while the animal with the large antennule span detects the spatial difference in fluid speed.

For example, Yen et al., [52] reported a velocity threshold of 20 μms^{-1} for an individual mechanosensor. If we assume that this threshold remains constant, an animal with a 1 mm antennal span can potentially detect a sheared flow of 0.02 s^{-1}. In contrast, an animal with a 5 mm antennal span and the same sensitivity could detect a shear of 0.006 s^{-1}. These shear levels would be created under an isotropic turbulence level (ε) of 0.5 and 0.04 $cm^2\ s^{-3}$, respectively [81]. Thus, the two hypothetical animals would be expected to have an order of magnitude difference in sensitivity to the same velocity gradient, despite equal sensitivity of the individual mechanosensors. A long antennule with a high sensitivity to velocity gradients benefits animals that live in the relatively quiescent environment of the deep ocean, but it could make them overly sensitive to ambient hydrodynamic disturbances in the upper water column. Since copepods must balance the risk of predation with the energetic cost of an unnecessary escape [14], acute sensitivity to shear may be disadvantageous where ambient turbulence is large. The trade-offs inherent in antennule architecture may result in correlations between morphology and turbulence level preferred by a copepod species. Since copepod species of similar size can have antennal spans that differ by 25 to 30% [82], knowledge of how copepod species are distributed may provide a quick and reliable measure of the instantaneous oceanic turbulence, particularly if distributions reflect active choices made by animals.

Behavioral assays quantifying the escape threshold of different developmental stages of a single species, *Acartia tonsa,* show a similar size-dependency [76] as behavioral sensitivity is markedly enhanced by increased antennule length [14,15]. This change in sensitivity does not appear to be the result of changes in the sensitivity of the individual mechanoreceptor. Boxshall et al., [83] suggested that sensory structures responsible for predator detection (namely the distal tips of the antennules) appear early in development and are conserved throughout ontogeny. Individual antennule segments lengthen as animals develop through the copepodid stages to adulthood. This gives rise to an increase in the overall length of the antennule with little change in morphology at the distal tips. Thus the inverse relationship between antennule size and escape threshold suggests that the mechanism underlying the higher predation risk of nauplii may be their decreased sensitivity to fluid signals created by potential predators.

THE ROLE OF FLUID VISCOSITY ON MECHANORECEPTION

Environmental conditions play a primary role in the transmission of sensory signals. Changes in the physical characteristics of the fluid can alter the detection distance and signal structure. For example, temperature is inversely correlated with kinematic viscosity of fluids. Animals in the tropics and regions surrounding the Mediterranean Sea experience viscosity values that are approximately 50% of those found in the polar region. The resonant frequency of an individual mechanoreceptor varies inversely with kinematic viscosity [48]. Similarly, the maximum displacement at its resonant frequency increases with the cube of viscosity. Thus mechanoreceptors of a polar copepod (living at -1°C) with a resonance frequency of 100 Hz would experience a shift in the frequency of maximum response to 200 Hz in equatorial regions (25°C). Furthermore, the maximum displacement of the seta at its resonant frequency decreases exponentially with viscosity, giving the equatorial copepods a maximum displacement that is $1/8^{th}$ of that in the polar copepod. This makes individual setae in the equatorial region much less sensitive to a given fluid speed. Under an assumption of constant frequency sensitivity, to maintain the same maximum displacement, equatorial animals would need to increase setal length by a factor of 4. However, in tropical and Antarctic congeners of *Euchaeta* (*E. rimana* and *E. antarctica*) setal length from segments 3, 7, 13 decrease (rather than increasing) by a factor of 1.8 to 2.0 (data from Yen & Nicoll [32]). Similar results are found for tropical and subpolar *Acartia* congeners (*A. tonsa,* living at 22°C, and *A. tumida*, at 3°C). This suggests that congeners living in different viscous regimes do not dynamically scale (to compensate for viscosity) the length of their seta to maintain spectral sensitivity. Instead, copepods may be regulating the size of their perceptive field. Higher viscosity (colder temperatures) increases the rate at which an initial disturbance is dampened; fluid mechanical signals in warm water maintain their velocity over a greater distance than in cold water. Thus, polar animals must have long setae that respond at lower signal amplitude in order to effectively perceive prey from a distance. This same level of sensitivity would increase the perceptive field of tropical copepods well beyond its ecologically relevant space.

The perceptive field must be large enough to allow an appropriate response, but not so large that it includes excessive sensory information. Signals emanating from a large distance may be so far away that they are irrelevant to an animal ability to effectively capture a prey or escape predators. As a result, copepods that live in low viscosity environments, where fluid signals travel much further, may have shorter setae to diminish their perceptive field.

SUMMARY

The ability to sense fluid motion is strongly influenced by morphological properties of setae, by the way in which they are organized into an ensemble along the mechanosensory organ (i.e., the antennule) and the physical charicteristics of the fluid in which they are embedded. Setal length and orientation affect how they encode basic properties such as velocity, frequency and direction, whereas the arraignment of setae along the body and antennules mediate perception of more complicated properties, such as shear. Morphological and physiological data indicate that the design of setae and antennules bias an organism towards detecting particular types of disturbances, or for efficient operation in certain environments. These structure–function relationships provide potential insight into the diets and the dominant predators of different copepod species and perhaps can explain the tremendous degree of variation in setal morphology and antennule architecture. However, structure–function predictions are premature because we generally lack complimentary data on both the design and ecological roles of the mechanosensory system in a particular organism. Thus, an important challenge is to use a comparative approach to determine whether design principles of mechanosensory systems can explain behavioral properties, and therefore provide insights into ecological interactions in the plankton.

REFERENCES

[1] Fields DM, Weissburg MJ & Browman HI (2007) Chemoreception in the salmon louse *Lepeophtheirus salmonis*: an electrophysiology approach. *Diseases of Aquatic Organisms*, 78, 161-168.

[2] Weissburg MJ, Doall MH & Yen J (1998) Following the invisible trail: Kinematic analysis of mate-tracking in the copepod *Temora longicornis*. *Philosophical Transactions of the Royal Society B: Biological Sciences*, 353, 701-712.

[3] Dehnhardt G, Mauck B, Hanke W & Bleckmann H (2001) Hydrodynamic Trail-Following in Harbor Seals (*Phoca vitulina*). *Science*, 293, 102-104.

[4] Pohlmann K, Grasso FW & Breithaupt T (2001) Tracking wakes: the nocturnal predatory strategy of piscivorous catfish. *Proceedings of the National Academy of Sciences the United States of America*, 98, 7371-7374.

[5] Doall MH, Colin SP, Strickler JR & Yen J (1998) Locating a mate in 3D: the case of *Temora longicornis*. *Philosophical Transactions of the Royal Society B: Biological Sciences*, 353, 681-689.

[6] Kiørboe T & Bagøien E (2005) Motility patterns and mate encounter rates in planktonic copepods. *Limnology and Oceanography*, 50, 1999-2007.

[7] Seuront L (2013) Chemical and hydromechanical components of mate-seeking behaviour in the calanoid copepod *Eurytemora affinis*. *Journal of Plankton Research*, 35, 724-743.

[8] Drenner RW & McComas SR (1980) The roles of zooplankter escape ability and fish size selectivity in the selective feeding and impact of planktivorous fish. *In*: Kerfoot WC (ed.), *Evolution and Ecology of Zooplankton Communities*. University Press of New England, Hanover, 587-593.

[9] Jonsson PR & Tiselius P (1990) Feeding behavior, prey detection, and capture efficiency of the copepod *Acartia tonsa* feeding on planktonic ciliates. *Marine Ecology Progress Series*, 60, 35-44.

[10] Saiz E & Kiørboe T (1995) Predatory and suspension feeding of the copepod *Acartia tonsa* in turbulent environments. *Marine Ecology Progress Series*, 122, 147-158.

[11] Fields DM & Yen J (2002) Fluid mechanosensory stimulation of behaviour from a planktonic marine copepod, *Euchaeta rimana* Bradford. *Journal of Plankton Research*, 24, 747-755.

[12] Jiang HS & Paffenhofer GA (2008) Hydrodynamic signal perception by the copepod *Oithona plumifera*. *Marine Ecology Progress Series*, 373, 37-52.

[13] Haury LR, Kenyon DE, & Brooks JR (1980) Experimental evaluation of the avoidance reaction of *Calanus finmarchicus*. *Journal of Plankton Research*, 2, 187-202.

[14] Fields DM & Yen J (1997) The escape behavior of marine copepods in response to a quantifiable fluid mechanical disturbance. *Journal of Plankton Research*, 19, 1289-1304.

[15] Kiørboe T, Saiz E, & Visser AW (1999) Hydrodynamic signal perception in the copepods *Acartia tonsa*. *Marine Ecology Progress Series*, 179, 97-111.

[16] Browman HI, Yen J, Fields DM et al., (2011) Fine-scale observations of the predatory behaviour of the carnivorous copepod *Paraeuchaeta norvegica* and the escape responses of their ichthyoplankton prey, Atlantic cod (*Gadus morhua*). *Marine Biology*, 158, 2653-2660.

[17] Fields DM, Durif, CM, Bjelland RM et al., (2011) Grazing rates of *Calanus finmarchicus* on *Thalassiosira weissflogii* cultured under different levels of ultraviolet radiation. *PLoS ONE*, 6, e26333.

[18] Fields DM & Yen J (1993) Outer limits and inner structure: the 3-dimensional flow field of *Pleuromamma xiphias* (Calanoida: Metridinidae). *Bulletin of Marine Science*, 53, 84-95.

[19] Yen J & Strickler JR (1996) Advertisement and concealment in the plankton: What makes a copepod hydrodynamically conspicuous? *Invertebrate Biology*, 115, 191-205.

[20] Breithaupt T & Ayers J (1998) Visualization and quantification of biological flow fields through video-based digital motion-analysis techniques. *Marine Freshwater Behavior* Physiology, 31, 55-61.

[21] Yen J, Sanderson B, Strickler JR & Okubo A (1991) Feeding currents and energy dissipation by *Euchaeta rimana*, a subtropical pelagic copepod. *Limnology and Oceanography*, 36, 362-369.

[22] Yen J & Fields D (1992) Escape responses of *Acartia hudsonica* nauplii from the flow field of *Temora longicornis*. *Archiv fur Hydrobiologie Beiheft*, 36, 123-134.

[23] Van Duren LA, Stamhuis EJ & Videler JJ (2003) Copepod feeding currents: flow patterns, filtration rates and energetics. *Journal of Experimental Biology*, 206, 255-267.
[24] Jiang H & Kiørboe T (2011) The fluid dynamics of swimming by jumping in copepods. *Journal of the Royal Society Interface*, 8, 1090-1103.
[25] Lenz PH & Hartline DK (1999) Reaction times and force production during escape behavior of a calanoid copepod, *Undinula vulgaris*. *Marine Biology*, 133, 249-258.
[26] Bundy MH, Gross TF, Vanderploeg HA & Strickler JR (1998) Perception of inert particles by calanoid copepods: behavioral observations and a numerical model. *Journal of Plankton Research*, 20, 2129-2152.
[27] Kalmijn AJ (1988) Hydrodynamic and acoustic field detection. *In*: Atema J, Fay RR, & Tavolga WN (eds.), *Sensory Biology of Aquatic Animals*. Springer Verlag, New York, 83-130.
[28] Gries T, Jöhnk K, Fields D & Strickler JR (1999) Size and structure of 'footprints' produced by *Daphnia*: impact of animal size and density gradients. *Journal of Plankton Research*, 21, 509-523.
[29] Strickler JR & Bal AK (1973) Setae of the first antennae of the copepod *Cyclops scutifer* (Sars): their structure and importance. *Proceedings of the National Academy of Sciences of the United States of America*, 70, 2656-2659.
[30] Friedman MM & Strickler JR (1975) Chemoreceptors and feeding in calanoid copepods (Arthropoda: Crustacea). *Proceedings of the National Academy of Sciences of the United States of America*, 72, 4185-4188.
[31] Weatherby TM & Lenz PH (2000) Mechanoreceptors in calanoid copepods: designed for high sensitivity. *Arthropod Structure and Development*, 29, 275-288.
[32] Yen J & Nicoll NT (1990) Setal array on the first antennae of a carnivorous marine copepod *Euchaeta norvegica*. *Journal of Crustacean Biology*, 10, 327-340.
[33] Weatherby TM, Wong KK & Lenz PH (1994) Fine structure of the distal sensory setae on the first antennae of *Pleuromamma xiphias* Giesbrecht (Copepoda). *Journal of Crustacean Biology*, 14, 670-685.
[34] Lenz PH, Weatherby TM, Weber W & Wong KK (1996) Sensory specialization along the first antenna of a calanoid copepod, *Pleuromamma xiphias* (crustacea). *Marine and Freshwater Behaviour and Physiology*, 27, 213-221.
[35] Fields DM, Shaeffer DS & Weissburg MJ (2002) Mechanical and neural responses from the mechanosensory hairs on the antennule of *Gaussia princeps*. *Marine Ecology Progress Series*, 227, 173-186.
[36] Keil TA & Steinbrecht RA (1984) *Mechanosensitive and Olfactory Sensilla of Insects.* Plenum Press, New York.
[37] Schmidt M & Gnatzy W (1984) Are the funnel-canal organs the 'campaniform sensilla' of the shore crab, *Carcinus maenas* (Decapoda, Crustacea)? II. Ultrastructure. *Cell Tissue Research*, 237, 81-93.
[38] Gaffal KP, Tichy J, Theiû J & Seelinger G (1975) Structural polarities in mechanosensitive sensilla and their influence on stimulus transmission (Arthropoda). *Zoomorphologie*, 72, 79-103.
[39] McIver SB (1985) Mechanoreception. *In*: Kerkut GA & Gilbert LI (eds.), *Comprehensive Insect Physiology Biochemistry and Pharmacology*. Pergamon Press, New York, 71-132.

[40] Keil TA (1997) Functional morphology of insect mechanoreceptors. *Microscopy Research and Technique,* 39, 506-531.

[41] Corey DP & Hudspeth AJ (1979) Ionic basis of the receptor potential in a vertebrate hair cell. *Nature*, 281, 675-677.

[42] Fields DM & Weissburg MJ (2004) Rapid firing rates from mechanosensory neurons in copepod antennules. *Journal of Comparative Physiology A: Neuroethology, Sensory, Neural, and Behavioral Physiology*, 190, 877-882.

[43] Shimozawa T, Kumagai T & Baba Y (1998) Structural scaling and functional design of the cercal wind-receptor hairs of cricket. *Journal of Comparative Physiology - A Sensory, Neural, and Behavioral Physiology*, 183, 171-186.

[44] Humphrey JAC, Devarakonda R, Iglesias I & Barth F (1993) Dynamics of arthropod filiform hairs I: mathematical modeling of the hair and air motions. *Philosophical Transactions of the Royal Society of London B: Biological Sciences*, 340, 423-444.

[45] Dechant HE, Rammerstorfer FG & Barth FG (2001) Arthropod touch reception: stimulus transformation and finite element model of spider tactile hairs. *Journal of Comparative Physiology A*, 187, 313-322.

[46] Laverack MS (1962) Responses of cuticular sense organs of the lobster Homarus vulgaris (Crustacea)-I. Hair-peg organs as water current receptors. *Comparative Biochemistry and Physiology*, 5, 319-335.

[47] Laverack MS (1962) Responses of cuticular sense organs of the lobster *Homarus vulgaris* (Crustacea)-II. Hair-fan organs as pressure receptors. *Comparative Biochemistry and Physiology*, 6, 137-145.

[48] Humphrey JAC & Barth FG (2008) Medium flow-sensing hairs: biomechanics and models. *In*: Casas J & Simpson SJ (eds.), *Advances in Insect Physiology*. Academic Press, Burlington, 1-80.

[49] Tautz J, Masters WM, Aicher B & Markl H (1981) A new type of water vibration receptor on the crayfish antenna - I. Sensory physiology. *Journal of Comparative Physiology A*, 144, 533-541.

[50] Huys R & Boxshall GA (1991) *Copepod Evolution*. The Ray Society, London.

[51] Hudspeth AJ & Logothetis NK (2000) Sensory systems. *Current Opinions in Neurobiology*, 10, 631-641.

[52] Yen J, Lenz PH, Gassie DV & Hartline DK (1992) Mechanoreception in marine copepods: electrophysiological studies on the first antennae. *Journal of Plankton Research*, 14, 495-512.

[53] Humphrey JAC, Barth FK & Voss K (2001) The motion-sensing hairs of arthropods: using physics to understand sensory ecology and adaptive evolution. *In*: Barth FG & Schmid A (eds.), *Ecology of Sensing*. Springer, New York, 105-125.

[54] Devarakonda D, Barth FG & Humphrey JAC (1996) Dynamic of arthropod filiform hairs. IV. Hair motion in air and water. *Philosophical Transactions of the Royal Society of London B: Biological Sciences*, 351, 933-946.

[55] Alcaraz M & Strickler JR (1998) Locomotion in copepods: patterns of movement and energetics of *Cyclops*. *Hydrobiologia*, 167/168, 404-414.

[56] Lenz PH (1993) Vibration sensitivity in marine copepods. *In*: Weise K, Gribakin FG, Popov NV & Renninger G (eds.), *Sensory Systems of Arthropods*. Birkhauser, Basel, 423-430.

[57] Bleckmann H, Breithaupt T, Blickhan R & Tautz J (1991) The time course and frequency content of hydrodynamic events caused by moving fish, frogs and crustaceans. *Journal of Comparative Physiology A*, 168, 387-398.

[58] Bush BMH & Pasztor VM (1983) Adaptation and sensory habituation in primary mechanoreceptive afferents of the lobster oval organ. *Journal of Physiology*, 343, 26-27.

[59] Alexander RM (1995) Leg design and jumping techniques for humans, other vertebrates and insects. *Philosophical Transactions of the Royal Society of London B: Biological Sciences*, 347, 235-248.

[60] Gronenberg W (1996) The trap-jaw mechanism in the dacetine ants *Daceton armigerum* and *Strumigenys sp*. *Journal of Experimental Biology*, 199, 2021-2033.

[61] Moore PA, Fields DM & Yen J (1999) Physical constraints of chemoreception in foraging copepods. *Limnology and Oceanography*, 44, 199-177.

[62] Adrian ED (1928) *The basis of Sensation*. Norton & Co., New York.

[63] Wiese K (1976) Mechanoreceptors for near field water displacements in crayfish. *Journal of Neurophysiology*, 39, 816-833.

[64] Young JZ (1939) Fused neurons and synaptic contacts in the giant nerve fibers of cephalopods. *Quarterly Journal of Microscopical Science*, 78, 367-386.

[65] Wiersma CA, Furshpan G & Florey E (1953) Physiological and pharmacological observations on muscle receptor organs of crayfish, *Cambarus clarkii* Girard. *Journal of Experimental Biology*, 30, 136-150.

[66] Palka J, Levine R & Schubiger M (1977) The cercus-to-giant interneuron system of crickets. I. Some attributes of the sensory cells. *Journal of Comparative Physiology* A, 119, 267-283.

[67] Davis AD, Weatherby TM, Hartline DK & Lenz PH (1999) Myelin-like sheaths in copepod axons. *Nature*, 398, 571.

[68] Ball EE & Cowan N (1977) Ultrastructure of the antennal sensilla of *Acetes* (Crustace, Decapod, Natantia, Sergestidae). *Philosophical Transactions of the Royal Society of London B: Biological Sciences*, 277, 429-456.

[69] Laverack MS (1976) *External Proprioceptors*. Chapman and Hall, London.

[70] Gnatzy W & Tautz J (1980) Ultrastructure and mechanical properties of an insect mechanoreceptor: stimulus-transmitting structures and sensory apparatus of the cercal filiform hairs of *Gryllu*. *Cell Tissue Research*, 213, 441-463.

[71] Tautz J (1979) Reception of particle oscillation in a medium. An unorthodox sensory capacity. *Naturwissenschaften,* 66, 452-461.

[72] Weissburg MJ (1997) Chemo- and mechanosensory orientation by crustaceans in laminar and turbulent flows: from odor trails to vortex streets. *EXS*, 84, 215-246.

[73] Tautz J (1977) Reception of medium vibration by thoracal hairs of caterpillars of *Barathra brassicae* L. (Leptoptera, Noctuidae) I. Mechanical properties of the receptor hairs. *Journal of Comparative Physiology*, 118, 13-31.

[74] Fletcher NH (1978) Acoustical response of hair receptors in insects. *Journal of Comparative Physiology A*, 127, 185-189.

[75] Kurbjeweit F & Buchholz C (1991) Structure and suspected functions of antennular sensilla and pores of three Arctic copepods *Calanus glacialis, Metridia longa, Paraeuchaeta norvegica*. *Meeresforsch*, 33, 168-182.

[76] Fields DM & Weissburg MJ (2005) Evolutionary and ecological significance of mechanosensory morphology: copepods as a model system. *Marine Ecology Progress Series*, 287, 269-274.

[77] Bundy MH & Paffenhofer GA (1993) Innervation of copepod antennules investigated using laser scanning confocal microscopy. *Marine Ecology Progress Series*, 102, 1-14.

[78] Fields DM (1996) *The Interaction of Calanoid Copepods with a Moving Fluid Environment: Implications for the Role of Feeding Current Morphology in Predator - Prey Interactions*. PhD thesis, State University of New York.

[79] Webster DR, Brathwaite A & Yen J (2004) A novel laboratory apparatus for simulating isotropic oceanic turbulence at low Reynolds number. *Limnology and Oceanography: Methods*, 2, 1-12.

[80] Fields DM (2010) Orientation affects the sensitivity of *Acartia tonsa* to fluid mechanical signals. *Marine Biology*, 157, 505-514.

[81] Hill PS, Nowell ARM, & Jumars PA (1992) Encounter rate by turbulent shear of particles similar in diameter to the Kolmogorov scale. *Journal of Marine Research,* 50, 643-668.

[82] Paffenhöfer GA (1998) On the relation of structure, perception and activity in marine planktonic copepods. *Journal of Marine Systems*, 15, 457-473.

[83] Boxshall GA, Yen J & Strickler JR (1997) Functional significance of the sexual dimorphism in the array of setation elements along the antennules of *Euchaeta rimana* Bradford. *Bulletin on Marine Sciences*, 61, 387-398.

[84] Holt JR & Corey DP (2000) Two mechanisms for transducer adaptation in vertebrate hair cells. *Proceedings of the National Academy of Sciences of the United States of America*, 97, 11730-11735.

In: Copepods: Diversity, Habitat and Behavior
Editor: Laurent Seuront
ISBN: 978-1-63117-846-7

Chapter 9

NANO- AND MICROFLUIDICS, RHEOLOGY, EXOPOLYMERIC SUBSTANCES AND FLUID DYNAMICS IN CALANOID COPEPODS

Ian R. Jenkinson*

Chinese Academy of Sciences Institute of Oceanology,
Key Laboratory for Marine Ecology and Environmental Sciences, Qingdao, China
Agence de Conseil et de Recherche Océanographiques, La Roche Canillac, France

ABSTRACT

During the last half century, much progress has been made on understanding feeding and encounter in copepods, including observation and modeling of fluid flow around their carapaces, appendages, setae and setules. This modeling has generally assumed a Newtonian medium, and laminar shearing at low Reynolds numbers, with zero velocity at the surface of both the copepod and its prey. This implies on the one hand no slippage of the fluid at surfaces, "*no slip*", and on the other no finite layer of no motion, "*no stick*". This conforms closely to the Derjaguin-Landau-Verwey-Overbeek (DLVO) model, widely used by engineers for over 60 years in nano/micrometer-scale fluidics problems involving surfaces. At the scales of pelagic calanoid copepods (< *ca.* 1 mm) the medium they live in is, at times and in places, functionally non-Newtonian. It presents excess viscosity and elasticity, imparted mainly by exopolymeric substances (EPS) secreted largely by algae and bacteria, and so may be most marked in blooms and chlorophyll maxima. Furthermore DVLO theory is increasingly being recognized as only a convenient assumption, now partly outdated by findings of slip layers up to several micrometers thick, notably on hydrophobic or omniphobic sculptured surfaces. This chapter briefly reviews DVLO theory, and then reviews some notable departures from this theory found in natural waters at scales applicable to copepods. The chapter reviews seawater rheology, and the possible effects of EPS producing stick layers of finite thickness, particularly on copepods' prey. It then considers possible slip layers associated with superhydrophobic drag reduction (SDR), omniphobic drag reduction (ODR) and slippery liquid-infused porous surfaces (SLIPS). Such possible effects include increase

* E-mail address: ianjenkinson@qdio.ac.cn/ian.jenkinson1@gmail.com.

(EPS layers) or decrease (SDR, ODR, SLIPS) in water velocity and shear rate close to the surface, as well as possible resistance to adhesion (slipperiness) and biofouling. Slipperiness in prey may hinder copepod feeding. Finally the differences are briefly treated between the relatively mucus-free calanoid copepods and more mucus rich harpacticoids, and a suggestion is made for interdisciplinary capacity building activities for rheology and nano/microfluidics to be understood better by teams investigating copepod dynamics and all aspects of plankton ecology and biogeochemistry.

INTRODUCTION

Advances in nano- and microfluidics since about 1990 have driven new surface science that shows that models of surface properties developed in the 19th and 20th centuries have ill-equipped biogeochemists and both microbial and plankton ecologists with engineering tools to understand many of the phenomena observed at the surfaces of living and non-living particles. This chapter exposes the importance of some recent advances in nano- and microfluidics, focusing on a particular area of biogeochemistry and plankton ecology, the feeding and hunting by planktonic, mostly calanoid, copepods.

In the 19th century Physics and Chemistry made enormous progress, driving engineering and the Industrial Revolution. Steam-powered, internal-combustion and electric engines were developed, together with chemical processes and purification, shaping the social, political, and ultimately the ecological and biogeochemical history of the following 20th century. However, the problems of electrical and turbulent boundary layers [1] remained mired in controversy. The 20th century saw arms races develop, first Anglo-US-German-Russian (1890-1945), then principally US-Soviet (1946-1990), driving huge motivation quickly to develop better and faster aircraft, missiles, ships, torpedoes and so on. These arms and machines develop huge physical pressures, so the small forces in soft, non-Newtonian materials tended to be largely neglected [2]. In the early 20th century, models of the tiny boundary layers were also greatly simplified and taught in the engineering schools principally in Europe and North America. In the 1940s, the simplified versions were drawn together to model the stability of colloidal systems in what is now known as the Derjaguin-Landau-Verwey-Overbeek (DLVO) model [3]. See below for further details.

The development of nano- and microfluidics, including nano- and microfiltration, at the end of the 20th century, with high added value in nano- and micro-machines, including lab-on-a-chip, reverse osmosis and desalination techniques, soon showed that DLVO did not suffice to model fluid dynamics at the nanometer, micrometer and even larger scales involved [4-9,19]. The published science has at times lagged behind innovation by trial and error, and the length scales happen to be at least in part those used by nauplii to adult copepods in swimming and prey capture. Measured in the sea at larger length scales (typically larger than 0.35 mm), the rheology, or visco-elasticity of natural waters is now known to be partly controlled by exopolymeric substances (EPS) exuded by algae and bacteria [10-15].

In this chapter, bearing in mind scales relevant to copepod hydrodynamics, I summarize the system of units considered, the length scales, particularly those relevant to nano- and microfluidics, the DVLO model, and observed departures from it, including superhydrophobic drag reduction (SDR) and wall slip as well as some effects that are unclear, or even contested. I subsequently consider the effects of EPS on copepod hydrodynamics and food handling, partly through its known effect on the rheology of the copepods' milieu, as well as by stickiness mechanisms. Copepods' ways of dealing with stickiness are also considered. Within this section, to minimize future confusion, I will list some different meanings of so-called "*sticky water*". The copepod surfaces will be considered, particularly the epicuticle of the feeding appendages, in relation to feeding, and that of the rest of the body in relation to swimming and ambush hunting. The surfaces of calanoid copepods' principal food, phytoplankton, are also considered. I finally treat some of the differences between harpacticoids and calanoids, particularly in relation to mucus.

UNITS

In this chapter I discuss mechanical stress quite explicitly, but mass very little. I follow Jenkinson & Sun [2] and rearrange the familiar framework of S.I. units, length [m], time [s] and mass [kg], into a framework of length [m], time [s] and pressure or stress [kg m^{-1} s^{-2}]. Further details are shown in Table 1.

Table 1. Units used in this chapter

Length	m
Time	s
Stress	Pa (pascals) = N m^{-2} (N = newtons)
Temperature	K (kelvins), °C (degrees Celsius) (0°C = -273.15K)
Salinity	Practical salinity units (psu)
Electric charge	C (Coulombs)
Molecular concentration	M (moles L^{-1}, where 1 mole = 6.02 $\times 10^{23}$ molecules)

NANO- AND MICROFLUIDICS IN RELATION TO THE DYNAMICS OF MICRO- AND MESOPLANKTON

The Length Scales of a Few Things

Nano- and microfluidics refer to flows at length scales below about 100 nm and 100 μm respectively. The discipline has recently progressed rapidly, and is being used to build entire micromachines, including microvalves, micro- and nanopumps, gears, cantilevers and other microdevices [16], driving intense research on processes at these scales.

Figure 1 has scale from 10^{-1} to 10^{7} nm. Table 2 also collects some of the scales mentioned in this chapter. At centimeter scale we have giant copepods such as *Bathycalanus* sp. directly subject to flow at the limits of turbulence. Here friction and leverage are important. *Bathycalanus*'s appendages and whiskery bits reach down to smaller scales, hence to different physical constraints. Its feeding parts are millimeter in scale, larger than the entire cephalosome of the adults of many small species and naupliar stages. Descending past the millimeter milestone, diffusion and viscosity become important. The next milestone, marked micrometer, includes the realms of cell biology. Between the micrometer to nanometer scales is where strong capillarity, molecular adhesion and recognition, as well as steric repulsion kick in. Here surface energy effects are important. Down at the nanometer scale we have chemistry, molecular structure, "*water-of-hydration shells*" or "*solvation shells*" [19] around ions, and the multimolecular shells of water itself (Tim Wyatt, *pers. com.).*

Table 2. Typical molecular and hydrated radii and diffusion coefficients in bulk water at 25°C. Summarized from Conlisk [3]

Solute	D_A ($cm^2\ s^{-1}$)	Molecular radius (nm)	Hydrated radius (nm)
Ions			
Na^+	1.33×10^{-9}	0.12	0.36
K^+	1.96×10^{-9}	0.15	0.33
Ca^{++}	2.1×10^{-9}	0.19	0.33
Cl^-	2.0×10^{-9}	0.16	0.33
Biomolecules			
Albumin	0.061×10^{-9}	-	7.2
Glucose	0.94×10^{-9}	-	1

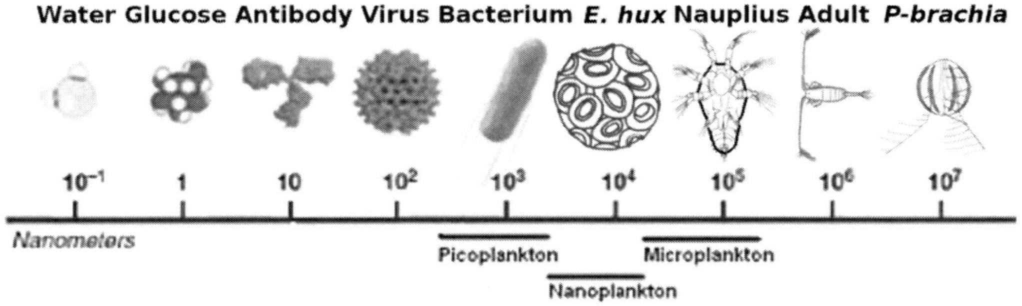

Figure 1. The length scales of a few things, ranging from water molecules to the ctenophore *Pleurobrachia pileus*. are compared to the nano- and microfluidics scales addressed in this chapter. *E. hux*: *Emiliana huxleyi*; *P-brachia*: *Pleurobrachia pileus*. Drawing of *E. huxleyi* and *P. pileus* from Jahnke [130] and Greeve [131], respectively.

Length Scales of Layers: The Derjaguin-Landau-Verwey-Overbeek (DLVO) Model

This section is a description of liquid-solid surface structures based on the Derjaguin-Landau-Verwey-Overbeek (DLVO) model, expanding on Conlisk [3].

Electrical Double Layer

Many solids acquire a negative charge when in contact with an aqueous medium. The development of electrical charge on a solid surface is a combination of physical and chemical processes. The surface charge is acquired through a variety of mechanisms such as ion adsorption, exposure to charged crystal surfaces, and ionization of surface groups. Among these, the ionization of surface groups plays a dominant role. For devices with surfaces made of glass or silicon, surface silanol groups (SiOH) undergo deprotonation, which results in development of negative charges on the walls.

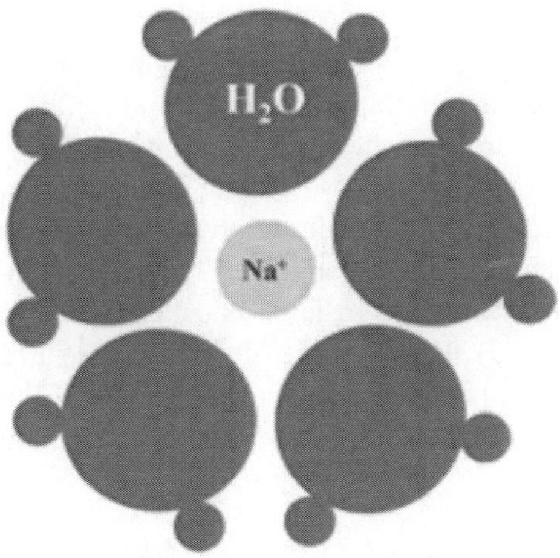

Figure 2. Sketch of the hydration of a sodium ion by water molecules. Values of the hydrated ionic radius a_H can range from 3 to 5 times the unhydrated ion radius a. For example the ratio $H = a_H/a$ is typically 1.83 for Cl^- and $H = 3.91$ for Na^+. Multivalent ions typically have hydration radii larger than monovalent ions because of the stronger electrical fields around them. Modified from Conlisk [3].

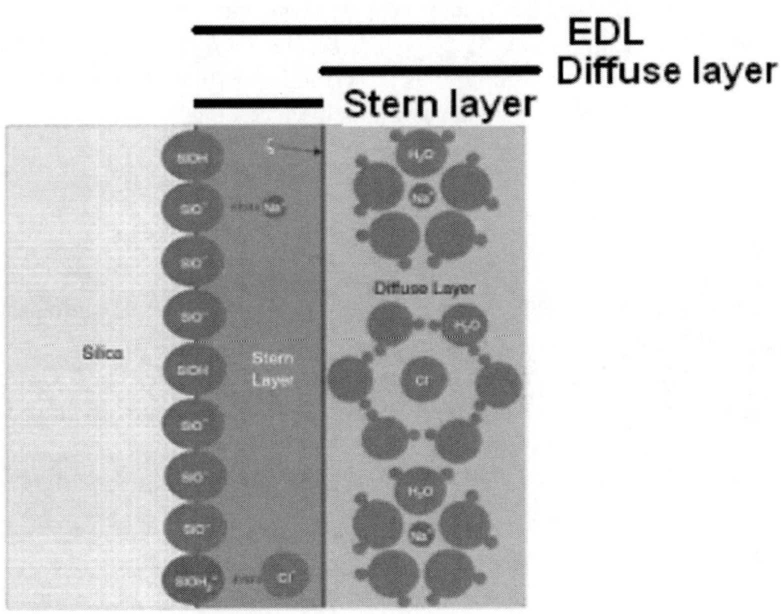

Figure 3. To show the hydration of ions by water molecules near a charged silica surface. The electric double layer (EDL) consists of a layer of counter ions pinned to the wall, the Stern layer, and a diffuse layer of mobile ions outside that layer. Here the wall is shown as being negatively charged, and the ξ-potential is defined as the Stern plane. Modified from Conlisk [3].

If the surface is negatively charged, the positively charged ions (called counterions) in the vicinity of the surface tend to collect near the charged surface, and an inner layer of counterions is attached to the surface. Outside this pinned layer, ions float within the electrolyte solution near the surface. The total system, including the pinned and floating ions or diffuse layer, is called the electrical double layer (EDL). On the basis of the analysis of the EDL surrounding an ion (Figures 2, 3), the local electric field can be in the order of 10^{10} V m^{-1}. Under the influence of this large electric field, the polar ends of water molecules can electrically bond to the ion, creating a "molecule" whose effective radius is larger than the unhydrated one. Several radii of solvated hydrated ions are shown in Table 2, but these are the radii considered to be time averages over a large number of ions.

The following treatment uses published values determined at 20°C unless otherwise stated. The layer outside the Stern layer consists of mobile ions and is called the diffuse layer, and the total system, including the Stern layer and the diffuse layer is called the Electrical Double Layer (EDL) (Figure 3). The presence of the charged surface and the ions thus creates an electrical field. Conlisk [3] presents parameters for simplified seawater, consisting of a solution of totally dissociated NaCl, i.e., Na^+ and Cl^- ions. The ionic strength of this solution

$$I = \sum_i z_i^2 c_i \tag{1}$$

where c_i is the molar concentrations of the electrolyte constituents of species i, z_i is their valence and here i is (Na^+, Cl^-). To obtain an approximate value of I for seawater, let us assume it to be a 3.5% (w/v) (35 g l^{-1}) solution of salt, NaCl. Since the atomic weights of Na^+ and Cl^- are, respectively, 23 and 35.5, their separate concentrations are, respectively, [(35 × 23)/(23 + 35.5), (35 × 35.5)/(23 + 35.5)] g l^{-1}, or (13.7606, 21.2393) g l^{-1}, (13.7606/23, 21.2393/35.5) M, or 0.598 M each. Since z_i is (1, 1), and c_i is (0.60, 0.60) M, $I = 1.2$ M.

The permittivity (or dielectric constant) of a medium, such as water

$$\varepsilon_r = \frac{\varepsilon_e}{\varepsilon_0} \tag{2}$$

where ε_0 is the permittivity of a vacuum, 8.85×10^{-12} C^2 N^{-1} m^{-2}, and ε_r is the permittivity of a medium (e.g., water) relative to ε_0.

Meissner & Wentz [17] determined ε_r at over a range of salinities at 20°C to be from 78.5 to 72.7 at salinities from 0 to 35, respectively. From Eq. 2, the respective values for ε_e are thus 6.95×10^{-12} and 6.47×10^{-12} C^2 N^{-1} m^{-2}.

The nominal length scale associated with the EDL is the *Debye length*, defined by

$$\lambda = \frac{\sqrt{\varepsilon_e RT}}{FI^{1/2}} = \frac{1}{\kappa} \tag{3}$$

where F is Faraday's constant (9.65×10^4 C mol^{-1}), ε_e is the electrical permittivity of the medium; R is the universal gas constant (8.31 J mol^{-1} K^{-1}), T is the temperature expressed in °K, I is the ionic strength, calculated above as 1.2 M.

For salinity 35, the calculated value of λ (Eq. 3) for our simplified seawater thus comes to 0.39 nm. The corresponding values for salinities 10, 1.0 and 0.10 are I = 0.34, 0.034 and 0.0034 mol L^{-1} and ε_r, = 75.2, 77.7 and 77.8 [17] and λ = 0.71, 3.0 and 7.3 nm.

Debye length is an important characteristic of interfacial electric double layers. It is a measure of the distance between charged interfaces where the electrostatic component of surface forces is important. Reduction of the Debye length due to increasing ionic strength leads to weaker electrostatic interactions and possible stability loss by dispersions and emulsions (an example of instability is shown when a liter of milk curdles in seconds following the addition of a drop of acidic lemon juice). Transition from the fresh to seawater in estuaries, with the formation of deltas of organic-rich sediment is a striking example illustrating importance of this parameter. As λ becomes smaller the voltage gradient (electric field) across the Debye layer on suspended and colloidal particles becomes greater, resulting in the particles aggregating into stronger and more compact flocs [3] of higher density [18]. This leads to more rapid sedimentation both in and downstream of the sharpest salinity gradients in estuaries with the frequent appearance of a turbidity maximum, mud banks and deltas, rich in organic matter. Some copepod species, particularly of the genera *Eurytemora* and *Acartia*, are specialized, having adapted to live in these stressful areas in and around estuarine turbidity maxima. Here, steep spatiotemporal gradients occur in salinity, and thus also in Debye lengths and in the electrical fields around mineral, non-living-organic and living particles. As a result, steep gradients will also frequently occur in the stickiness, density and size distribution of marine snow, marine organic aggregates and transparent exopolymer particles (TEPs), often giving high loads of recently aggregated and rapidly sinking or rising [18] suspended particulate matter.

Turbulent Boundary Layers on Copepod Surfaces

The Reynolds number is the main dimensionless parameter associated with viscous flow, and for boundary layer flow on a flat plate of length L, it is defined by

$$Re = \frac{\rho U_{\infty} L}{\mu} \tag{4}$$

where ρ is the fluid density, μ is the viscosity, U_{∞} is the velocity far from the plate, and L is its length [1].

At low values of Re (< *ca.* 1), a copepod lives in an immediate fluid environment free of turbulence, while at high values of Re (> *ca.* 100) it lives in a fluid that is turbulent at body-length scales, but a shearing, non-turbulent boundary layer covers all its body surfaces.

Slip Layers and Non-Stick Layers

Wall Slip

Rothstein [8] reviews the development of the notion of wall slip. The concepts of wall slip and slip length were formalized by Navier in 1823 [3] as shown in Figure 5.

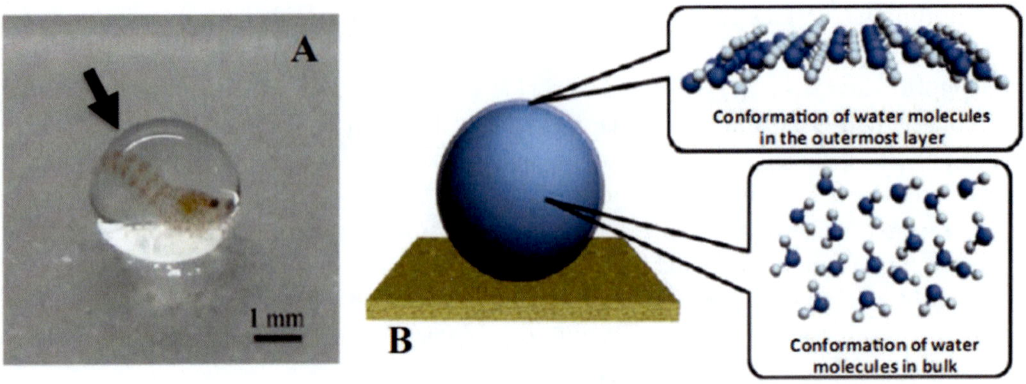

Figure 4. (A) Optical photograph of a several-mm^3 droplet, in which a *Caridea serrata* shrimp was trapped, placed on a superhydrophobic pedestal. (B) Schematic illustration of a droplet on a superhydrophobic plate, showing the H-bonding net around the droplet consisting of a regular array of water molecules. The conformation of water molecules closest to the water/solid surface and water/atmosphere interface is different from that in the bulk. Interfacial molecules form regular structure in order to maximize their H-bonding partners and thereby form a compact H-bonding net. Modified from Su et al. [134], and reproduced with kind permission.

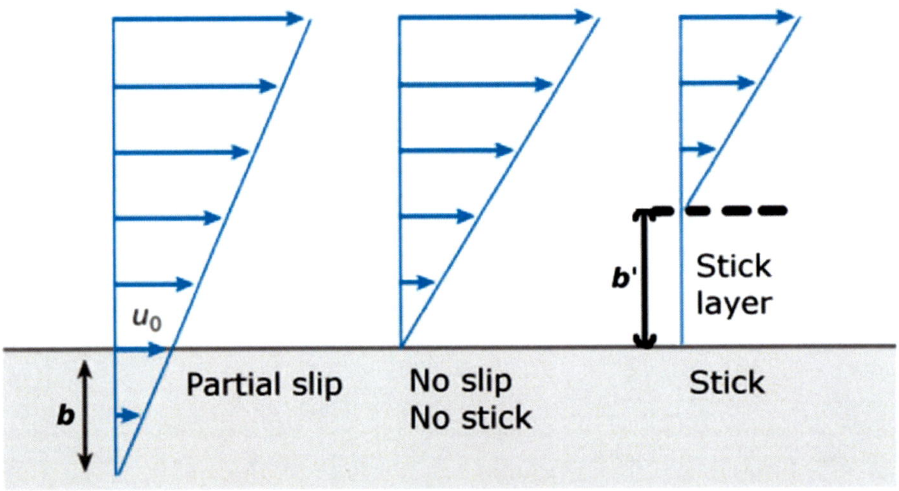

Figure 5. Schematic diagram of slipping and sticking at a fluid-solid interface by a fluid subject to simple shearing. *b* is slip length; *b'* is sticking length; u_0 is slip velocity. Modified from Rothstein [8].

In Navier's model, the magnitude of the slip velocity, u_0, is proportional to the magnitude of the shear rate experienced by the fluid at the wall:

$$u_0 = b\dot{\gamma} \tag{5}$$

where *b* is the slip length and $\dot{\gamma}$ is shear rate such that

$$\dot{\gamma} = \left|\frac{\partial u}{\partial y}\right| \tag{6}$$

where *u* is the fluid velocity at any point and *y* is the distance of that point from the surface. When the value of *b* is negative (i.e., notionally within the wall) it is a slip length, but when *b* is positive (i.e., in the "fluid") it becomes a stick length (Figure 5) (see "Stick layer" section below).

There followed much debate on the subject during the 19th century. During the 20th century, however, published experimental results generally confirmed that no slip existed at liquid-solid boundaries, so the functional "*no-slip*" idea came to be universally accepted as the normal condition [8,20,21], apart from minor exceptions and dissident voices such as that of Grasso et al., (2002) [22] and Drost-Hansen (2006) [23].

The Lotus Leaf Effect and Superhydrophobic Drag Reduction

Better experimental techniques and the advance of requirements by the micro- and nanofluidics industry, however, led to a wealth of papers in the 1990s finding violations of the no-slip law mainly at hydrophobic walls. More dramatically, the 2000s have seen the discovery of superhydrophobic drag reduction (SDR), and by 2010 Rothstein [8] could authoritatively state that, "*The no-slip boundary condition* [was] *no more than a convenient approximation found to hold under most normal flow conditions*". Slip lengths up to 25 μm have now been found [5].

Until recently, it was considered that SDR could exist when the superhydrophobic surface was in either of two states, the Wenzel state (i.e., when liquid occupies the spaces between the surface irregularities) or the Cassie state, in which gas is trapped between the irregularities [8]. The Cassie state tends to show more powerful SDR than the Wenzel state, but a disadvantage to pelagic organisms might be the susceptibility of SDR to compression and dissolution of the trapped gas by hydrostatic pressure. Barthlott & Neinhuis's [4] much cited paper represents a landmark. They investigated the water-repellence and self-cleaning of lotus leaves. Quickly, workers found such surfaces to show SDR, with slip lengths of many nanometers in tubes with hydrophobic coatings [7].

Slip Layers and Hierarchical Microstructuring

As well as the Cassie and Wenzel states, there is a recently reported third drag-reducing state. Cottin-Bizonne et al., [24] modeled that, while regularly spaced and sized nano- and microstructures can give the biggest slip lengths and hence the most drag reduction (DR), the amount of DR can be very sensitive to small changes in structure shape and size. A fractal hierarchy of structure sizes gives the most stable DR. This may explain the generally hierarchical distribution of structure sizes on flowering plants [25]. If microstructuring on copepods also generally occurs at a hierarchy of sizes, this too would be compatible with a function evolved for stable drag reduction. The setules on setae and other parts of the copepod body may function as part of such hierarchical structuring.

The Flower Petal Effect

Water drops stick to many flower petals rather than rolling off. Feng et al., [6] show that these petals bear a hierarchy of nanometer- to micrometer- scale, hydrophobic irregularities. Like on lotus leaves, water drops round up due to repulsion, but because of the particular sizes and shapes of the irregularities, the drops remain pinned to such petals. After the Lotus Leaf effect (which can incorporate SDR and self-cleaning of the surface whether in the Cassie state or the Wenzel state), I class the Flower Petal effect as a second superhydrophobic state.

Slippery Liquid-Infused Porous Surfaces (SLIPS)

A third new state has now been identified on the pitchers of the insect-catching pitcher plant *Nepenthes alata*. Insects are attracted to sweet, scented nectar on the rim (peristome) of the pitcher. But it is a trap. The nectar is slippery and the insects aquaplane on it and slip into digesting fluid in the pitcher. Bauer & Federle [26] showed that the peristome bears *hydrophilic* microstructures, and is lubricated by the aqueous nectar. The nectar prevents the insects' feet getting a grip by resistance to dewetting, or resistance to draining between the foot and the peristome structures, or both. This system inspired Wong et al., [27] to propose microstructures with water-immiscible fluid trapped in the spaces between the microstructures. They are termed slippery liquid-infused porous surfaces (SLIPS). The authors then tested such SLIPS for omniphobic action. To be functionally useful, either in evolved organisms or in manufactured objects, the fluid would have to be resistant to erosion by the ambient shear field, or replenished by reservoirs. Unlike Cassie state surfaces (with their compressible and soluble gas), SLIPS are predicted to be omniphobic, and insensitive to increased hydrostatic pressure. They are now being intensively investigated industrially for commercial applications in marine antifouling [135]. SLIPS properties are potentially useful to aquatic organisms, particularly those covered by replenishable mucus, such as algae, mollusks, tunicates, hydrozoans, and these organisms should be investigated for SLIPS-mediated anifouling properties. Such surfaces might also produce surface laminar and/or turbulent drag reduction. Pelagic copepods seem unlikely candidates to bear SLIPS, as they secrete little or no mucus. As many of their prey do, however, the predatory copepod then has to deal with it. We further consider a hunting copepod below.

Possible SDR on Plankton Surfaces

Jenkinson & Sun [15] found that flow of cultures of harmful algae species and of a bacterium in non-turbulent flow in capillaries of different radii, and at different values of shear stress could be increased (drag reduction) or decreased (drag increase) relative to that of pure seawater or culture medium. Drag increase was probably caused by EPS-mediated increased viscosity now well known in seawater [11-14,28-32] as well as in cultures of both phytoplankton [10] and marine bacteria [33]. Jenkinson & Sun [15] suggested the drag reduction, however, may have been SDR associated irregular hydrophobic surfaces of the cultured organisms and/or on the EPS they were producing.

Stick Layers

Stick layers occur when there is a layer of non-deforming material adhering to the surface, even though such a layer may consist mainly of water. Some examples of stick layers are described hereafter.

Electrical Double Layers

EDLs are treated above. They are of thickness about 0.7 to 7 nm.

Layers of Adhering Materials Such as Polymers, or Sculpturing

In the sea most surfaces, both solid and gaseous, rapidly accumulate a skin of organic material, particularly surfactant matter. Such layers on the outside of copepods seem not to be frequent in pelagic copepods, although a limited layer of what appears to be a 1- to 2-μm layer of bacterial "plaque" was shown by scanning electron microscopy on copepod cuticles [34].

Layers of Epibionts Adhering to the Cuticle

Various epibionts, including diatoms, protozoa and bacteria inhabit the outer surfaces of copepods, details of which follow.

Diatoms

Studies have indicated that high infestations of epibiotic diatoms on copepods fail to change respiration rate or excretion rate of ammonia or phosphate relative to non-infested copepods, yet infested copepods were found to have higher mortality rates. Epibionts are considered to benefit from excretion of nutrients by their hosts, especially those on the uropods, near the anus, or near the mouthparts, where they may benefit from sloppy feeding. The host's swimming would also bring nutrients past the epibiont, and thin the microlayer/boundary layer around it, allowing faster nutrient uptake. There is no record of hosts feeding on algae attached to their exoskeleton. Pelagic copepods appear to be infested more generally than benthic ones, although the reason is not known [34]. Epiphytic diatoms may not only affect swimming speed but may increase the host's visibility to predators by increasing or changing their hydrodynamic signal [35].

Protozoa

Carman & Dobbs [34] review protozoan epibionts of the Crustacea, including ciliates of the sub-classes Hypostomata, Suctoria, Hymenostomata, Peritricha and Spirotricha, as well as foraminiferans, naked rhizopods, choanoflagellates and ellobiopsids. The number of epibiont protozoa can range from none (very common) up to 250 per host. The authors cite 13 references (not counting older reports, cited elsewhere) referring to different patterns of distribution on the body that are characteristic of different epibionts as well as of different host species, development stages and sometimes the sex of the host. Infestation intensity may also be a function of time of year. As well as the Suctoria and Peritricha, apostome ciliates (subclass Hypostomata) are widespread epibionts [35]. Ciliates range in the symbiotic relationship from exuviotrophic ectocommensals to parasites, see Carman & Dobbs [34] and references therein.

Bacteria (*sensu lato*)

Bacterial epibiont infestation of copepods can be divided (according to the classification of the authors) into that by *Vibrio* and that by other bacteria. Details follow.

Bacteria other than Vibrio. Non-*Vibrio* bacteria encountered on copepods include *Acinetobacter*, *Aeromonas*, *Alteromonas*, *Bacillus*, *Chromobacterium*, *Cytophaga*, Flavobacterium, *Leucothrix mucor*, *Pseudomonas* and other unidentified rods and cocci (classification following [34]). The incidence of copepods with bacteria is reported to be unrelated to the abundance of free-living bacteria, and may involve the physiological stage of the host copepod. Numbers of epibiotic bacteria are reported to vary from none to 10^5 or even 6×10^5 per copepod [34]. Epibiotic bacteria are generally unevenly distributed on copepods, favoring areas between segments, around mouthparts, near the anus, and on legs.

On pelagic copepods, filamentous bacteria are not found abundantly [34]. In *Labidocera* sp., a planktonic copepod, filamentous bacteria occur only at the bases of feeding appendages and legs, where they may contribute little to surface drag. Among benthic copepods, however, filamentous bacteria are relatively common. Indeed, fouling by the cyanobacterium *Leucothrix mucor* on the harpacticoid *Tigriopus brevicornis* has been shown to change swimming patterns [37].

Vibrio. Along with other *Vibrio* species, *V. cholerae* has been much studied because it causes diseases and the strain O1 (VCO1) causes cholera in humans [38-40]. Copepod cuticles are comprised essentially of chitin [39,40]. The VCO1 connection with chitin is an excellent example of a successful bacterium-substrate interaction, with a detailed influence on its lifestyle in the different environments it colonizes [38,40]. It has a variety of life forms, which can succeed each other. They show various forms, including (i) a free-living flagellated form and (ii) a multicellular form, that produces layers of bacteria or biofilms on surfaces. This multicellular form can colonize the chitinous cuticles of copepods, other crustacea, and chitin-containing phytoplankton and macrophytes in marine, brackish and fresh waters. Vibrios are attracted to and attach to both chitin and copepods. However, attachment of vibrios to copepods is less efficient than their attachment to chitin particles. Pruzzo et al., [39] suggest that this is because the copepod exoskeleton bears a wax epicuticle that prevents close contact with the chitin until bacterial enzymes (e.g., lipase) have digested it. Scars and pits on copepod exoskeletons occur in *Acartia* species [34,41], which result from the activity of chitinoclastic or wax-degrading epibiotic bacteria (possibly *Vibrio*). Based on observations, however, Carman & Dobbs [34] consider that chitinoclastic activity by epibiotic bacteria is rather exceptional.

In contrast to the lack of relationship between abundance of epiphytic (probably non-*Vibrio*) bacteria and that of free-living bacteria in the surrounding water, abundance of *Vibrio alginolyticus* on the rock-pool copepod *Tigriopus fulvus*, was positively correlated with the abundance of *V. alginolyticus* in the water [42].

Glycocalyxes

Glycocalyxes are considered here as they cover many of the prokaryotic and eukaryotic prey of copepods. Most cells, including bacteria [43], protists [44], and the cells of multicellular organisms [45] bear a glycocalyx (also known by terms including, "cell coat", "extracellular matrix" and "spore mucilage"). It consists of a system of polymer fibers, largely of polysaccharides. They are generally arranged in an intertwining network. These membrane carbohydrates occur usually in combination with proteins or lipids as glycoproteins or glycolipids. Many other carbohydrate compounds, e.g., proteoglycans, are loosely attached to the outer surface of the cell as well. One can say that the entire outside surface of the cell often has a loose carbohydrate coat, the glycocalyx.

Glycocalyx molecules manage the recognition of and adhesion to surrounding cells and other structures [46]. Functions of the glycocalyx and its different components are multiple and probably still not well elucidated Many of the carbohydrates moieties of glycocalyx glycoproteins act as *receptor sites* for binding pheromones, and they can communicate such information into the cell as part of two-way cross-talk [47]. When bound to external antigens, this combination activates attached proteins inside the cell that, in turn, activate a cascade of intracellular enzymes. Some carbohydrate moieties enter into immune reactions, based on lock-and-key recognition mechanisms. Some harmful algae, moreover, use their glycocalyx to effect short-range or contact allelopathic action, to harm other members of the plankton community [48], including multicellular potential predators [49], such as copepods. While the glycocalyx serves the protist to detect potentially useful and harmful molecules outside the cell [50], it may also serve potential predators, including Crustacea such as copepods, to better identify their prey. In both prokaryotes and eukaryotes, the glycocalyx also mediates adhesion between their cells in colonies, as well as the initial adhesion of epibionts to their substrates and hosts. Typically, the thickness of glycocalyxes can range from 0.1 μm or less to several μm [2].

Vicinal Water (VW)

Drost-Hansen [23] reviews considerable published evidence for layers of changed properties of water close to surfaces. The causes of "vicinal water" (VW) are often poorly understood, and, Drost-Hansen's ideas notwithstanding could be several. Manifestations of VW seem to vary, sometimes without obvious cause, leading some to doubt their existence. The existence of VW is thus rarely mentioned in physics or engineering textbooks. In the present chapter we are interested in what happens around copepods, not the physico-chemical (molecular) reasons for it. Some discussion of mechanisms is unavoidable, however, in order to consider the evidence.

Viscosity of VW

A succinct summary of properties that have been reported for water close to surfaces is shown in Table 3. In particular, the viscosity has been reported to be increased by a factor of 2 to 10. No characteristic deformation rates or stresses are given for the measurements of such viscosity. A relevant length scale, however, may be the layer thickness. Published values of long-range ordering effects extend from 0.05 to 5 μm from the surface (Table 4) while Grasso et al., [22] lists publications reporting H-bonding effects from a few hundreds of nm.

Evidence that this VW viscosity is fundamentally different from that of the bulk phase is provided by measurements of viscosity of water squeezed between flat horizontal plates 30 to 90 nm apart. Peschel & Adlfinger [51] found that the viscosity not only showed a strong negative relationship with inter-plate separation, but also goes through a number of maxima and minima, corresponding with temperatures 15, 30, 45 and 60ºC (the so-called "Drost-Hansen temperatures"). Other workers cited by Drost-Hansen [23] have found similar increases in viscosity, as well as in other properties (Table 3), in small capillaries at the same temperatures. These anomalies do not occur far from surfaces.

Thermal anomalies measured by differential scanning calorimetry, are reported close to 60°C, a "Drost-Hansen temperature" [23] (see last paragraph) in highly dilute (~0.01%) aqueous solutions of the quaternary amine, cetyl-trimethyl ammonium salicylate. Such solutions form large worm-like micelles, and the author suggests that these results indicate that the micelles are vicinally hydrated, which may explain their unusual rheological properties.

Table 3. Summary of some of the properties of vicinal water. From Drost-Hansen [23]

Property	Bulk water	Vicinal water
Density (kg m^{-3})	1.00 x 10^3	0.96-0.97 x 10^3
Specific heat (cal kg^{-1})	1.00 x 10^3	1.25 x 10^3
Thermal expansion coefficient (°C^{-1})	250 × 10^{-6}	300-700 × 10^{-6}
Adiabatic compressibility (Atm^{-1})	7 × 10^{-17}	35 × 10^{-17}
Heat conductivity (cal s^{-1} °C^{-1} m^{-1})	1.4 × 10^{-3}	10 × 10^{-3} to 50 × 10^{-3}
Viscosity (mPa s)	0.89	2-10
Activation energy, ionic conduction (kcal $mole^{-1}$)	4	5-8
Dielectric relaxation frequency (Hz)	19 × 10^9	2 × 10^9

For references, see Drost-Hansen [23].

Table 4. Experimentally determined values for "long range ordering effects" within interfacial water layers at the boundary of solid surface and water. From Drost-Hansen (2006) [23], in which please find the references

Reference	Method	Solid boundary	Penetration depth (μm)
Etzler and Lilies (1986)	Dielectric constant	Sheets of mica	2-5
Drost-Hansen (1976)	Adhesion at glass	Glass	1.5
Henniker (1949)	Disjoining pressure	Mica or steel plates	< 1
Mastro and Hurley (1985)	Surface conductivity	Glass tube	0.3-0.4
Peschel and Adlfinger (1969, 1976)	Conductivity	Quartz particles	0.2-0.3
Falk and Kell (1966)	Viscosity	Glass plates	0.25
Montejano et al (1983)	Conductivity	Pyrex glass	0.05-0.2
Drost-Hansen (1969)	Viscosity	Pyrex glass	*ca.* 0.2
Steveninck et al (1991)	Viscosity	Convex quartz plates	0.16
Braun and Drost-Hansen (1981)	Rigidity modulus	Convex glass plates	0.15
Bailey and Koleske (1976)	Disjoining pressure	Convex fused-silica plates	*ca.* 0.1
Nir and Stein (1971)	Disjoining pressure	Quartz plates	0.1
Derjaguin (1933)	Rigidity modulus	Glass	0.1
Antonsen and Hoffman (1992)	Viscosity	Glass plates	*ca.* 0.1
Clifford (1975)	Air-bubble flow	Glass tubes	*ca.* 0.1

Assuming that VW really does form close to surfaces, what about the surfaces of other suspended particles and dissolved macromolecules? Reviewing considerable evidence, Drost-Hansen [23] concludes that there is a critical size for macromolecules below which no effects of VW are detected, and above which they are. This size corresponds to molecular mass of 1 to 5 kDa. The increased viscosity suggests that other rheological properties might also be changed in such solutions or suspensions. Furthermore VW-type properties are destroyed by fluid shearing, and conditions without shearing for a critical time t_c are required for them to re-establish. Values of t_c are found to range from about 1 s to about 1 day.

Other Properties of VW

Similar manifestations of VW are often reported from the vicinity of quite different types of surface, for example those of polystyrene and silica. Furthermore, the density of VW is reported up to 3 to 4% less than that of bulk water, and the specific heat up to 25% higher. (Table 3). Drost-Hansen [23] considers that VW can extend from surfaces, out to a "*penetration distance*" of *ca.* 0.1 to 5 μm (Table 4), and that they are unaffected by ionic strength, pH, or any surface-specific epitaxial ordering.

VW effects may also be particularly important in intracellular processes, but that is outside the scope of this chapter.

Objections to the Theory of VW

Yaminsky et al., [52] carefully showed that at surfaces of glass or of fused silica, in contact with water, the silica can interact with the water to form a gel. This gel formation is variable and difficult to repeat, and can show complex dependence on time, temperature and other properties. This may lead to variable physical properties being measured in thin water layers that are ascribed to fundamental properties of the water, but are in reality due to the presence of silica gel. Many of the experiments cited to justify the existence of VW were conducted in glass or fused-silica apparatus. So findings used to oppose the now highly unfashionable "polywater" idea [53], appear to weaken the concept of VW as well. More recently, however, new structuring phenomena at surfaces have been proposed [22], and much work is required to deepen understanding of the phenomena that may be responsible for different manifestations of VW.

THALASSORHEOLOGY AND EXOPOLYMERIC SUBSTANCES IN COPEPODS' ENVIRONMENTS

Thalassorheology

The Viscosity and Elasticity of Seawater and Other Natural Waters

Thalassorheology, the rheology of seawater and other natural waters, has recently been reviewed [2], so only a brief outline is given here in relation to copepods. In natural water, the liquid phase consists of water and, particularly in seawater, its dissolved small molecules, mainly salts more or less dissociated into cations and anions. The viscosity of this phase is controlled by temperature and salinity. Its elastic modulus G' is zero, which means that, apart from effects due to gravity waves in surface and subsurface density discontinuities, it has no memory of past conformation and thus no tendency to spring back after deformation.

While it has previously been written [2, and references therein] that the viscosity of the aquatic phase is independent of length-scale, this is strictly true only down to a length scale often considered to be close to the Debye length (Eq. 3), roughly the size of water shells (*ca.* 0.2 – 0.3 nm) [54] (Figures 2 and 3). Increase in viscosity of pure water at larger scales due to VW effects at particle surfaces (nm- to μm-scale, Table 4) may also occur.

Thalassorheology, Lumpiness and EPS

The intra-sample variability found in the rheological properties of seawater led me to suggest centimeter-scale heterogeneity [11]. Excess viscosity was similarly found to be variable at 5-10-cm scale in the sea [13]. This rheological variability has been termed "lumpiness" [2,11,55]. It may be due to physical, chemical and biological mechanisms of aggregation by EPS, as well as direct formation of EPS structures by the producer organisms. An extensive and advanced literature on aggregation by sludge EPS [56] is available to inspire further research on marine and freshwater EPS.

The following types of EPS suspended or colloidal in seawater can all be sticky: marine snow [57], transparent exopolymeric particles (TEP) [58,59], as well as phytoplankton [60-65] and bacterioplankton [66]. Partly through inter-particle bridging by fibers and fibrils of EPS or other matter, much of this stickiness is transferred to length scales larger than the aggregates themselves, and therefore imparts viscoelastic properties to the medium at length scales exceeding aggregate size.

Attraction to and Avoidance of EPS and Its Effects on Swimming

Savage [67] described patches in the North Sea of *Phaeocystis* sp., a sometimes-colonial haptophyte that sticks to fishermen's and plankton nets. The fishermen knew that these patches are avoided by herring and are thus associated with low catches. Later, Seuront and Vincent [32] showed that *Phaeocystis globosa* blooms increase seawater viscosity up to 4-fold, and give it centimeter-scale heterogeneity, even when the colonies had been strained out. These authors showed furthermore that the swimming trajectories of the copepod *Temora longicornis* were more convoluted in *P. globosa* bloom water than in clear water, suggesting that the copepods were avoiding zones of high EPS viscosity. The *Phaeocystis* EPS was constraining copepods' swimming trajectories *in situ*. On the other hand Bochdansky & Bollens [68] found that *Acartia hudsonica* slightly favored patches of the diatom *Skeletonema* sp., but this is nutritious food for them.

Zooplankton sometimes associate with phytoplankton patches and thin layers (TLs), but sometimes avoid them. In a study involving a mucus-rich TL of marine snow, Alldredge et al., [69] found that most zooplankton avoided it. The authors suggested that the abundant TEP within the aggregates may have clogged delicate feeding appendages. In particular, the diatom *Odontella longicruris* occurred abundantly in the aggregates, and appeared to be especially avoided by herbivores, which preferred *Thalassiosira* sp. McManus et al., [70], in contrast, found that zooplankton could form TLs associated with phytoplankton TLs, or else their own TLs, separate from the phytoplankton TLs. In particular, zooplankton preferentially avoided TLs rich in toxic diatoms *Pseudo-nitzschia* spp. It would be expected that, as a general rule, zooplankton including copepods would like to feed in phytoplankton TLs, where the concentration of food is higher than in most parts of the water column. Little is still known, however, about the effects on zooplankton in TLs of toxic and non-toxic algae and their specific rheological properties [71,72].

Various Meanings of "Sticky Water"

The different meanings of "sticky water" found in the literature relevant to plankton are briefly reviewed as follows.

1. DVLO-type EDL of water in contact with a solid surface (Figure 3); with thickness the Debye length $\lambda = O(0.1 \text{ to } 1 \text{ nm})$ at salinity 10 to 0.1 respectively.
2. The "sticky water" of Strickler [73]. This is DVLO-type boundary-layer water where the streamlines hug a rigid body moving in a stationary fluid, such as that of a sinking copepod [74], or a stationary body in a uniformly moving fluid. This meaning of "sticky" is a synonym for what engineers confusingly call the "viscous regime", and it is in opposition to turbulent water or the engineers' "inviscid" regime.
3. Non-DVLO-type layers of vicinal water (VW), with ordering up to ca. 5 µm from a solid surface (Table 4). Such VW may be due to several phenomena, difficult to study because of intermittence or time-dependence [22,23].
4. The "sticky water" referred to by Kiørboe [75]. Such "sticky water" is due to EPS, including mucus, made by organisms such as *Phaeocystis*, certain dinoflagellates and diatoms as well as algal spores and many macroalgae [76]. This EPS structures the water [77], making it more viscous and sticky, and at least in some cases also lumpy. It appears also to have made swimming trajectories by *Temora longicornis* more compact [32].
5. Water trapped in eddies for example between structures in coral reefs has been described as "sticky" [78,79]. This is an example of a more general phenomenon of water trapped in eddies, for example along the bank of a river, or along an irregular coastline bordered by an alongshore current. This "stickiness" is a function of eddying.
6. Finally we should not forget Maxwell's definition of viscosity [80] (in a Newtonian fluid) as "internal friction", where friction itself is a form or sticking that resists movement between two parallel surfaces in contact with the fluid. So in a sense all fluids are "sticky" internally in that they resist deformation, with a stress, the viscous modulus, G'' [Pa] proportional to the shear rate [s^{-1}]. Elastic materials, on the other hand are internally "sticky" in that they resist deformation with a stress, the elastic modulus, G' [Pa] proportional to the absolute amount of shear [$m\ m^{-1}$ or dimensionless] the material is showing relative to its shape in the absence of external force. This elasticity is a memory of past shape. Ideal liquids have no memory, therefore no elasticity. Kiørboe's [75] "sticky water" (item 4 above) is viscoelastic seawater that will show both increased G'' and finite G'.

Effects of EPS on Feeding by Copepods and *Daphnia*

Prieto et al., [81] investigated copepods feeding on the diatom *Thalassiosira weissfloggii* and its associated abundant TEP. The copepods ingested the diatoms but not the TEP. However, their presence affected the TEP so that it formed larger particles that sank faster.

Malej & Harris [82] found that the presence of TEP reduced feeding on diatoms by copepods, and they suggested that the TEP mechanically interferes with the handling of food and its ingestion. Later, however, Ling & Alldredge [83] suggested that this finding might have been because the copepods were feeding on TEP (with its associated bacteria and algae) instead of on the diatoms. It might be that both effects are at work, and that the dominant one in any particular circumstance might depend on the real or "perceived" (by the copepod) nutritional and toxicological status of the diatoms present [84].

In further support of feeding hindered by mucus and filaments, Young et al., [62] compared *Daphnia* feeding on algae considered "edible" with that on "inedible" algae. The "edible" algae were *Scenedesmus*, while the "inedible" ones were the filamentous *Ulothrix* sp. and *Phormidium tenue* as well as algae that aggregate into lumps, including *Achnathes* and *Palmella. Daphnia* did eat the so-called "inedible" algae, but more slowly than the "edible" ones. It was observed microscopically that the physical structure of mucus and filaments impeded food manipulation. Moreover, when given a mixture of "edible" and "inedible" algae, the *Daphnia* ate even the "edible" algae more slowly than in the absence of the "inedible" ones. They suggested that the presence of the mucus and filaments of the "inedible" algae hindered feeding even on the "edible" algae.

While EPS and mucus thus often hinder copepod feeding, copepods sometimes actually feed on it beneficially. Richman et al., [85] found that *Acartia negligens* fed actively on mucus produced by several species of coral, and assimilated around 50% of its organic matter. Coral mucus, however, may be more attractive than that produced by plankton because its C:N ratio is lower, while it also contains energy-rich wax-ester products of coral metabolism. Feeding is specifically considered in the next section. Effects of EPS on hunting are considered below in the section, "Ambush feeding".

THE COPEPOD SURFACE IN RELATION TO FEEDING AND SWIMMING

Feeding

In zooplankton, "Smaller predators hunt individual cells, whereas larger ones use feeding currents, mucous nets or elaborate filters to collect them en masse" [86]. Copepods fall between the smaller and the larger categories. Pelagic copepods are among the quickest feeders in the plankton, able to clear around 10^6 of their body volume per day [75]. Like most other crustaceans their bodies, including the appendages and mouthparts they use for feeding are made of chitin, with calcareous inclusions, covered by an exocuticle, the composition of which is discussed below.

Much remarkable, painstaking observation has been made of feeding-appendage morphology and filter-feeding between setules at small length (ca. 5-10 μm) and time (ca. 10 ms) scales, e.g., [73,87-93]. For an example, see Figure 6. From these and similar observations, flow around the appendages and setules was considered laminar, with Reynolds number estimated at 10^{-2} to 10^{-1} [94]. Descriptions and physical modeling of suspension feeding have always been based on implicit assumptions of DLVO-theory with no slipping and no sticking at walls. Jørgensen [92] concluded that such filtering between setules is physically impossible because the cells would be prevented from touching the setules by their surrounding non-turbulent boundary layer. This may have influenced subsequent workers to conclude that the observations of filtering must have been wrong. Recent ideas, however, are coming back to the notion that these observations cannot have been mostly wrong, and that filtering between setules may be important after all [75]. After all, seeing is believing. These observations need to be reassessed theoretically in the light of Lotus-leaf theory and other non-DLVO models of surface slipping and sticking. If the setules are superhydrophobic and self-cleaning, water will pass between them more readily, and even sticky cells will have less tendency to stick to them. A "stickiness arms race" and a "fouling-antifouling arms race" may thus be parts of Smetacek's [86] "watery arms race".

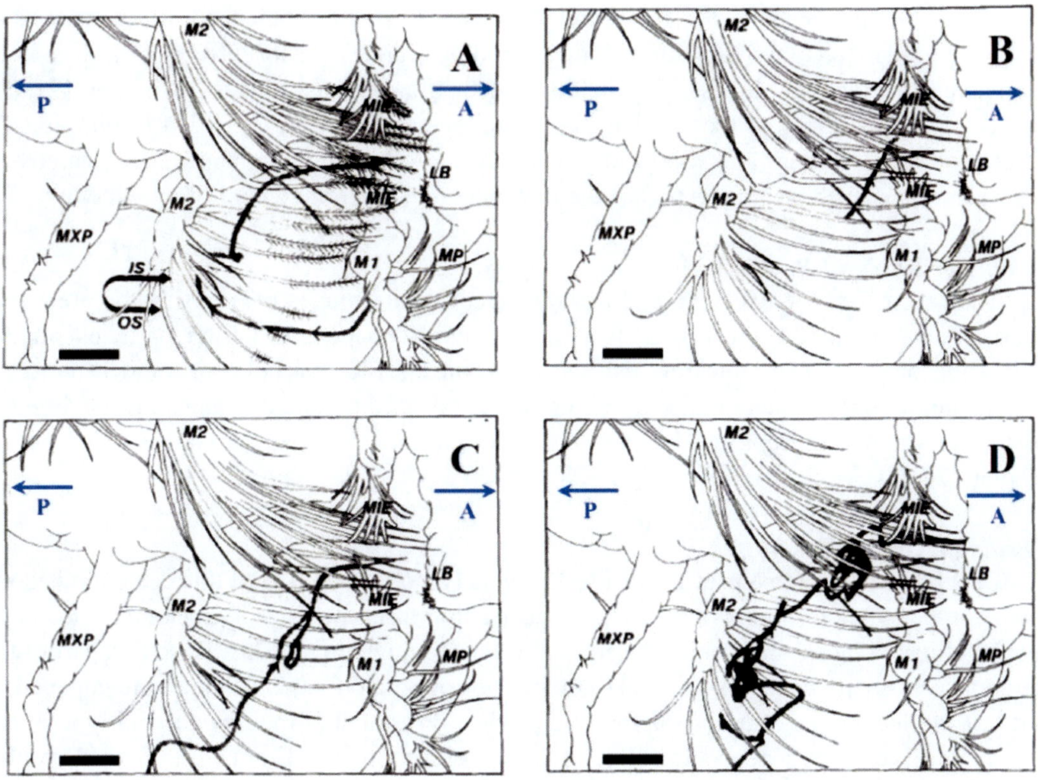

Figure 6. Representative digitized cell paths across the 2nd maxillae during passive capture of *Thalassiosira weissfloggii* cells by *Eucalanus elongatus*. The low amplitude motions of the M2 were omitted for clarity, and the position of the appendage was fixed at midcycle. (A) Cell approaches from outer side of M2, passes between setae at the proximal-distal joint, then is swept across inner surface of the M2 toward the setal tips and MIE without sticking to the setae. (B) Cell path similar to view (A), but cell passes between M2 setae near MIE. (C) Cell approaches M2 from the inner side through the space between the left and right M2. No passage between setae. d. Cell approaches from outer side of M2 and passes between setae of the distal segment, but sticks to the appendage briefly at several locations as it is swept across toward the MIE. A: anterior; P: posterior; LB: labrum; MP: mandibular palp; MI: first maxilla; M1E: first maxilla endite; M2: second maxilla; MXP: maxilliped; OS: outer surface; IS: inner surface. View is focused on the inner surface of the left M2, which is held at an angle ventro-lateral to the body wall. Not all of the setules are in focus in view A, and all setules were omitted in views B, C, and D for clarity. The scale bar is 200 μm. Modified from Price & Paffenhöfer [91]. Reproduced with kind permission.

The set of physical parameters associated with solid-liquid boundaries is much richer than just DLVO and Newtonian-water ideas, useful though this simplified version of Physics was for 20th-century product-innovation and engineering. Unfortunately many biologists believed such straightjacketing ideas too literally. From now on, models of small-scale, low-force interactions in aquatic biological systems should be continually reassessed as knowledge of physics progresses, driven by the current revolution in industrial micro- and nanofluidics.

Previously, feeding in plankton copepods was generally classified into: (i) filter feeding and (ii) raptorial feeding [95]. Based on high-resolution (ca. 1 μm) and high frequency (ca. 500 Hz) filming [89], it was concluded, perhaps under the influence of Jørgensen [92], that they were not "filter feeders" in the sense of using the setae on their feeding appendages to filter phytoplankton passively out of their feeding currents. More recently it has been considered they actively use sensory inputs to detect, capture and ingest their prey [91,96]. Kiørboe [75] has reviewed the problem still more recently, and has classified feeding into (i) feeding-current feeding and (ii) ambush feeding. As mentioned above, he tentatively reinstates some filter-feeding as part of the first category, an arrangement I follow below.

Filter Feeding and Raptorial Feeding

The categories, filter feeding and raptorial feeding, are modes of handling the prey. In a short review, Boyd [88] considered that few calanoid copepods can filter particles smaller than 5 μm or larger than 100 μm, and that most grazers filter particles of a much narrower size range. It follows that larger particles are handled individually, and may be broken, pierced or crushed by raptorial feeding. All food particles >100 μm are subject to raptorial feeding [75,88].

Feeding-Current Feeding

Copepods detect some prey in feeding currents [73]. At least part of this detection is done by sensing a cloud, or microzone, of chemicals leaking from the cells. This cloud is elongated into an ellipsoid by shearing and acceleration of the feeding current [75,97], so that part of the cloud arrives at the copepod's sensors before they prey itself. This prompts the copepod to redirect the feeding current and capture the particle. Presumably the amount and type of the "scent" will help the copepod to select or reject as appropriate. From experimental results, Frost [98] concluded that, below saturating food concentrations, *Calanus pacificus* feeds on both large (*ca.* 80 μm) and small (*ca.* 4 μm) cells, but handles and eats larger cells with greater effectiveness than it handles and eats small cells. Moreover, other things being equal, larger prey should secrete more chemicals, allowing detection at a greater distance. The clearance rate on "large" prey is indeed larger than on "small" prey, which has been deduced from observed size spectra successfully preyed upon by copepods, relative to the size spectra present in the water [75,88,90].

Furthermore, Hong et al., [99] found that *Acartia tonsa* fed on *Karenia breve* (18-45 μm in length and width) and *Karlodinium veneficum* (< 20 μm long; sizes from [100]) by feeding-current feeding (i.e., "*the copepod generates a feeding current to attract and capture prey*") and not by raptorial feeding.

Ambush Feeding

General Considerations

Kiørboe [75] points out that ambush feeding is mandatory for pelagic cyclopoid copepods, and is also found amongst some calanoids, including *Acartia.* Ambush-feeding copepods wait motionless, then attack prey that move through their perceptive sphere. The jump attack is accomplished in only 2-10 ms. The rapidity of the attack reduces the thickness of boundary layers around the copepod, while enabling it to penetrate the prey's boundary layers more. If the attack were slower, the thicker boundary layers would push the prey away. Ambush feeding with such active attack jumps is restricted to powerful, large (> 250 µm) plankton, with a streamlined body shape, including chaetognaths and some copepods. This streamlining together with the pointed front end with projecting grasping processes will further help the predator to penetrate the boundary layer [75].

EPS As a Defense Against Ambush Feeding

Approach of the predator to the prey will subject the boundary layer to squeezing deformation. The relevant viscosity for squeezing is the 2D squeezing/elongational (SE) viscosity, which is larger than the shearing viscosity by the Trouton ratio, *T*. In Newtonian liquids *T* is always 3, but in different polymer solutions *T* can be as high as 10^3 [102]. Incorporation of loosely bound EPS (LB-EPS) in the boundary layer around mucous phytoplankton (thickness around 10-100 µm) may thus be a powerful protection against ambush feeding. Modeling combined with rheological characterization of such LB-EPS is required. Furthermore, LB-EPS may be difficult for copepods to feed on because their appendages, setae and setules will easily pass through it. Sufficiently strong mucus, however, such as tightly bound EPS (TB-EPS) or some TEP and organic aggregates, would be graspable or filterable.

Thus even small amounts of polymeric thickening in the water, whether diffuse or associated with the prey surface, might make ambush feeding much more difficult, depending on the rheological properties of the polymer, including *T*, at the strain rates and scales involved in the deformation. In principle, this effect could be tested experimentally. However, pointed appendages that could penetrate the boundary layer and grasp the prey would be an additional advantage to the predator.

SDR As a Possible Aid to Ambush Feeding

Asmolov et al., [9] have modeled the hydrodynamic forces (in Newtonian liquid) impeding the approach of a sphere with a hydrophilic surface to a flat superhydrophobic surface. The authors found reduction in the forces impeding approach (SDR). Approach would thus be facilitated. The amount of the reduction is sensitive to details in the parameters, particularly the different length scales involved in SDR. Asmolov et al., [9] point out that for real situations, atomic force microscopy could be used to measure the forces developed. Such forces therefore need to be modeled and measured in future investigations of the nano- and microfluidics of copepod feeding, along with millisecond and micrometer-scale videomicroscopy from the scale of whole-copepod ambush feeding to that of a setule approaching a picoplankton cell [103]. Such studies would provide insight into whether such a type of predation would provide evolutionary pressure for copepods to develop a superhydrophobic surface. Approach may be hindered, however, and some length scales greatly increased by the rheological effects of any tightly or loosely-bound EPS around the prey.

Rheological and Microfluidics Considerations on Feeding

Because of the boundary-layer constraints in ambush feeding, it is not surprising that ambush-feeding copepods are notably clean of mucus, since any mucus would tend to thicken boundary layers, either by increasing squeezing/elongational viscosity or by coating surfaces, or both. Similarly, copepods that use filter feeding (in many cases the same ones) should avoid producing mucus as loosely-bound mucus would tend to increase viscosity generally, and tightly-bound mucus might stick to the setae and setules of the filtering appendages, in both cases increasing boundary-layer thickness. Copepods finding themselves in rheologically thickened water may then feed as best they can by converting from filter-feeding to "fling-and-clap" feeding [73]. Benthic copepods, particularly harpacticoids (see below), are likely not affected by these constraints, and thus have less need to keep clean of mucus. The importance of mucus and EPS in feeding is also considered above in the "Ambush feeding" section.

Copepod Feeding Appendages

For eight species of *Calanus* and *Neocalanus*, Yamaguchi et al., [93] review the inter-seta spacing on the feeding appendage, the second maxilla (Mx 2), as they consider that the inter-seta spacing is indicative of the size of particles that can be removed from the water. Spacing ranges from *ca.* 3 µm in *Neocalanus plumchrus* naupliar stage-1 to -3 to *ca.* 10 µm in *Calanus hyperboreus* stage 6 females (pre-adults) (Figure 7).

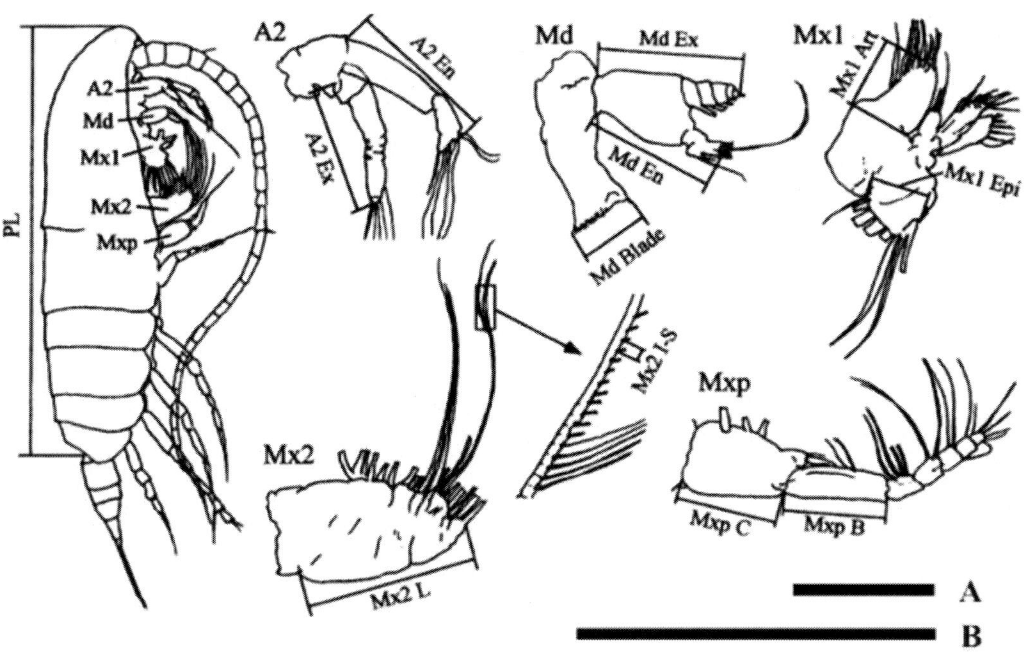

Figure 7. Diagrams showing the prosome length (PL) and 11 parts of feeding appendages measured for interspecific comparisons (abbreviations shown in the parentheses): A2 endopods (A2 En), A2 exopods (A2 Ex), Md cutting blade (Md Blade), Md endopods (Md En), Md exopods (Md Ex), Mx1 arthrite (Mx1 Art), Mx1 epipodite (Mx1 Epi), length of Mx2 (Mx2 L), inter-seta spacing of near the tip of strong spine on Mx2 basis (Mx2 1-S), Mxp coxa (Mxp C), and Mxp basis (Mxp B). Scale bars are both 1 mm, A for whole body (*Neocalanus plumchrus* C5) and B for feeding appendages (except for strong spine on Mx2 basis). Modified from Yamaguchi et al. [93]. Reproduced with kind permission.

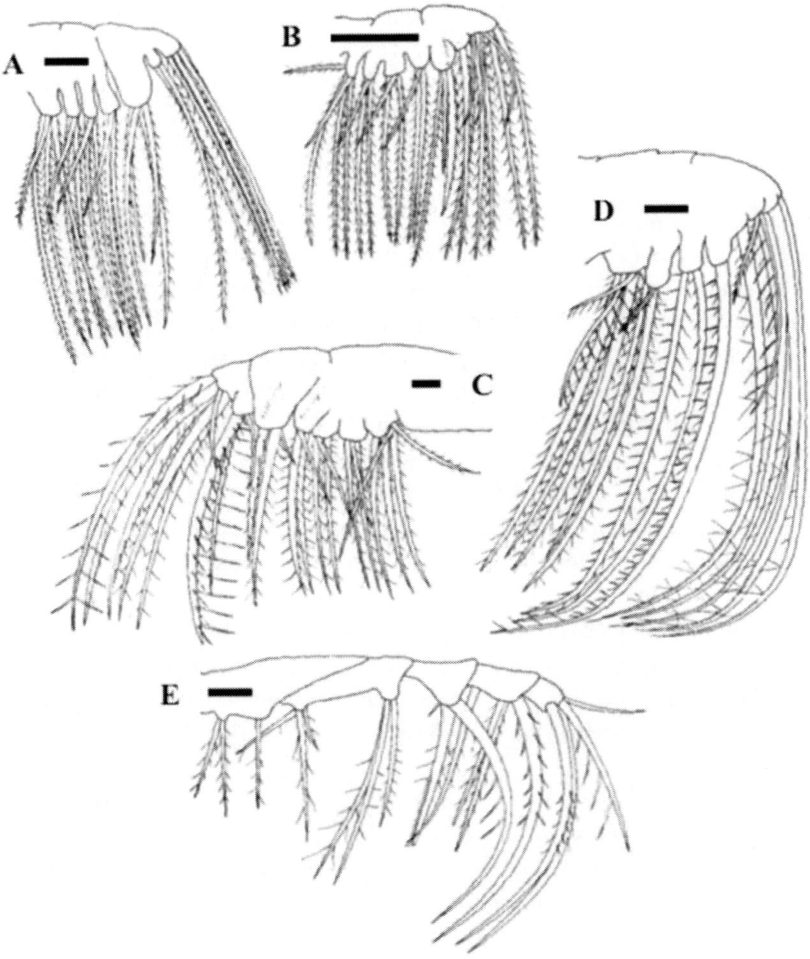

Figure 8. Setulation of maxillae. A, *Pseudocalanus*; B, *Temora*; C, *Centropages*; D, *Acartia*; E, *Oithona*. Modified from Marshall & Orr [87]. Copyright (1986) by Association for the Sciences of Limnology and Oceanography, Inc. Reproduced with kind permission.

Marshall & Orr [86], however, show that each seta bears two rows of setules (Figure 8), and these authors consider that the inter-*setule* distance determines the size of food taken. For smaller copepods of the genera *Pseudocalanus*, *Temora*, *Centropages*, *Acartia* and *Oithona*, Marshall & Orr [87] found this distance to range mostly from 2 to 6 μm, in some cases up to 17 μm. In many copepods, the setae of the maxilliped are longer, thicker, and often pointed, and may be used almost exclusively for grasping. Anraku and Omori [104] also give very useful drawings of marine plankton copepods (not shown here).

The Crustacean Epicuticle

Most free-living crustaceans are covered by a cuticle, which is hard, except just after moulting. The external layer in contact with the water is the epicuticle, and we shall consider it because its surface properties may change the way the copepod interacts with ambient water and other materials. The cuticle of large decapods (crabs) has been extensively studied, but that of copepods much less so. In crabs and copepods, the mass of the cuticle consists of chitin. This is partly calcified in the crabs studied, but apparently not in copepods. The outer layer, or epicuticle, may be rich in hydrophobic wax in plankton copepods. The evidence for this is considered hereafter and discussed in comparison to decapod epicuticle.

The Decapod Epicuticle

The interface between the copepod and its milieu is its carapace, which covers the whole body including setae and setules, although it is thin and flexible at the limb joints. It is thickest where it covers the cephalothorax dorsally, and this part has been the most studied [105]. Much work has been done on larger crustaceans. Beneath the cuticle (or carapace) in the shore crab *Carcinus maenas* is an epidermis. This epidermis is overlain by a cuticular "principal layer", itself overlain by a "pigmented layer", with a final thin wrapping, the epicuticle [106,107]. The organic matrix of the principal and pigmented layers of the cuticle is composed mainly of units of chitin-protein fibers. In crabs, the units are spatially organized according to a twisted-plywood model, characterized by curved "arcs", that may resist predators [108] by providing toughness and protection against fracture propagation. These twisted structures are clearly multiscalar [106-108]. Interfibrillar lacunae start to become calcified with hours of ecdysis. Penetrating the cuticle from the underlying epidermis to the outside are "pore canals", apparently containing cellular material, that likely allow communication to the outside. The edible crab *Cancer pagurus* presents similar structures, in which Fabritius et al., [108] describe the chitin structure and development as "*Molecules of the sugar N-acetylglucosamine polymerize to form anti-parallel chains of a-chitin. Crystallized chitin chains coated with a protein matrix form about 5-7 nm thick nanofibrils. The nanofibrils further assemble to chitin-protein fibers with diameters of 50–250 nm. These fibers arrange with parallel long axes to form horizontal planes. In stacks of these planes, the fiber orientation changes helically and generates a twisted plywood structure, similar to that in* Carcinas maenas, *that forms the three main layers of the cuticle.*" In Fabritius's [108] terminology, from proximally to distally, overlying the epidermis is the membranous layer, overlain successively by the endocuticle, the exocuticle and finally the external, thin epicuticle.

In *C. pagurus* [108], the thin epicuticle that covers the whole integument contains no helicoidally arranged fibers and consists of a compact layer and a subjacent layer with numerous pore canals as described earlier. Raman spectroscopy has shown [108] that the compact epicuticle consists mainly of waxes containing C—H stretching bands that appear in hydrocarbons, but little or no proteins or chitin. The pore canals crossing the inner layer, and considered by Girard-Guille [107] to contain cellular material, were found by Fabritius [108] to contain mineral particles that are not calcite. Ultrastructural investigation of the soft membranous layer, which separates the proximal surface of the cuticle from the underlying epithelium, shows regularly layered fibrous planes that consist of chitin and proteins.

Becker et al., [109] measured the surface properties of carapaces from 42 decapod and 3 squillid species. Properties varied from extremely hydrophilic (wetting angle 0°) to moderately hydrophobic (91°). Colonization by fouling organisms was also investigated, but no relationship between fouling and wetting angle was discovered. It would be interesting to make a similar investigation for different copepod species.

The Copepod Epicuticle, Compared with Decapods

Copepod Epicuticle Structure

Like the edible and shore crabs, copepods have a chitinous exoskeleton covered by an epicuticle thought by Pruzzo et al., [39] to be of wax This epicuticle probably protects the cuticle from attack by chitinolytic bacteria, which are ubiquitous in the ocean. Chitinolytic bacteria of the genera *Enterococcus* and *Vibrio*, including *E. faecalis* and *V. cholerae* of human origin, have been found largely associated with copepods in both fresh- and seawater [39]. These same *E. faecalis* continued to be active and culturable much longer when grown with chitin particles, (*ca.* 30 days), than with live freshwater copepods (9-15 days) [39], suggesting that the epicuticle does indeed protect the chitin exoskeleton of copepods from bacterial attack.

In copepods, Raymont et al., [105] showed sections of the cuticle ("integument") of *Calanus finmarchicus* by transmission electron microscopy (TEM), combined with staining. From this they concluded that the outer zone consists of about six 1-μm layers, and contains basic protein, some mucoprotein, mucopolysaccharides and lipoproteins. The inner zone is also layered and is of similar composition except for a paucity of mucoproteins. It is underlain by an epithelium.

Parts of the cuticle, in particular that part covering appendages, bear setae covered with a similar outer and inner cuticle, underlain not by an epithelium but by, what may be homologous, a cell-containing lumen. Setae in turn bear setules composed of cuticle, but setules have no lumen [105].

Copepod Epicuticle Composition

Because the epicuticle is the copepod's interface with its environment, its surface properties are vitally important to its life style and survival. Organisms as diverse as flowering plants, insects and humans bear waxy epicuticles [110]. Many copepods in salt lakes [111] and in the sea [112,113] produce large amounts of wax for energy storage and buoyancy regulation. Like the cuticle of decapods, that of copepods and their setae consists of chitin [105]. Chitin is a highly hydrophobic polysaccharide but it is easily transformed into a variety of soft, hydrophilic derivatives. In crustacean carapaces it generally occurs as nanofibres encased in protein sheaths [108]. But is the thin epicuticle surface hydrophilic or hydrophobic, and does it really contain any wax as Pruzzo et al., [39] believe? The surface properties must be important not only in harboring or excluding pathogenic bacteria, but also in determining feeding and encounter dynamics. I have been unable to find any clear information as to whether copepod epicuticle is hydrophobic or hydrophilic, or if it is waxy. In natural waters, however, surfaces generally accumulate a coating of hydrophobic or amphiphilic algal or bacterial exopolymers [114,115], which in both cases would generally be hydrophobic to the outside.

Copepod Epicuticle Function

Boxhall [116] remarks that most planktonic copepods have a generally smooth body surface externally, but that some, such as *Centraugaptilus horridus*, are covered with spinules, and reticulate ornamentation occurs on the planktonic harpacticoid *Aegisthus aculeatus*. Spinules and other microsculpturing are borne also on the integument of some planktonic harpacticoid misophrioid copepods, particularly *Benthomisophria palliata*, but not in others. A system of grooves, seen by SEM, decorates the uropods of the freshwater cyclopoid *Acanthocyclops venustus*. The functional significance of these markings is poorly understood, but that the adaptive advantages are presumed [116] to outweigh any disadvantage of increasing hydrodynamic drag in the free-swimming forms. Nano- and microsculpturing on hydrophobic surfaces are now known, however, to sometimes *reduce* drag by superhydrophobic drag reduction (SDR), at both low and high Reynolds-number flow [8]. If hydrophobic, as well as producing SDR, this nano- and microsculpturing may further have an anti-stick or self cleaning role, like that on the petals and leaves of some flowering plants [4,6,8], and thus help to keep the copepod surface clean and free of fouling organisms. Superhydrophobic and sculptured amiphilic surfaces are now being developed industrially to reduce sticking by organisms fouling marine structures [27]. Similar techniques should be used to investigate whether copepods and other plankton use similar mechanisms to reduce biofouling.

Swimming

Some effects of EPS on copepod swimming aspects of EPS on hunting (ambush feeding) have been considered above. Detailed models of copepod swimming and hovering in relation to energy consumption, feeding currents and chemoreception have been made [117-120]. The effect of excess copepod density has also been modeled on feeding and swimming fields as well as suspension feeding currents [121]. The detection of hydrodynamic signals by copepods has also been modeled [122]. The hydrodynamic modeling was done by Computational Fluid Dynamics (CFD) using FLUENT software. This work represents powerful advances to understand the phenomenology and energetics of pelagic copepod life. Yet at that time this work was done, techniques of CFD could not yet model the effects of non-Newtonian fluids, and neither were non-DLVO effects taken into account. This work thus cannot tell us about the subject of the present chapter, but it deserves to provide the basis for future work that will. Already a new version of FLUENT is available that can take into account non-Newtonian liquid [123].

Harpacticoids Are Different

The free-living copepods, calanoids and cyclopoid, abundant in the plankton, show remarkably little mucus secretion to the outside, except for the hardened mucus of egg cases and spermothecae. On the other hand, harpacticoids produce mucus more frequently. Benthic and meiobenthic harpacticoids, including *Diarthrodes nobilis* [124] and *Pseudostenhelia wellsi* [125] build mucous burrows. This mucus may be used "garden" microflora, with the mucus and associated microflora being subsequently ingested [124,125]. Two other harpacticoid species, *Nitocra lacustris* [126] and *Laophonte* sp. [127] do indeed ingest mucus with bacteria. In another mucous-tube-building meiofaunal harpacticoid *Stenhelia palustris*, secretory products, glands and secretory pores have been described [128]. Furthermore, dense concentrations of mucous tubes made on seagrass blades by *Dactylopodia tisboides* attract meiofauna [129].

How is the abundant mucus production in harpacticoids compared to the "clean" lifestyle of most calanoids and cyclopoids related to their respective lifestyles, mostly benthic (or particle-associated) and planktonic? Work is needed to explore this question further.

Conclusion and Recommendations

Currently, most aquatic zooplankton and microbe ecologists as well as biogeochemists are poorly informed about many modern engineering models that can help explain nano- and micrometer-scale processes, particularly those observed at surfaces. I hope that this review of rheology, nano- and microfluidics and biofouling will be useful to improve understanding other processes in the lives of pelagic calanoid copepods, including swimming, feeding, protection against biofouling, parasitism and infection. This approach should also be extended to wider areas of pelagic ecology and biogeochemistry.

The disciplines of oceanography, rheology and surface science are all composite, and have of necessity been built from parts of Physics, Chemistry and Biology. I suggest there is now a need to form a multidisciplinary group of experts in pelagic ecology, biogeochemistry, rheology, nano/microfluidics, biofouling and allied fields. Such a group should associate to build capacity by building capacity in this new branch of aquatic science. The first step should be to teach each other what they know. This discipline might be called, *Rheology micro/nanoFluidics and bioFOuling* (RheFFO). The next step would be for the trained experts to build capacity in younger researchers, to make new discoveries.

Acknowledgments

My thanks go to Tim Wyatt for vastly improving this chapter. Project supported by the State Key Program of National Natural Science of China (Grant No. 41230963).

References

[1] Davidson PA, Kaneds Y, Moffatt K & Sreenivasan KR (eds., 2011) *A Voyage through Turbulence*, Cambridge University Press, Cambridge.

[2] Jenkinson IR & Sun J (2010) Rheological properties of natural waters with regard to plankton thin layers. A short review. *Journal of Marine Systems*, 83, 287-297.

[3] Conlisk AT (2013) *Essentials of Micro- and Nanofluidics,* Cambridge University Press, Cambridge.

[4] Barthlott W & Neinhuis C (1997) Purity of the sacred lotus, or escape from contamination in biological surfaces. *Planta*, 202, 1-8.

[5] Ou J, Perot B & Rothstein JP (2004) Laminar drag reduction in microchannels using ultrahydrophobic surfaces. *Physics of Fluids*, 16, 4635 (9 p.).

[6] Feng L, Zhang Y, Xi J et al., (2008) Petal effect: a superhydrophobic state with high adhesive force. *Langmuir*, 24, 4114-4119.

[7] Shirtcliffe NJ, McHale G, Newton MI & Zhang Y (2009) Superhydrophobic copper tubes with possible flow enhancement and drag reduction. *Applied Materials & Interfaces*, 1, 1316-1323.

[8] Rothstein JP (2010) Slip on superhydrophobic surfaces. *Annual Review of Fluid Mechanics*, 42, 89-209.

[9] Asmolov ES, Belyaev AV & Vinogradova OI (2011) Drag force on a sphere moving toward an anisotropic superhydrophobic plane. *Physics Reviews E*, 84, 026330 (11 p.)

[10] Jenkinson IR (1986) Oceanographic implications of non-newtonian properties found in phytoplankton cultures. *Nature*, 323, 435-437.

[11] Jenkinson IR (1993) Bulk-phase viscoelastic properties of seawater. *Oceanologica Acta*, 16, 317-334.

[12] Jenkinson IR & Biddanda BA (1995) Bulk-phase viscoelastic properties of seawater: relationship with plankton components. *Journal of Plankton Research*, 17, 2251-2274.

[13] Seuront L, Lacheze C, Doubell MJ et al., (2007) The influence of *Phaeocystis globosa* on microscale spatial patterns of chlorophyll a and bulk-phase seawater viscosity. *Biogeochemistry*, 83, 173-188.

[14] Kesaulya I, Leterme SC, Mitchell JG & Seuront L (2008) The impact of turbulence and phytoplankton dynamics on foam formation, seawater viscosity and chlorophyll concentration in the eastern English Channel. *Oceanologia*, 50, 167-182.

[15] Jenkinson IR & Sun J (2014) Drag increase and drag reduction found in phytoplankton and bacterial cultures in laminar flow: are cell surfaces and EPS producing rheological thickening and a Lotus-leaf Effect? *Deep Sea Research Part II*, 101, 216-230.

[16] Gad-el-Hak, M. (ed., 2002). *The MEMS Handbook (Mechanical Engineering Handbook Series)*. CRC Press, Boca Raton.

[17] Meissner T & Wenz FJ (2004) The complex dielectric constant of pure and sea water from microwave satellite observations. *IEEE Transactions on Geoscience and Remote Sensing*, 42, 1836-1849.

[18] Mari X, Torréton JP, Trinh, CBT et al., (2012) Aggregation dynamics along a salinity gradient in the Bach Dang estuary, North Vietnam. *Estuarine and Coastal Marine Science*, 96, 151-158.

[19] Alemán JV, Pelegrí J & Sangrà P (2006) Ocean rheology. *Journal of Non-Newtonian Fluid Mechanics*, 133, 121-131.

[20] Richardson (1973) On the no-slip boundary condition. *Journal of Fluid Mechanics*, 59, 707-719.

[21] Pedley TJ (1997) Introduction to fluid dynamics. *Scientia Marina*, 61 (Suppl. 1), 7-24.

[22] Grasso D, Subramaniam K, Butkus M et al., (2002) A review of non-DLVO interactions in environmental colloidal systems. *Reviews in Environmental Science and Bio/Technology*, 1, 17-38.

[23] Drost-Hansen W (2006) Vicinal hydration of biopolymers: cell biological consequences. *In*: Pollack GH, Cameron IL & Wheatley, DN (eds.), *Water and the Cell*, Springer, New York, 175-217.

[24] Cottin-Bizonne C, Barentin C & Bocquet L (2012) Scaling laws for slippage on superhydrophobic fractal surfaces. *Physics of Fluids*, 24, 012001 (13 p.).

[25] Koch K, Bohn HF & Barthlott W (2009) Hierarchically sculptured plant surfaces and superhydrophobicity. *Langmuir*, 29, 14116-14120.

[26] Bauer U & Federle W (2009) The insect-trapping rim of *Nepenthes* pitchers: surface structure and function. *Plant Signaling and Behavior*, 4, 1019-1023.

[27] Wong T-S, Kang SH, Tang, SKY et al., (2011) Bioinspired self-repairing slippery surfaces with pressure-stable omniphobicity. *Nature*, 477, 443-447.

[28] Carlson DJ (1987) Viscosity of sea-surface slicks. *Nature*, 329, 823-825.

[29] Zhang Z, Zhang A, Liu L et al., (2003) Viscosity of sea surface microlayer in Jiaozhou Bay and adjacent sea area. *Chinese Journal of Oceanology Limnology*, 21, 351-357.

[30] Seuront L, Vincent D & Mitchell JG (2006) Biologically induced modification of seawater viscosity in the Eastern English Channel during a *Phaeocystis globosa* bloom. *Journal of Marine Systems*, 61, 118-133.

[31] Seuront L, Leterme SC, Seymour JR et al., (2010). Role of microbial and phytoplanktonic communities in the control of seawater viscosity off East Antarctica (30-80° E). *Deep-Sea Research II*, 57, 877-886.

[32] Seuront L & Vincent D (2008) Increased seawater viscosity, *Phaeocystis globosa* spring bloom and *Temora longicornis* feeding and swimming behaviours. *Marine Ecology Progress Series*, 363, 131-145.

[33] Básaca-Loya A, Burboa MG, Valdez MA et al., (2008). Aggregation behavior and rheology of culture broths of *Rhodosorus marinus*. *Revista Mexicana de Física*, 54 (Suppl. 2), 119-126.

[34] Carman KR & Dobbs FC (1997) Epibiotic microorganisms on copepods and other marine crustaceans. *Microscopy Research and Technique*, 37, 116-135.

[35] Bagøien E & Kiørboe T (2005) Blind dating - mate finding in planktonic copepods. III. Hydromechanical communication in *Acartia tonsa*. *Marine Ecology Progress Series*, 300, 129-133.

[36] Ohtsuka S, Hora M, Suzaki T et al., (2004) Morphology and host-specificity of the apostome ciliate *Vampyrophrya pelagica* infecting pelagic copepods in the Seto Inland Sea, Japan. *Marine Ecology Progress Series*, 282, 129-142.

[37] McAllen R & Scott GW (2000) Behavioural effects of biofouling in a marine copepod. *Journal of the Marine Biological Association of the United Kingdom*, 80, 369-370.

[38] Colwell R (1996). Global climate and infectious disease: the cholera paradigm. *Science*, 274, 2025-2031.

[39] Pruzzo C, Vezzulli L & Colwell RR (2008) Global impact of *Vibrio cholerae* interactions with chitin. *Environmental Microbiology*, 10, 1400-1410.

[40] Signoretto C, Burlacchini G, Pruzzo C & Canepari P (2005) Persistence of *Enterococcus faecalis* in aquatic environments via surface interactions with copepods. *Applied Environmental Microbiology*, 71, 2756-2761.

[41] Nagasawa S (1987) Exoskeletal scars caused by bacterial attachment to copepods. *Journal of Plankton Research*, 9, 749-753.

[42] Carli A, Pane L, Casareto L et al., (1993) Occurrence of *Vibrio alginolyticus* in Ligurian Coast rock pools (Tyrrhenian Sea, Italy) and its association with the copepod *Tigriopus fulvus* (Fisher 1860). *Applied Environmental Microbiology*, 59, 1960-1962.

[43] Costerton JW, Irvin RT & Cheng KJ (1981) The bacterial glycocalyx in nature and disease. *Annual Review of Microbiology*, 35, 299-324.

[44] Preisig HR, Anderson OR, Corliss JO et al., (1994) Terminology and nomenclature of protist cell surface structures. *Protoplasma*, 181, 1-28.

[45] Hall JE (2010) *Guyton and Hall Textbook of Medical Physiology: Enhanced E-book.* Elsevier Health Sciences, New York.

[46] Sansone F & Casnati A (2013) Multivalent glycocalixarenes for recognition of biological macromolecules: glycocalyx mimics capable of multitasking. *Chemical Society Reviews*, 42, 4623-4639.

[47] Falciatore A & Bowler C (2002) Revealing the molecular secrets of marine diatoms. *Annual Review of Plant Biology*, 53, 109-130.

[48] Yamasaki Y, Shikata T, Nukata A et al., (2009) Extracellular polysaccharide protein complexes of a harmful alga mediate the allelopathic control within the phytoplankton community. *International Society for Microbial Ecology Journal*, 3, 808-817.

[49] Kim D, Kumamoto O, Lee KS et al., (2004) Deleterious effect of *Chattonella marina* on short-necked clam (*Ruditapes philippinarum*); possible involvement of reactive oxygen species. *Journal of Plankton Research*, 26, 967-971.

[50] Hauton C (2012) The scope of the crustacean immune system for disease control. *Journal of Invertebrate Pathology*, 110, 2451-260.

[51] Peschel G & Adlfinger K, (1970) Viscosity anomalies in liquid surface zones: IV. The apparent viscosity of water in thin layers adjacent to hydroxylated fused silica surfaces. *Journal of Colloid and Interface Science*, 34, 505-510.

[52] Yaminsky VV, Ninham BW & Pashley RM (1998) Interaction between surfaces of fused silica in water. evidence of cold fusion and effects of cold plasma treatment. *Langmuir*, 14, 3223-3235.

[53] Malkina AD & Derjaguin BV (1950) *Kolloidnyi Zhurnal*, 12, 431.

[54] Wernet P, Nordlund D, Bergmann U et al., (2004) The structure of the first coordination shell in liquid water. *Science*, 304, 995-999.

[55] Jenkinson IR & Wyatt T (2008) Rheological properties of exopolymeric secretions in HABs may be functions of length scales. *In*: Moestrup, Ø. et al (Eds.) *Proceedings of the Twelfth International Conference on Harmful Algae*, International Society for the Study of Harmful Algae and Intergovernmental Oceanographic Commission of UNESCO, Copenhagen and Paris, 126-128.

[56] Liu XM, Sheng GP, Luo HW et al., (2010) Contribution of extracellular polymeric substances (EPS) to the sludge aggregation. *Environmental Science & Technology*, 44, 4355-4360.

[57] Heissenberger A, Leppard G & Herndl G (1996) Ultrastructure of marine snow. II. Microbial considerations. *Marine Ecology Progress Series*, 135, 299-308.

[58] Thornton DCO (2004) Formation of transparent exopolymeric particles (TEP) from macroalgal detritus. *Marine Ecology Progress Series*, 282, 1-12.

[59] Rochelle-Newall EJ, Mari X & Pringault O (2010) Sticking properties of transparent exopolymeric particles (TEP) during aging and biodegradation. *Journal of Plankton Research*, 32, 1433-1442.

[60] Kiørboe T, Anderson KP & Dam HG (1990) Coagulation efficiency and aggregate formation in marine phytoplankton. *Marine Biology*, 107, 235-245.

[61] Kiørboe T & Hansen JLS (1993) Phytoplankton aggregate formation: observations of patterns and mechanisms of cell sticking and the significance of exopolymeric material. *Journal of Plankton Research*, 15, 993-1018.

[62] Young S, Palm M, Grover J & McKee D (1997) How *Daphnia* cope with algae selected for inedibility in long-running experiments. *Journal of Plankton Research*, 19, 391-397.

[63] Wotton R (1996) Colloids, bubbles, and aggregates - a perspective on their role in suspension feeding. *Journal of the North American Benthological Society*, 15, 127-135.

[64] Thornton DCO (2002) Diatom aggregation in the sea: mechanisms and ecological implications. *European Journal of Phycology*, 37, 149-161.

[65] Yamaguchi H, Sakamoto S & Yamaguchi M (2007) Nutrition and growth kinetics in nitrogen- and phosphorus-limited cultures of the novel red tide flagellate, *Chattonella ovata* (Raphidophyceae). *Harmful Algae*, 7, 26-32.

[66] Biddanda B (1986) Structure and function of marine microbial aggregates. *Oceanologia Acta*, 9, 209-211.

[67] Savage RE (1932) *Phaeocystis* and herring shoals. *Journal of Ecology*, 20, 326-340.

[68] Bochdansky AB, Bollens, SM (2004) Relevant scales in zooplankton ecology: Distribution, feeding, and reproduction of the copepod *Acartia hudsonica* in response to thin layers of the diatom *Skeletonema costatum*. *Limnology and Oceanography*, 49, 625-636.

[69] Alldredge AL, Cowles TJ, MacIntyre S, Ruines JEB, Donaghay PL, Greenlaw CF.; Holliday DV, Dekshenieks MM, Sullivan JM, Zaneveld J R (2002) Occurrence and mechanisms of formation of a dramatic thin layer of marine snow in a shallow Pacific fjord. *Marine Ecology Progress Series*, 233, 1-12.

[70] McManus MA, Alldredge AL, Barnard AH et al., (2003) Characteristics, distribution and persistence of thin layers over a 48 hour period. *Marine Ecology Progress Series*, 261, 1-19.

[71] GEOHAB Global Ecology and Oceanography of Harmful Algal Blooms (2013). McManus MA, Berdalet E, Ryan J, Yamazaki H, Jaffe JS, Ross ON, Burchard H, Jenkinson I, Chavez FP (Eds), *Algal Blooms, GEOHAB Core Research Project: HABs in Stratified Environments.* Intergovernmental Oceanographic Commission of UNESCO and Science Council for Ocean Research, Paris and Newark, Delaware, USA.

[72] Berdalet E, McManus M, Ross O et al., (2014) Understanding harmful algae in stratified systems: review of progress and identification of gaps in knowledge. *Deep-Sea Research Part II*, 101, 4-20.

[73] Strickler RD (1984) Sticky water: a selective force in copepod evolution. In: Meyers DG, Strickler JR (eds) *Trophic Interactions within Aquatic Ecosystems*, American Association for the Advancement of Science, Washington DC, pp. 187-239.

[74] Yen J (2000) Life in transition: balancing inertial and viscous forces by planktonic copepods. *Biological Bulletin, Woods Hole*, 198, 213-224.

[75] Kiørboe T (2011) What makes pelagic copepods so successful? *Journal of Plankton Research*, 33, 677-685.

[76] Boney AD (1981) Mucilage: The ubiquitous algal attribute. *British Phycological Journal*, 16, 115-132.

[77] Decho AW (1990) Microbial exopolymer secretions in ocean environments: their role(s) in food webs and marine processes. *Oceanography and Marine Biology Annual Review*, 28, 73-153.

[78] Wolanski E & Spagnol S (2000) Sticky waters in the Great Barrier Reef. *Estuarine, Coastal and Shelf Science*, 50, 27-32.

[79] Zeidberg L & Hamner W (2002) Distribution of squid paralarvae, *Loligo opalescens* (Cephalopoda: Myopsida), in the Southern California Bight in the three years following the 1997–1998 El Niño. *Marine Biology*, 141, 111-122.

[80] Wikipedia contributors (2014) 'Viscosity'. *Wikipedia, The Free Encyclopedia*, http://en.wikipedia.org/wiki/Viscosity [accessed 05 February 2014].

[81] Prieto L, Sommer F, Stibor H & Koeve W (2001) Effects of planktonic copepods on transparent exopolymeric particles (TEP) abundance and size spectra. *Journal of Plankton Research*, 23, 515-525.

[82] Malej A & Harris R (1993) Inhibition of copepod grazing by diatom exudates: a factor in the development of mucus aggregates? *Marine Ecology Progress Series*, 96, 33-42.

[83] Ling SC & Alldredge A (2003) Does the marine copepod *Calanus pacificus* consume transparent exopolymer particles (TEP)? *Journal of Plankton Research*, 25, 507-515.

[84] Ianora A, Miralto A, Poulet S et al., (2004) Aldehyde suppression of copepod recruitment in blooms of a ubiquitous planktonic diatom. *Nature*, 429, 403-407.

[85] Richman S, Loya Y & Slobodkin LB (1975) The rate of mucus production by corals and its assimilation by the coral reef copepod *Acartia negligens*. *Limnology and Oceanography*, 20, 918-923.

[86] Smetacek V (2001) A watery arms race. *Nature*, 411, 745.

[87] Marshall SM & Orr AP (1966) Respiration and feeding in some small copepods. *Journal of the Marine Biological Association of the United Kingdom*, 46, 513-530.

[88] Boyd CM (1976) Selection of particle sizes by filter-feeding copepods: A plea for reason. *Limnology and Oceanography*, 21, 175-180.

[89] Paffenhöfer GA, Strickler J & Alcaraz M (1982) Suspension-feeding by herbivorous calanoid copepods: a cinematographic study. *Marine Biology*, 67, 193-199.

[90] Nival P & Nival S (1976) Particle retention efficiencies of an herbivorous copepod, *Acartia clausi* (adult and copepodite stages): Effects on grazing. *Limnology and Oceanography*, 21, 24-38.

[91] Price HJ & Paffenhöfer GA (1986) Capture of small cells by the copepod *Eucalanus elongatus Limnology and Oceanography*, 31, 189-194.

[92] Jørgensen CB (1983) Fluid mechanical aspects of suspension feeding. *Marine Ecology Progress Series*, 11, 89-103.

[93] Yamaguchi A, Dalpadado P & Ikeda T (2003) Morphological comparisons of feeding appendages of *Calanus* and *Neocalanus* copepods. *Bulletin of the Fisheries Society of Hokkaido University*, 54, 59-65.

[94] Koehl MAR & Strickler JR (1981) Copepod feeding currents: food capture at low Reynolds number. *Limnology and Oceanography*, 26, 1068-1073.

[95] Mann KH & Lazier JRN (1996) *Dynamics of Marine Ecosystems* (2nd edition), Blackwell, Cambridge.

[96] Mann KH & Lazier JRN (2006) *Dynamics of Marine Ecosystems* (3rd edition), Blackwell, Cambridge.

[97] Jenkinson IR & Wyatt T (1992) Selection and control of Deborah numbers in plankton ecology. *Journal of Plankton Research*, 14, 1697-1721.

[98] Frost BW (1972) Effects of size and concentration of food particles on the feeding behavior of the marine planktonic copepod *Calanus pacificus*. *Limnology and Oceanography*, 17, 805-815.

[99] Hong J, Talapatra S, Katz J et al., (2012) Algal toxins alter copepod feeding behavior. *PLOS One*, 7, e36845.

[100] Steidinger KA, Wolny JL & Haywood AJ (2008) Identification of Kareniaceae (Dinophyceae) in the Gulf of Mexico. *Nova Hedwigia*, 133, 269-284.

[101] Bruno E, Andersen Borg CM & Kiørboe T (2012) Prey detection and prey capture in copepod nauplii. *PLoS ONE*, 7, e47906.

[102] Barnes H, Hutton J & Walters K (1989) *An Introduction to Rheology*. Elsevier, Amsterdam.

[103] Tiselius P, Saiz E & Kiørboe T (2013) Sensory capabilities and food capture of two small copepods, *Paracalanus parvus* and *Pseudocalanus* sp. *Limnology and Oceanography*, 58, 1657-1666.

[104] Anraku M & Omori M (1963) Preliminary survey of the relationship between the feeding habit and the structure of the mouth-parts of marine copepods. *Limnology and Oceanography*, 8, 116-126.

[105] Raymont JEG, Krishnaswamy S, Woodhouse MA & Griffin RL (1974) Studies on the fine structure of Copepoda: observations on *Calanus finmarchicus* (Gunnerus). *Proceedings of the Royal Society of London. Series B: Biological Sciences*, 185, 409-424.

[106] Giraud-Guille MM (1984a) Fine structure of the chitin-protein system in the crab cuticle. *Tissue and Cell*, 16, 75-92.

[107] Giraud-Guille MM (1984b) Calcification initiation sites in the crab cuticle: The interprismatic septa: An ultrastructural cytochemical study. *Cell Tissue Research*, 236, 413-420.

[108] Fabritius HO, Karsten ES, Balasundaram K et al., (2012) Correlation of structure, composition and local mechanical properties in the dorsal carapace of the edible crab *Cancer pagurus*. *Zeitschrift für Kristallographie*, 227, 766-776.

[109] Becker K, Hormchong T & Wah M (2000) Relevance of crustacean carapace wettability for fouling. *Hydrobiologia*, 426, 193-201.

[110] Oshima Y, Shikata M, Koyama T et al., (2013) MIXTA-like transcription factors and WAX INDUCER1/SHINE1 coordinately regulate cuticle development in *Arabidopsis* and *Torenia fournieri*. *Plant Cell*, 25, 1609-1624.

[111] Kopprio GA, Kattner G, Graeve M et al., (2012) Exceptional lipid storage mode of the copepod *Boeckella poopoensis* in a pampean salt lake, Argentina. *Aquatic Biology*, 15, 275-281.

[112] Jónasdóttir S (1999) Lipid content of *Calanus finmarchicus* during overwintering in the Faroe–Shetland Channel. *Fisheries Oceanography*, 8, 61-72.

[113] Teerawanichpan P & Qiu X (2012) Molecular and functional analysis of three fatty acyl-CoA reductases with distinct substrate specificities in copepod *Calanus finmarchicus*. *Marine Biotechnology*, 14, 227-236.

[114] Vojvodić V & Ćosović B (1996) Fractionation of surface active substances on the XAD-8 resin: Adriatic Sea samples and phytoplankton culture media. *Marine Chemistry*, 54, 119-133.

[115] Verdugo P & Santschi PH (2010) Polymer dynamics of DOC networks and gel formation in seawater. *Deep-Sea Research II*, 57, 1486-1493.

[116] Boxhall GA (1982) On the anatomy of the misophrioid copepods, with special reference to *Benthomisophria palliata* Sars. *Philosophical Transactions of the Royal Society, B: Biological Sciences*, 297, 125-181.

[117] Jiang H, Osborn TR & Meneveau C (2002a) The flow field around a freely swimming copepod in steady motion. Part I: Theoretical analysis. *Journal of Plankton Research*, 24, 167-189.

[118] Jiang H, Meneveau C & Osborn TR (2002b) The flow field around a freely swimming copepod in steady motion. Part II: Numerical simulation. *Journal of Plankton Research*, 24, 191-213.

[119] Jiang H, Osborn TR & Meneveau C (2002c). Hydrodynamic interaction between two copepods: a numerical study. *Journal of Plankton Research*, 24, 235-253.

[120] Jiang H, Osborn TR & Meneveau C (2002d) Chemoreception and the deformation of the active space in freely swimming copepods: a numerical study. *Journal of Plankton Research*, 24, 495-510.

[121] Jiang H & Strickler JR (2005) Mass density contrast in relation to the feeding currents in calanoid copepods. *Journal of Plankton Research*, 27, 1003-1012.

[122] Jiang H & Paffenhöfer GA (2008) Hydrodynamic signal perception by the copepod *Oithona plumifera*. *Marine Ecology Progress Series*, 373, 37-52.

[123] Tsukahara T, Kawase T & Kawaguchi Y (2011) DNS of viscoelastic turbulent channel flow with rectangular orifice at low Reynolds number. *International Journal of Heat and Fluid Flow*, 32, 529–538.

[124] Hicks GRF & Grahame J (1979) Mucus production and its role in the feeding behaviour of *Diarthrodes nobilis* (Copepoda: Harpacticoida). *Journal of the Marine Biological Association of the United Kingdom*, 59, 321-330.

[125] Chandler G & Fleeger J (1984) Tube-building by a marine meiobenthic harpacticoid copepod. *Marine Biology*, 82, 15-19.

[126] Decho AW & Fleeger JW (1988) Ontogenetic feeding shifts in the meiobenthic harpacticoid copepod *Nitocra lacustris*. *Marine Biology*, 97, 191-197.

[127] Decho AW & Moriarty DJW (1990) Bacterial exopolymer utilization by a harpacticoid copepod: A methodology and results. *Limnology and Oceanography*, 35, 1039-1049.

[128] Nehring S (1993) Tube-dwelling meiofauna in marine sediments. *Internationale Revue der gesamten Hydrobiologie und Hydrographie*, 78, 521-534.

[129] Peachey RL & Bell SS (1997) The effects of mucous tubes on the distribution, behavior and recruitment of seagrass meiofauna. *Journal of Experimental Marine Biology and Ecology*, 209, 279-291.

[130] Janke J (1992) Dominant species in phytoplankton blooms. 1. *Emiliana huxleyi* (Lohmann) Hay and Mohler. ICES identification leaflet for plankton, No. 178, 4 p.

[131] Greeve W (1975) Ctenophora. CIEM, Fiche d'identification du zooplancton, No. 146, 6 p.

[132] Su B, Wang S, Song Y & Jiang L (2011) A miniature droplet reactor built on nanoparticle-derived superhydrophobic pedestals. *Nano Research*, 4, 266-273.

[133] Epstein AK, Wong TS, Belisle RA et al., (2012) Liquid-infused structured surfaces with exceptional anti-biofouling performance. *Proceedings of the National Academy of Sciences of the United States of America*, 109, 13182-13187.

In: Copepods: Diversity, Habitat and Behavior
Editor: Laurent Seuront
ISBN: 978-1-63117-846-7

Chapter 10

COPEPOD BEHAVIOR: OCEANOGRAPHIC CUES, DISTRIBUTIONS AND TROPHIC INTERACTIONS

***C. B. Woodson*[1*], *D. R. Webster*[2] *and A. C. True*[2]**
[1]COBIA Lab, College of Engineering, University of Georgia, Athens, GA, US
[2]School of Civil and Environmental Engineering, Georgia Institute of Technology, Atlanta, GA, US

ABSTRACT

In order to fully understand the role copepods play in marine ecosystems, we delve into the sensory world of copepods, and build on this foundation to show how copepods utilize their environment to bridge the gap between dispersed prey and fast-moving predators. We begin by describing in detail the copepod's environment and the cues that we know are employed based on available literature. We then explore the implications of these behaviors on copepod distributions, trophic interactions, and overall marine ecosystem production. As we learn more about copepod sensory capabilities, it is becoming clearer, that copepods provide a critical link between physical and behavioral determination of organism distributions in the ocean. This link has profound effects on higher trophic level production and top-down *versus* bottom-up control in marine ecosystems.

INTRODUCTION

Copepods have evolved behavioral responses to their environment over millions of years that can have profound effects on their distributions, trophic interactions, and ultimately marine fishery production [1,2]. This is largely because copepods play a unique role in many marine food webs due to their position as secondary consumers [3,4], where they provide a critical intermediary link between photosynthesis (primary production) and higher trophic levels. Copepods are also prey for a wide variety of tertiary consumers such as predatory

* Corresponding author: Email: laurent.seuront@flinders.edu.au.

zooplankton, larval and adult fishes, and filter-feeding marine mammals [5-8]. Consequently, copepod behavior can have dramatic direct and indirect consequences through marine ecosystems. For this reason, there has been intense effort by oceanographers to understand the dynamics of copepod behavior relative to their environment for many decades.

Early work on copepod behavior focused on explaining the role of strong diel vertical migration (DVM) patterns that many species exhibit [9-12]. Researchers explained this behavior as the interplay between feeding and predation where copepods would ascend from depth to the surface at night to feed and back down at dawn to avoid visual predators, all cued predominantly by light intensity. The motivation behind this behavior, predator avoidance, was a catalyst to the next few decades of research on copepod behavior [4,13,14].

As experiments and observations became more sophisticated, it appeared that copepods utilize the full suite of environmental stimuli available to them, and that while predation may be the evolutionary driving force for individual behavior, it may not be, as recent studies suggest, the most important factor in driving population structure and trophic dynamics [1,8,15]. Responses to cues such as light, fluid density (changes in temperature or salinity), gradients in velocity, turbulence, chemicals from prey, and prey presence in combination with ocean circulation patterns across a range of scales can lead to aggregation in patches defined by vertical or horizontal orientations [7,16-18]. Direct responses to mating cues and predator disturbances can lead to changes in population distributions as a function of depth [5,19-21]. In many cases, we are just beginning to understand how copepod behavior can influence the productivity of both lower and higher trophic levels, the organization of food webs, and pelagic ecosystem functioning [1,8,22,23].

Our understanding of copepod behavior has grown dramatically over the past few decades through critical collaborations between copepod ecologists, oceanographers, and fluid dynamists and through application of new technology to observation, experimentation, and analysis [24-26] and through the aid of increasingly sophisticated individual-based models [22,23,27]. From the fluid physics standpoint, we now understand how turbulence can affect both copepod feeding and predation rates [28-30], how fluid velocity gradients can elicit escape responses from predators [31,32], and how copepods utilize the properties of flow to develop feeding currents and minimize energetic expenditure [33]. Application of three-dimensional imaging systems and fractal analysis of trajectories has provided new insights into subtle responses of copepods to environmental cues, and how these responses persist through time [25,34]. These advances have led to a variety of new questions about how the marine environment functions. How do copepods translate environmental cues to particular behaviors? How do these responses directly affect distributions of copepods, their predators, and their prey? How do these changes affect production of higher trophic levels and the maintenance of biodiversity? Increasingly sophisticated computer models of pelagic ecosystems [27,35] are beginning to address these questions and new technologies for imaging copepod behavior [36-38] will enable oceanographers to test hypotheses developed by these models in the near future.

In this chapter, we review the basic cues that an individual copepod is exposed to from changes in light intensity and duration to ephemeral disturbances due to predator movements. For each cue, we begin with the basic physics behind the cue generation and propagation through the environment and reception by an individual copepod. We then discuss observed behavioral responses to the cue, and the potential implications for ecological function. Once we have reviewed the state-of-knowledge of copepod sensory and behavioral ecology, we delve into the how these behaviors affect distributions and predator-prey interactions. We then discuss, based on available literature, the cascading effects of physics, behavior, and consequent distributions on trophic cascades and ecosystem functioning. Finally, we discuss the current state-of-the-field and future directions for understanding the role copepods play in the marine environment. Our intention is to give the reader a bottom-up view of the ocean in order to catalyze interest and research in pelagic ecosystem functioning as it relates to copepods and their behaviors especially in the light of new studies that suggest a much different role of copepods in marine systems that is not driven by predation response behaviors, but more closely tied to the spatial structure of prey in the ocean [1,2,8].

Oceanographic Cues and Behavioral Responses

In the ocean, an individual copepod is exposed to a myriad of environmental, chemical, and biological cues, which it must receive and respond to in order to survive. Copepods have a number of sensory capabilities that facilitate behavioral responses to cues and structure in the ocean. These include sensitivity to light, temperature, salinity, fluid density, velocity gradients, and chemicals. This section will review our knowledge of copepod behavioral response to each of these cues in order to set the stage for the subsequent discussion of organism distributions and interactions.

Before discussing particular cues, it is important to discuss the scales over which a copepod interacts with its environment. A single individual copepod can typically swim at *ca.* 5 mm s^{-1} allowing it to cover at most *ca.* 400 m in a day. The relative scales of horizontal (km's, m s^{-1}) and vertical (m's, mm s^{-1}) gradients and flows in the ocean therefore largely constrain copepod behavior to sensing vertical structure and moving vertically. However, this is not always the case especially in the vicinity of fluid density fronts (common in estuaries, river plumes, and in upwelling regimes) where horizontal gradients can be of the same magnitude as vertical gradients. For now, we consider predominantly vertical structure in the ocean, with the caveat that there are potential situations where copepods may be able to directly respond and interact along horizontal gradients [39].

Environmental Cues: Light

Light in the ocean plays important roles in both primary and secondary production. In this section, we briefly cover the role that light plays in copepod behavior in order to describe the 'world through a copepod's eyespot'. The treatment here is not meant to be exhaustive, but simply to familiarize and remind the reader that light plays a unique role in copepod behavior that can have large effects on their sensory response, distributions, and productivity. Copepods use light in many ways because it can affect prey distributions, predation risk, and habitat quality [40–42]. Light as a cue in the ocean is an omnipresent factor that provides copepods with a mechanism to locate prey and avoid predators. By using light cues, copepods can deal with the wide range of light attenuation in different water masses easily and utilize light to accomplish many basic life processes.

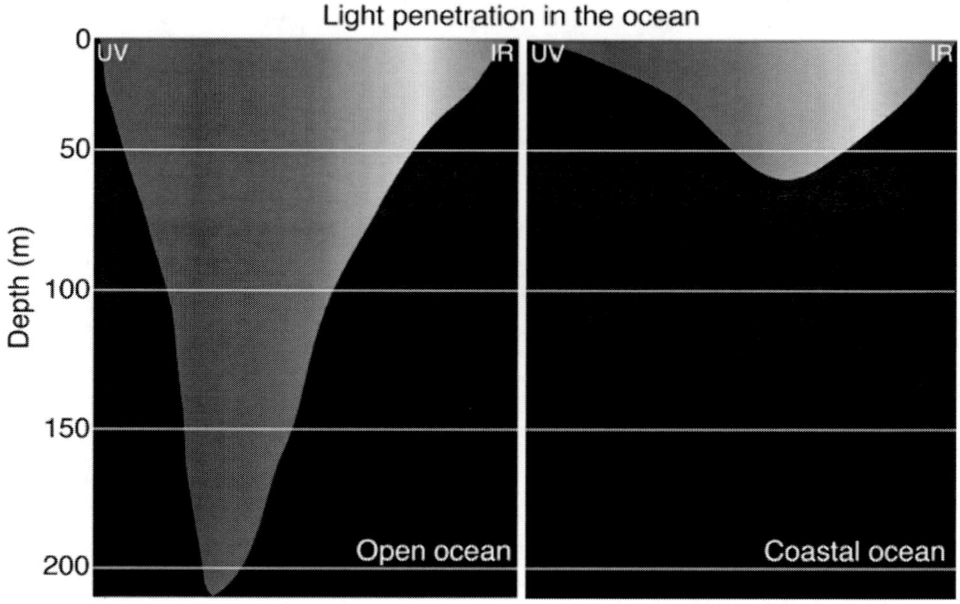

Figure 1. Light attenuation in the ocean. Attenuation is much higher in coastal waters due to increased turbidity associated with runoff, increased primary productivity, and increased sediments. Copepods are most sensitive to blue-green light that can penetrate much deeper into the ocean.

Light is absorbed rapidly over depth in the ocean (Figure 1). Consequently, light provides an ideal cue for sensing susceptibility to visual predators with gradients that scale reasonably well to copepod behavior and swimming capabilities. However, other factors such as turbidity and colored dissolved organic matter (CDOM) can have dramatic effects on the attenuation of light in the ocean. Regardless of the particular decay constant (dependent on water clarity and content), copepods can utilize light gradients to maintain position within the water column relative to predator visual acuity, and they may use different aspects of the light distribution to cue particular behaviors as will be discussed below.

Copepods are generally sensitive to light and numerous copepod species, such as *Acartia tonsa*, demonstrate a broad range of sensitivity and are positively phototactic [43]. Copepods sense light through a single eyespot located centrally on the main body or head of the animal. The eyespot senses light intensity generally in the range of visual light (400-700 nm), and can provide weak infrared and UV sensitivity depending on species [4]. Light intensity can be used in a variety of behaviors and can also mediate responses to other environmental cues [42]. Perhaps the most fundamental use of light by copepods is in either changes in photoperiod inducing dormancy [44] or in the use of light intensity to initiate DVM and depth maintenance [9,12].

Light is generally considered the primary proximate cue controlling DVM [41,45]. Lampert [46] observed that DVM is intrinsically linked to light-dependent mortality in surface waters, and typically occurs only in the presence of such predation pressure. Further, DVM is a flexible behavior that allows for balancing various constraints including effectiveness of protection, energy input, and other environmental variables [47]. Four major hypotheses have been suggested to explain DVM. First, the preferendum or isolume hypothesis corresponds to zooplankton following a preferred light level during migration.

Buskey et al. [48], for example, observed copepod behavior consistent with this hypothesis. Second, the absolute intensity threshold hypothesis suggests that DVM is at least partly controlled by an absolute threshold. *Calanopia americana* is an example of a copepod that follows behavior consistent with this hypothesis [49,50]. Third, the rate of change hypothesis suggests that the relative rate and direction of light intensity change from ambient level mediates DVM. This behavior has been observed in *A. tonsa* [51] among other species [41]. Fourth, the transparency-regulator hypothesis emphasizes the role of water transparency in regulating structural drivers and dynamic drivers [52]. Williamson et al. [52] argue that this hypothesis allows for more accurate explanation of the broad range of observed DVM behaviors.

Species that undergo DVM are most sensitive to wavelengths (blue/green) that are prevalent at sunset when migration often occurs [53]. While light is the primary proximate cue, exposure to predators can additionally influence migrations by affecting the magnitude of movement and inducing migrations in previously non-migrating species. For many species, the behavioral response is commonly activated from chemical or (possibly) mechanical cues from fish predators [41]. Recent studies have also found that some copepods are more sensitive to predator flow disturbances when exposed to higher illumination, which suggests that copepods are sensitive to their visual susceptibility to predation [42]. In this study, the copepod, *Calanus finmarchicus*, tunes its response to predator-like flow disturbances based on the ambient light intensity, which illustrates individual copepods awareness of their particular environment.

Many species of copepods are known to modulate their depth based on the phase of the moon, again a response to ambient light levels that appears related to predator avoidance. Avoidance behavior (rigorous downward swimming) to ultraviolet light has also been observed in some species likely to avoid ultraviolet radiation damage [54]. Responses to light have also been shown to induce swarming behavior in many estuarine and coastal species of copepods [55]. Swarming in response to light has been proposed as a mechanism for mating aggregation [20]. In the light shafts of mangroves, Buskey [56] observed intense swarms of the copepod, *Dioithona oculata*, and illustrated increased mating success within the swarms. However, whether swarming behavior is a true response to environmental changes or an artifact associated with intense light beams is still unclear [45].

Finally, a potential example of the use of light affecting horizontal distributions is the increased abundance of estuarine copepods, such as *Eurytemora affinis*, within the estuarine turbidity maxima (ETM) region [40,57,58]. Although ETMs are also often regions of other convergent oceanographic processes, they can be sensed using light by copepods. In shallow estuaries, light can penetrate the entire water column with minimal attenuation. However, changes in turbidity often occur at scales (10's m) similar to that of vertical structure in the open ocean. However, whether copepods actively employ light as a cue to maintain position within these regions as a mechanism for reducing predation risk is still an open question.

Environmental Cues: Temperature, Salinity, and Fluid Density

Gradients in temperature, salinity, and fluid density in many ways form the most consistent environmental cues available to copepods in the ocean. In the ocean, temperature and salinity follow distinct profiles that lead to the development of gradients in fluid density called 'fronts' in the horizontal, and 'clines' in the vertical [2]. Regions of strong vertical gradients or clines are common throughout the ocean, are often associated with subsurface chlorophyll maxima, and occur at scales (m's) that can be exploited by individual copepods [59,60]. Due to the strong associations between temperature, salinity and prey resources, it is expected that copepods employ these cues in foraging behavior. However, recent studies have suggested that this is not necessarily the case, at least for several common coastal species [15,60,61].

The presence of a salinity gradient has long been recognized as affecting copepod swimming behavior with the organism consistently recoiling from the salinity gradient layer [62]. The most important factor in creating patchiness of copepods in this context is modified swimming behavior [63]. Harder [64] observed species-dependent distributions in the presence of a fluid density gradient and documented minimum threshold levels for several species to accumulate at salinity gradient layers. Most copepods tended to accumulate near the fluid density gradient layer. Interestingly, *Temora longicornis* showed no accumulation at salinity gradient layers in the absence of a coincident fluid density gradient. *T. longicornis* did, however, show aggregation at a fluid density gradient layer in absence of a coincident salinity gradient. Thus, fluid density appears to be the most important property controlling behavior that leads to aggregation, at least for this species. Lougee et al. [65] observed similar influence of haloclines on copepod distributions and argued that in estuaries, salinity and fluid density gradients are generally coincident. In that scenario, the results of the proximate cue being salinity or fluid density would ultimately be the same. Many estuarine copepods are suspected of using fluid density cues, often in the form of salinity, to maintain position within the estuary that can trigger tidally-driven vertical migration [66-68]. These responses enable copepods to undergo small-scale (2-6 m) vertical movements triggered by changes in salinity. Migrations at these scales enable the copepods to maintain position within the ETM [57,69].

There is also similar, but not as exhaustive, evidence that copepod swimming behavior is influenced by temperature, however the response may be linked to changes in water viscosity associated with temperature changes. *T. longicornis*, for instance, demonstrates higher swimming speed, fewer break periods, and more frequent jumps for higher fluid temperature [70]. Temperature can also directly affect metabolic rates and lead to dormancy [44]. Changes in temperature have also been shown to affect responses of estuarine copepods to salinity [71].

Regarding the mechanism of aggregating, Woodson et al. [60] observed that individual copepods rarely cross a fluid density gradient layer, when strength is at supra-threshold levels, and rather swim along a layer boundary or turn around. They reported behavior threshold levels for *A. tonsa* and *T. longicornis* crossing the fluid density gradient layer. The behavior around salinity gradient and fluid density gradient layers is also species-dependent. Woodson et al. [15] observed that *E. affinis* behaved differently than the other copepod species tested. *E. affinis* often crossed the fluid density gradient layer, changed swimming patterns, and increase residence time in the layer region. For this species, Seuront [68] observed increased complexity of swimming path with increasing salinity and concluded that *E. affinis* employs salinity gradients as a proximate cue to maintain position near regions of high productivity in estuaries. The state of the adult may also be a factor for the behavioral response. Michalec et al. [69] found an optimal salinity range for swimming activity in the copepod *Pseudodiaptomus annandalei*, which differed for ovigerous females compared to other females and males.

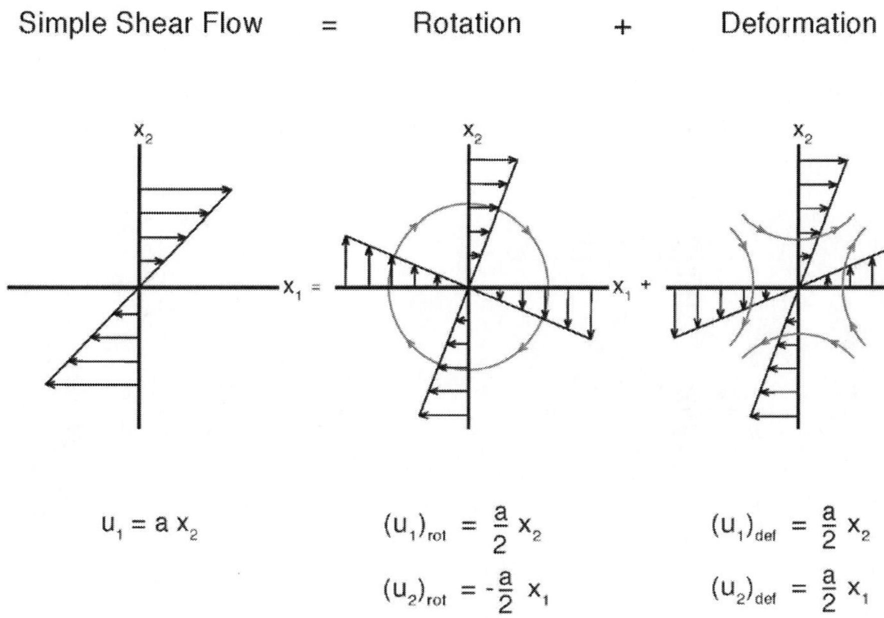

Figure 2. Shearing motion will rotate and stretch any straight material line as demonstrated by this example illustration. The decomposition of this simple example can be generalized to three-dimensions via Eqn. (1).

The ecological result of aggregations around steep salinity gradient or fluid density gradient layers may vary as well. For instance, Heuch [72] observed salmon louse copepodid aggregations at step salinity gradients and suggested that the aggregations may increase the frequency of host/parasite encounters. The response for this copepodid seems to be influenced by both the absolute salinity and the change in salinity with depth. In another example, Bochdansky and Bollens [73] suggested that it may be advantageous for copepods to respond to physical cues rather than searching for ephemeral food patches. The observations of Woodson et al. [15] support this suggestion, at least for *E. affinis*, for whom the fluid density gradient layer appears to act as the initiation of a cue hierarchy in the search for food and high resource patches.

Environmental Cues: Fluid Motion

In any flow field there are multiple fluid mechanical cues that a copepod could sense, namely fluid acceleration, vorticity, and deformation rate. A general three-dimensional velocity gradient, can be decomposed into its irrotational and rotational components defined by the deformation rate (or strain rate) tensor, e_{ij}, and the rotation tensor, r_{ij}, as (Figure 2):

$$\frac{\partial u_i}{\partial x_j} = e_{ij} + \frac{1}{2} r_{ij} \tag{1}$$

where u_i is the velocity vector and the indices i and j indicate the three coordinate directions [74].

The deformation rate tensor can be thought of as the relative velocity gradient due to pure deformation and can itself be decomposed into the linear deformation rate (diagonal terms) and the shear deformation rate (off-diagonal terms):

$$e_{ij} = \begin{vmatrix} \frac{\partial u_1}{\partial x_1} & \frac{1}{2}\left(\frac{\partial u_2}{\partial x_1}+\frac{\partial u_1}{\partial x_2}\right) & \frac{1}{2}\left(\frac{\partial u_3}{\partial x_1}+\frac{\partial u_1}{\partial x_3}\right) \\ \frac{1}{2}\left(\frac{\partial u_2}{\partial x_1}+\frac{\partial u_1}{\partial x_2}\right) & \frac{\partial u_2}{\partial x_2} & \frac{1}{2}\left(\frac{\partial u_3}{\partial x_2}+\frac{\partial u_2}{\partial x_3}\right) \\ \frac{1}{2}\left(\frac{\partial u_3}{\partial x_1}+\frac{\partial u_1}{\partial x_3}\right) & \frac{1}{2}\left(\frac{\partial u_3}{\partial x_2}+\frac{\partial u_2}{\partial x_3}\right) & \frac{\partial u_3}{\partial x_3} \end{vmatrix} \tag{2}$$

The rotation tensor can be thought of as the relative velocity gradient due to fluid rotation:

$$r_{ij} = \frac{\partial u_i}{\partial x_j} - \frac{\partial u_i}{\partial x_j} = -\varepsilon_{ij}\omega_k \tag{3}$$

hence,

$$r_{ij} = \begin{vmatrix} 0 & -\left(\frac{\partial u_2}{\partial x_1}-\frac{\partial u_1}{\partial x_2}\right) & \left(\frac{\partial u_1}{\partial x_3}-\frac{\partial u_3}{\partial x_1}\right) \\ \left(\frac{\partial u_2}{\partial x_1}-\frac{\partial u_1}{\partial x_2}\right) & 0 & -\left(\frac{\partial u_3}{\partial x_2}-\frac{\partial u_2}{\partial x_3}\right) \\ -\left(\frac{\partial u_1}{\partial x_3}+\frac{\partial u_3}{\partial x_1}\right) & \left(\frac{\partial u_3}{\partial x_2}-\frac{\partial u_2}{\partial x_3}\right) & 0 \end{vmatrix} \tag{4}$$

and

$$r_{ij} = \begin{vmatrix} 0 & -\omega_3 & \omega_2 \\ \omega_3 & 0 & -\omega_1 \\ -\omega_2 & \omega_1 & 0 \end{vmatrix} \tag{5}$$

where ω_k is the vorticity vector and ε_{ijk} is the alternating tensor. For the example shear flow shown in Figure 2 (i.e., $u_1 = ax_2$), the deformation rate and rotation tensors are, respectively, given by

$$e_{ij} = \begin{vmatrix} 0 & \frac{1}{2}a & 0 \\ \frac{1}{2}a & 0 & 0 \\ 0 & 0 & 0 \end{vmatrix} \quad (6)$$

and

$$r_{ij} = \begin{vmatrix} 0 & a & 0 \\ -a & 0 & 0 \\ 0 & 0 & 0 \end{vmatrix} \quad (7)$$

These results, and Figure 2, demonstrate that there is simple shear deformation aligned at 45° relative to the x_1 and x_2 axes and rotation aligned with the x_3 axis only.

Thus, any three-dimensional velocity field can be thought of as the superposition of rotational and deformational flow components, and the ability of a copepod to sense the magnitude and orientation of the flow disturbance allows it to distinguish, for instance, the suction force from a predator [75] versus the fluid mechanical signature of a phytoplankton cell entrained in its own feeding current [76]. Some authors suggest that copepod size has evolved to utilize specific flow characteristics [77], and many copepods have developed adaptive behaviors to remove signal noise from their immediate surroundings [77,78].

Copepods sense fluid motion using setae (bristle- or hair-like structures) that are mechanically sensitive along the first antenna. Deflection of the setae triggers neural signals (i.e., spikes) that possess amplitudes and sensitivity that suggest the capability for very rapid responses such as prey capture and escapes from predators [79,80]. Copepod setae are sensitive to fluid velocities as small as 20 μm s^{-1} and demonstrate response to stimuli in the range of 40 to 1000 Hz [81]. Copepod setae possess a wide range of lengths. Shorter setae are sensitive to rapid fluid motion, whereas long setae function as low velocity sensors [82]. The morphology (i.e., setal shapes, orientations, and degree of ornamentation) is spectacularly diverse among copepods, which indicates a vast range of mechanosensory capability [83,84]. This suggests that copepod setal morphology may be tuned to specific functionality. For instance, predatory copepods should have numerous short setae, with fewer long setae, in order to effectively detect the flow disturbance of rapidly moving prey [83].

A majority of studies of hydromechanical cues have focused on predator-prey detection interactions [13,85]. In fact, Kiørboe [1] mentions mechanoreception of predators and of prey as key factors in the success of copepods. Kiørboe and Visser [14] argue that predator perception of prey depends on the absolute magnitude of the velocity disturbance created by moving prey, whereas prey perception of predators is a function of the spatial gradients of velocity and the sensor span; see also [86]). In the prey-perception-of-predators scenario, the signal strength is proportional to the spatial gradient of velocity as well as the size of the prey (i.e., the spatial span of the sensors). Although it should be noted that observations of *A. tonsa* do not fully support the hypothesis that threshold for behavioral response solely depends on the velocity difference (i.e., the product of velocity gradient and sensor span; [87]).

Table 1. Thresholds for escape responses to fluid motion in several copepod species

Species	Fluid Stimulus	Threshold Deformation Rate (s^{-1})	Reference
Acartia hudsonica	Siphon	2.2	[90]
Acartia hudsonica	Oscillating bar	1.7	[204]
Acartia hudsonica, nauplii	Siphon	0.80	[85]
Acartia tonsa	Siphon	0.38	[19]
Acartia tonsa	Siphon	1.19-2.49	[87]
Acartia tonsa, nauplii	Siphon	0.38	[19]
Calanus finmarchicus	Rotating Cylinder	0.40	[205]
Calanus helgolandicus, nauplii	Siphon	0.52-3.24	[206]
Centropages hamatus	Siphon	1.2	[90]
Centropages hamatus	Oscillating bar	1.4	[90]
Centropages typicus, nauplii	Siphon	2.60-2.79	[206]
Euchaeta rimana	Siphon	2.4	[19]
Euchaeta rimana	Pipette jet	3.3	[32]
Eurytemora affinis	Pipette jet	1.9	[13]
Eurytemora affinis, nauplii	Siphon	1.88-2.65	[206]
Euterpina acuftifrons, nauplii	Siphon	1.92-4.25	[206]
Labidocera madurae	Siphon	6.3	[19]
Oithona spp.	Siphon	3.8	[19]
Paracalanus parvus	Siphon	6.2	[204]
Pleuromamma xiphias	Siphon	4.6	[19]
Temora longicornis	Pipette jet	6.5	[13]
Temora longicornis	Siphon	2.7	[90]
Temora longicornis	Oscillating bar	> 15	[90]
Temora longicornis, copepodid	Siphon	0.73-0.81	[31]
Temora longicornis, nauplii	Siphon	2.78-3.96	[31]
Temora turbinata	Siphon	3.9	[204]
Tortanus discadatus	Siphon	0.3	[90]
Tortanus discadatus	Oscillating bar	1.2	[90]

The most significant flow disturbance is quantified by the deformation rate (Eq. 2; [14,19]. Copepods appear to be capable of perceiving acceleration, but do not behaviorally respond to acceleration [87]. Similarly, the presence of fluid vorticity (Eqs. 3-5) did not elicit escape response [87]. The behavioral response does not depend on pure linear deformation rate versus pure shear deformation rate, but rather is mediated by the magnitude of the deformation rate. Numerous studies have used siphons [88], moving spheres [89], and other flow disturbance generators [87,90] to document threshold levels of deformation rate to evoke an escape response (Table 1). There is substantial variability among the reported threshold values. Much of the variability may be due to sensory differences (and setal morphology differences) among species, although there may also be an influence of measurement approach among various researchers. The escape response of *A. tonsa* also depends on the relative location of the flow disturbance, which indicates a directional bias and creates effective '*blinds spots*' of weaker perceptible ability [91]. The greatest susceptibility for *A. tonsa* is when attacked from the side. Thus, documenting a threshold deformation rate for escape response requires flow measurements in a plane relevant to the copepod's orientation. In addition, the threshold deformation rate to elicit escape response appears to be mediated by ambient illumination level with a lower threshold in lighted conditions compared to dark conditions [42].

Known sensitivity of copepod setae appear to be consistent with the ability for remote detection of inert particles by feeding-current-generating and cruising copepods [86], although sensing of prey is strongly biased toward larger prey particles [14]. Consistent with the morphological diversity mentioned above, the combination of numerous mechanosensors facilitates the ability to spatially locate the flow disturbance of moving prey [92]. *Oithona plumifera*, for instance, show an ability to accurately project the prey's instantaneous location and velocity prior to attack with adaptable attack behavior based on the perceived signal patterns [93]. Sensing of prey within the capture region is facilitated by mechanoreceptive setae on the basal sector of the first antenna, whereas setae on the mid and distal regions are utilized to orient to prey located outside the capture region [84]. Detection of prey depends on the relative spatial location of the prey and the prey's motility [94]. Motile prey are detected in front of the copepod, whereas non-motile prey are detected ventrally or lateral-ventrally to the copepod.

Mechanoreception in copepods is important in contexts beyond predator-prey interactions as well. For instance, Jiang and Osborn [95] discuss the hydrodynamic signal between copepods of similar size. The signals in that case may not be symmetric with one copepod sensing another who is unaware of the first. Consistent with this scenario, Bagøien and Kiørboe [96] observed that *A. tonsa* use hydrodynamic signals to locate mates. Males and females appear unaware of the presence of a potential mate until one individual creates a flow disturbance by hopping. Copepods are capable of distinguishing the source of a hydrodynamic disturbance (prey, predator, mate, etc.) based on the information contained in the signal [76]. This suggests that different types of behavioral responses to different types and magnitudes of various hydrodynamic signals are likely. In another context example, Woodson et al. [61] observed copepods orienting to thin layers of velocity gradients with increased swimming speed and turning frequency that result in increased local residence time near the thin layer structure. The threshold deformation rate in this case was substantially smaller (roughly one to two orders of magnitude smaller) than the threshold to evoke an escape response. Woodson et al. [60] hypothesized that the response to the velocity gradient layer indicated the initiation of a cue hierarchy in the search for resource rich zones. Finally, sensing of turbulent fluid motion has been hypothesized to induce copepod behavioral responses such as sinking or swimming into less turbulent regions [97-99]. While the relevant mechanosensory signal in fluctuating turbulent velocity fields has been speculated [100,101], direct measurements have not yet been performed. Further, turbulent flow conditions may significantly affect the availability and reliability of hydrodynamic signals from predators, prey, and conspecifics.

Environmental Cues: Chemical and Biological

Chemosensing and chemical signals play important roles in copepod ecology. Chemical cues in the marine environment typically consist of highly soluble (molecular diffusivity ~ 10^{-5} cm^2 s^{-1}) compounds including amino acids, proteinaceous compounds, nucleic-acid related compounds, glycerolipids, and pheromones [102-105]. The means of release and the hydrodynamically-regulated advection and diffusion of a chemical signal determine the amount and specificity of chemical information conveyed [104,106]. The focus of this section is on biophysical coupling concerning the release of chemical cues, subsequent hydrodynamic advection and diffusion, and finally cue perception and implications for behavioral responses.

The two most important parameters that define the appropriate regimes of fluid flow and chemical transport in any aquatic environment are the Reynolds number *Re*

$$Re = \frac{\rho UL}{\mu} = \frac{UL}{\nu} \tag{8}$$

and the Péclet number *Pe*

$$Pe = \frac{UL}{\Gamma} \tag{9}$$

where U is the characteristic velocity, L is the characteristic length scale, ρ is fluid density, μ is dynamic viscosity, ν is kinematic viscosity, and Γ is molecular diffusivity of the chemical. *Re* represents the ratio of inertial to viscous forces acting in a flow, and *Pe* describes the ratio of advective to diffusive effects acting on a passive scalar, such as chemical concentration, in a flow field.

Low and intermediate values of *Re* indicate laminar flow (i.e., less than the order of 100) with very low values (i.e., less than 1) indicating viscosity-dominated flow. Larger values of *Re* indicated turbulent flow in which random velocity fluctuations lead to enhanced chemical transport. Similarly, low values of *Pe* indicate a regime dominated by diffusive transport, whereas larger values of *Pe* indicate an advection-dominated regime. Thus, these parameters form the basis for a quantitative understanding of the relevant transport processes as well as the chemical concentration cue likely to be experienced by receivers [106].

The spread of chemicals emitted into a stagnant fluid environment is governed by Fickian diffusion in which chemical flux $\vec{q}$ is proportional to the concentration gradient as:

$$\vec{q} = -\Gamma \vec{\nabla} c \tag{10}$$

where c is chemical concentration, and the diffusion transport equation is given as:

$$\frac{\partial c}{\partial t} = \Gamma \nabla^2 c \tag{11}$$

Time scales of molecular diffusion are extremely long and gradients of chemical solute develop slowly. At low *Re*, the addition of flow advection is included in the chemical flux:

$$\vec{q} = \vec{u}c - \Gamma \vec{\nabla} c \tag{12}$$

where $\vec{u}$ is the velocity vector, and transport is described by the advection-diffusion equation:

$$\frac{\partial c}{\partial t}+\vec{u}\cdot\vec{\nabla}c=\Gamma\nabla^2 c \qquad (13)$$

Finally, chemical flux in high *Re* turbulent flows is additionally influenced by random velocity fluctuations. The turbulent time-averaged flux of a chemical solute, $\bar{\vec{q}}$, is given as:

$$\bar{\vec{q}}=\bar{\vec{u}}\bar{c}-\Gamma\vec{\nabla}\bar{c}+\overline{\vec{u}'c'} \qquad (14)$$

where the overbar notation indicates time-averaged and the prime notation indicates fluctuation away from the time-averaged value. Equation (11) reveals the additional turbulent flux term corresponding to the covariance of the velocity and concentration fluctuations. The time-averaged concentration field is now governed by the Reynolds-averaged advection-diffusion equation as:

$$\frac{\partial \bar{c}}{\partial t}+\bar{\vec{u}}\cdot\vec{\nabla}\bar{c}=\Gamma\nabla^2\bar{c}-\vec{\nabla}\cdot\left(\overline{\vec{u}'c'}\right) \qquad (15)$$

In many cases, the covariance term is replaced by a 'turbulent diffusivity' term that models the effect of turbulence as a diffusive process.

Copepods demonstrate sensory capability to detect dissolved chemicals and use the cues to respond to food, mates, and other items [107]. Food particle detection *via* chemosensation can be more efficient than hydrodynamic detection [108]. Copepod cruising strategy is more effective for smaller particles and chemical plume sensing is more effective for larger particles, although this conclusion is highly sensitive to the assumed detection threshold levels [108]. *T. longicornis*, for instance, can trail follow sinking marine snow particles from a distance of greater than 25 particle radii *via* chemosensation [109]. Copepod distributions are also influenced by chemical interactions with phytoplankton food particles [110]. When near to patches of food particles, *A. tonsa* demonstrated the ability to detect the patch edge and to maintain position precisely enough to take advantage of food presence [111]. In this context, Woodson et al. [60] reported threshold concentrations of chemical exudates of phytoplankton for *A. tonsa* and *T. longicornis* to initiate excited area-restricted search behavior (i.e., increased swimming speed and turning frequency). Similar behavior is observed in several other species as well [15].

Numerous studies have demonstrated mate detection *via* chemosensation, and the impressive mate locating ability may help explain copepod success in the ocean [112]. Males have the ability to chemically distinguish virgin females compared with mated females [113]. Copepods also appear to be unique in their ability to track chemical information in a three-dimensional space to locate a mate. In the trail following scenario, the female leaves a trail of pheromone that the male randomly intersects. Once the male intersects the trail, he accurately follows it and even corrects for heading in the wrong direction [104]. By leaving a pheromone trail, the female increases the probability of encounter with a mate by up to 100 times [114]. Male *T. longicornis* locate females by following chemical trails for up to 10 seconds and track over distances exceeding 13 cm or 130 body lengths [115,116]. *Centropages typicus* demonstrate similar behavior and track trails up to 31 second old and 17 cm long [96]. The freshwater species *Hesperodiaptomus shoshone* demonstrate shorter (3.8 cm maximum) and briefer (2.4 s maximum) tracks, which suggests either weaker chemical sensitivity or less persistence of the chemical cue [117]. Alternate to the trail following scenario, female *Pseudocalanus elongatus* create a nearly spherical cloud of pheromone with heterogeneous chemical distribution around their hovering position [118]. A male swims in the cloud surrounding the female with intermittent physical contact with the female. In another alternate behavior, female *Oithona davisae* distributes pheromone in a trail intermittently due to their hop-sink swimming behavior [119]. Males use local search strategies to bridge the gap between neighboring patches of pheromone.

Copepods also demonstrate chemosensory ability beyond sensing food or mates. For instance, female *T. longicornis* show a chemosensory response to dimethyl sulfide (DMS) suggesting that plumes of this gas may provide information about the nearby presence of algae food particles [120]. In another example, *T. longicornis* and *E. affinis* sense water-soluble diesel oil and avoid the contaminated region [121]. Further, chemosensation may be used in combination with other sensory modes including flow sensing [84]. The response to combined cues in this case is species dependent.

Environmental Cues: Multiple Stimuli and Interactions

The way in which behavioral responses change when multiple cues are combined leads to the idea of a cue hierarchy and provides a basic tool for quantifying the ecological implications of individual behavior. A central theme in sensory ecology is the idea of a cue hierarchy in triggering changes in behavior [122]. A cue hierarchy explains the relative importance of various types of sensory cues in triggering behavioral responses in relation to one another. For example, the presence of chemical cues induces *T. longicornis* to strengthen or cancel responses to velocity gradient cues, whereas combined chemical and velocity gradient cues evoke the same response in *A. tonsa* as the isolated cues [60]. The velocity gradient acts as an initial cue to restrict search area in a foraging copepod, and chemical cues then allow the copepod to refine the search area for a resource patch with greater effectiveness [60]. Finally, mechanical contact with individual prey cells initiates feeding behaviors [60]. Each cue allows an individual copepod to systematically and effectively search their local environment for prey or resources. Similarly, as mentioned above for predator-prey interactions, light can modulate the hydromechanical threshold for escape response among copepods [42].

BEHAVIOR AND DISTRIBUTIONS

In this section, we focus on how individual copepod behaviors when viewed collectively affect their distributions over the depth of the water column. As discussed previously, due to the disparity in vertical (meters) and horizontal (kilometers) scales in the ocean, most copepods interact with their surroundings in the vertical direction where scales are better matched to the foraging and swimming capabilities of an individual (e.g., *ca.* 5 mm s^{-1}, covering 10's of m in a few hours). Two recent articles review the effects of behavior on zooplankton distributions and the implications on higher trophic levels [2,22].

Foraging and feeding behavior are driven by both mechano- and chemoreception in an effort to locate resource patches. Field studies by Mullin and Brooks [123] found that the copepod *Calanus pacificus* would not be able to survive in the water column except at a preferential depth with concentrated food resources. Similarly, Daro [124] found that a dominant copepod species was food limited at most depths and that minimum concentrations necessary for survival (200 mg cm^{-3} chl-*a*) were only found in one narrow depth range between 20 and 30 m. Furthermore, *in situ* evidence for microscale patchiness [17,125,126] shows that limited food resources inhibit copepod growth and thus copepods must be able to actively locate and exploit isolated resource patches in order to survive. Indeed, the idea of area-restricted search behavior [15,60,127] in individuals through altered swimming speed and increased turn frequency can often produce population scale aggregations. Many species of copepods including *Pseudocalanus minutus* [128], *T. longicornis* and *P. elongans* [129], and *A. tonsa* [111], are observed to decrease swimming speed (or hop frequency) when encountering food patches. Leising et al. [130] also note predation risks for a foraging copepod exploiting different portions of the water column, and modeling studies by Leising and Franks [127,131] examined the effect of food concentrations and patchiness on copepod swimming behavior and found behavior consistent with area-restricted foraging.

Individual Behavior and Depth Distributions

Individual copepod behaviors, such as predator avoidance, foraging, and mating, have been shown to have dramatic effects on the distribution of copepods in relation to oceanographic properties. Some clear examples of this are in estuaries as described previously in the section on responses to salinity, and in the '*thin layers*' literature; see e.g., [132-134]. Thin layers are a common feature of the ocean and represent fine-scale (m's) distributions of both phytoplankton and zooplankton (Figure 3). These layers are often associated with similar scale gradients in temperature, salinity, fluid density, velocity, and turbulence. Copepods, through their behavioral adaptations, appear to be cued into the formation and maintenance of these high resource patches [61,135,136]. Simple individual-based models that simulate behaviors observed in the laboratory over longer time scales have further illustrated that these patches can lead to large-scale distribution patterns in copepods [15,131].

Habitat partitioning is thought to be a critical component in the ability of the pelagic environment to maintain observed levels of biodiversity especially in the plankton. Differences in DVM along with behavioral responses to fluid mechanical, chemical, and phototactic cues can have significant trophic consequences. Many calanoid copepods, like many other zooplankters, are diel vertical migrators, i.e., they exhibit large changes in depth on a diel cycle driven by phototactic cues [137]. When vertical migratory behavior is superimposed on the physical structure of the water column and coupled with individual behavioral responses to patchy cues, population-scale, high-population-density aggregations can result [18]. Experiments on a variety of zooplankton showed that almost every species tested aggregated at salinity and/or temperature interfaces (i.e., the pycnocline; [64]). Fieldwork also showed multiple species of copepods aggregating in the layer just above the permanent halocline [97]. Interestingly, depth distribution was species-segregated. Bollens et al. [138] tested the effects of a variety of mechanical, chemical, and visual cues in producing patchy aggregations of the copepod *A. hudsonica*. Genin et al. [37] acoustically tracked over 375,000 zooplankters over a coral reef in the Red Sea and observed active swimming against vertical flows resulting in depth maintenance that on a population scale could lead to dense aggregations. Finally, Pierson et al. [139] documented that copepods make periodic forays into deeper waters throughout the night, suggesting a higher frequency component to classical conceptualization of DVM.

Foraging behaviors in response to interacting cues in the form of a cue hierarchy can also lead to dense aggregations of copepods in vertical layers based on laboratory experiments and numerical modeling [15,127]. Acoustic data suggest these aggregations can form during a combination of DVM and foraging behavior for organisms in the copepod size classes (Figure 4; [18]). But to date, direct observation of copepod foraging behavior over time scales that could lead to dense aggregations of copepods have not been made.

Direct responses to predators can also affect the distribution of copepods. Bollens and Frost [5] documented the initiation of DVM in response to predators. In fact, recent evidence suggests that some copepod species make repeated migrations between the surface and depth in order to minimize predation risk and maximize feeding opportunity [139]. Whether these migrations are cued by local changes in light intensity or are intrinsic to copepod behavior after food saturation is an on-going question for biological oceanographers that would benefit greatly from advanced *in situ* imaging systems [36,38,126,140].

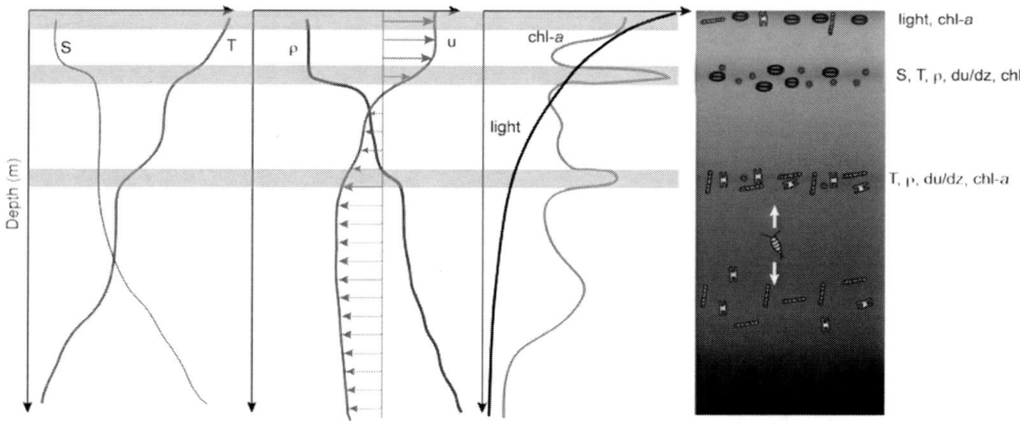

Figure 3. Gradients in the ocean at scales relevant for copepod behavior. Salinity (S), temperature (T), fluid density (ρ), velocity gradients (du/dz), Chlorophyll a (chl-a), and light all contribute to the copepod sensory environment, but not all will be associated with preferred prey items.

It is expected that convergent swimming behavior likely plays the largest role in the depth maintenance and distributions of copepods. Whether individual depth preferences develop into large-scale aggregations at the population level is often determined by an interaction between physics and biology [22]. Several mechanisms have been proposed for the development of patches of phytoplankton (e.g., gyro-tactic trapping, shearing, *in situ* growth, buoyancy, convergent behavior). However, at the scale of copepods, swimming speeds are greater than vertical velocities for most of the ocean [2]. Fine-scale cues associated with the formation of phytoplankton patches can differ depending on the mechanism causing the patchiness. Therefore, it is expected that some patches may lead to copepod aggregations, while others may not. For example, *in situ* growth of phytoplankton can create a patch in the absence of a fluid density gradient or velocity gradient. Copepods therefore may not be able to sense the patch until much later when phytoplankton achieve sufficient density to elicit a chemical response.

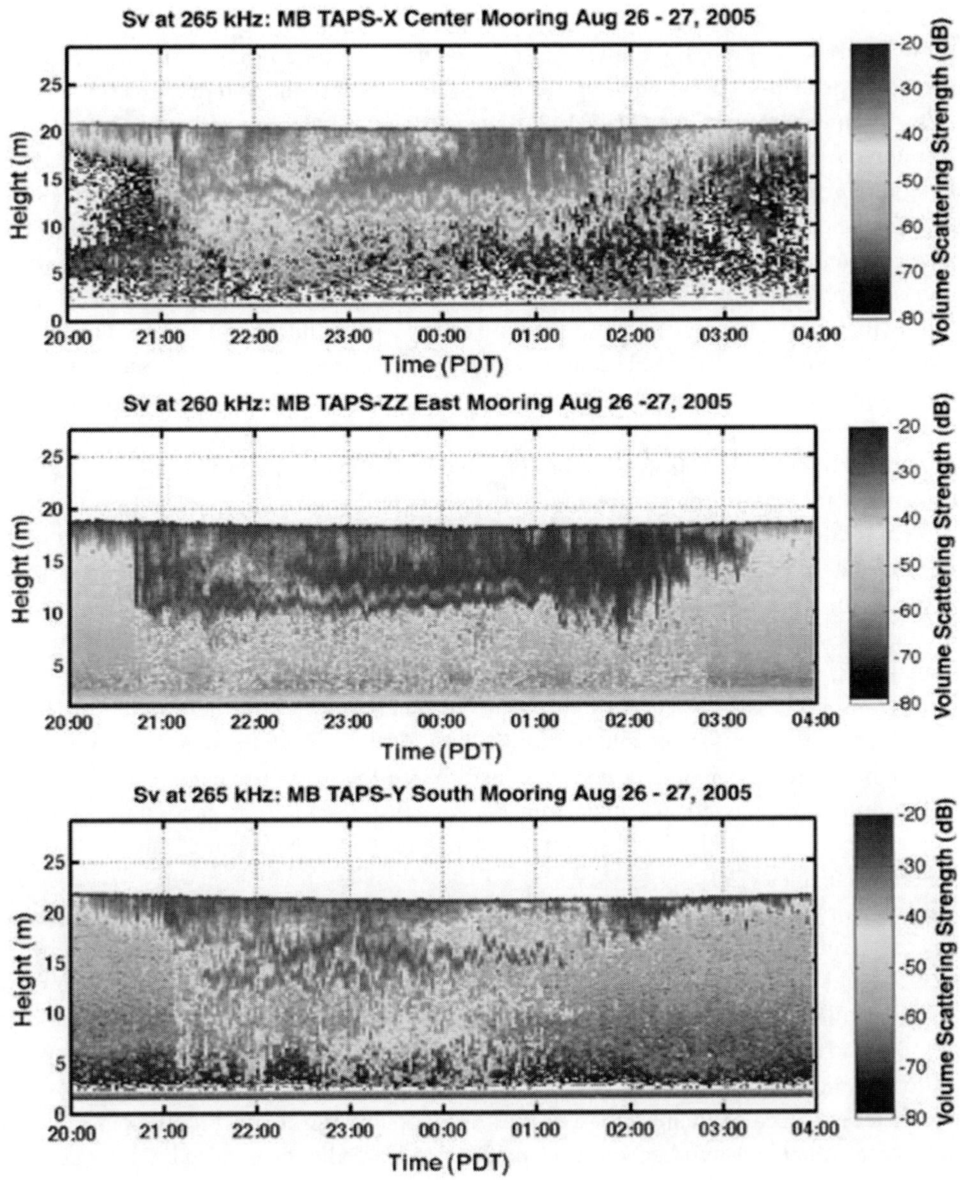

Figure 4. Three volume scattering strength records are displayed, centered on midnight on 26–27 August 2005. These patterns illustrate the similarities and differences in acoustic scattering at three sites in Monterey Bay showing aggregation in layers at night (adapted from [18]).

Aggregations: Physical and Behavioral Drivers

Aggregative behavior in zooplankton is often credited either to physical processes or to behavior [141]. Physical processes such as internal waves, currents, stratification, and turbulence, may affect aggregations, but only at relatively large spatial and temporal scales. However, behavioral interactions that promote aggregations and swarming act at small scales that may not be observed by typical oceanographic or biogeography studies. Behavioral responses that lead to aggregations are attributed to either social interactions, or individual responses to environmental cues [142,143]. These types of behavior differ in that the former is a response to signals/cues that are the product of animals at certain population density; whereas the latter is the result of attraction to a surroundings cue [141,144]. Copepods may exhibit social behavior during mating, but responses to environmental cues, such as fluid mechanical, chemical, and phototactic cues, appear to be the primary driving forces behind copepod swarms [144,145]. Copepod interaction with environmental conditions is directly affected by their ability to perceive and respond to fluid mechanical and chemical information in their surroundings [143,146]. The interaction of these processes at relevant scales affects observed community structure and, therefore, is an important aspect of plankton ecology [1,15,145].

Phytoplankton closely follow distributions of passive scalar quantities, whereas copepods show intense variability even at the smallest scales. On large scales, advective motions of currents can act to transport organisms into general areas [3]. On smaller scales, turbulent motions mix passive particles. Turbulent advection can explain plankton patchiness at large scales (order of kilometers), but variation at small scales is also present [147,148]. The patchiness, described at the inertial scales (scales of Coriolis parameter and Rossby radius) by Abraham [148], is the result of turbulent transport, or behavioral responses [97,149]. The strength of this process is often defined by the turbulence intensity or turbulent diffusivity. Surface winds play an important role in the variation of turbulence through the water column, with the highest turbulent intensity near the surface and lowest in the pycnocline [150]. Fluid density stratification acts to attenuate vertical velocity fluctuations, and thus may aid in creating vertical aggregations of passive particles, such as marine snow, and many phytoplankton species [133].

Fluid density and velocity gradients are more common and easily detectable cues for foraging at larger scales than chemical exudates or phytoplankton presence alone. The presence of fluid density and velocity gradients warrants brief searches for other information, thus focusing the search region, reducing energetic expenditure, and improving individual fitness [15,59,151,152]. Physical gradients are more common than chemical exudate patches, and therefore meet the criteria for detectability at larger scales. Additionally, momentum diffuses faster than scalar quantities (exudates, phytoplankton cells), thus velocity gradients often surround chemical exudate or food patches [59]. Gallager et al. [152] proposed that aggregations at these layers are a result of copepods using this information to narrow search regions. In contrast, waterborne chemical exudates provide more reliable, but possibly less available cues that also induce swarming and area-restricted search behavior [107,110]. Finally, contact cues (mechanical or chemical) with individual prey cells are likely to provide the most reliable, close-range cues that initiate feeding response [111]. This detectability-reliability issue has been thoroughly addressed for specialist insects in terrestrial systems, and suggests that foraging individuals employ a host of cues to improve search success and foraging efficiency [122].

TROPHIC INTERACTIONS AND PRODUCTION

If the ocean were well mixed, copepods would be solely dependent on their sensory capabilities and sensory space, which extends at most a few millimeters from the organism itself. In this case, an individual copepod would need to locate prey items (typically single phytoplankton cells) that would be located several meters apart with no other information to locate them. This situation would be equivalent to human beings being required to locate food that is similar in size to a cookie, and each cookie is located several kilometers apart with no information on where the next cookie might be if one is found. In such a situation, copepods would be largely extinct or rare. As it stands, copepods are the most abundant multi-celled organism on earth, largely due to the unique structure of the marine environment that is driven by flow physics and accentuated by biological processes [2].

Convergence and Aggregation

The importance of physical drivers in developing such structure to the marine environment is clearly portrayed in an example illustrated by McManus and Woodson [2] using an allegorical comparison between the scales described above through which copepods and humans interact with their world. The ocean is structured across multiple scales ranging from basin scale circulation to mesoscale eddies and filaments to fronts and clines to fine-scale distributions of phytoplankton [153,154]. These patterns are the result of flow, fluid dynamic instabilities, and stabilizing processes associated with fluid density discontinuities that are primary subjects of the field of physical oceanography. Here, we want to illustrate that the ocean is not well-mixed and is quite structured even at scales relevant to copepod foraging, and it is this structure that determines the production of copepods in the marine environment and how copepods fill a critical gap between primary and tertiary consumers.

Some of the clearest examples of the role copepods play in marine food webs comes from examining thin layers and fronts (e.g., [6,8,16,140]. The term '*thin layer*' in the context of biological and physical oceanography is more or less synonymous with biophysical coupling and has come to be defined as spatially coherent plankton patches (marine snow, bacteria, phytoplankton, zooplankton, icthyoplankton; [59,155]) in which biomass can be several orders of magnitude greater than the water column immediately above or below the layer (Figure 5). Thin layers range in thickness from a few centimeters to a few meters vertically, span up to square kilometers horizontally, and can persist for days [132,156]. They are a widespread phenomenon occurring in virtually all marine environments where conditions are favorable including fjords, river mouths, continental shelves, and shelf basins [157]. Copepods are known to aggregate in or near thin phytoplankton layers, and aggregation can occur rapidly within a few hours [18,132]. In these regions of vertical aggregations of biological species, the population densities can be enormous (up to several thousand per m^3; [158]). The extremely dense aggregation of biomass in thin layers has profound implications on the health and vitality of marine ecosystems, specifically through its linkage of fine scale (< 1 m) individual behavioral processes and submesoscale (tens to hundreds of meters) processes that underlay the spatiotemporal distribution of marine fishery productivity [8].

As McManus & Woodson [2] point out, fronts are surface outcroppings of vertical fluid density gradients and consequently have similar characteristics to thin layers. Fronts are regions of horizontal gradients in oceanographic properties that have long been known as biological hotspots in the ocean [7,151,159]. Flows in fronts are often convergent with a frontal jet developing along one side of the front [3]. In these regions, phytoplankton are often aggregated and found in densities that can support large populations of copepods similar to those reported for thin layers [160-162].

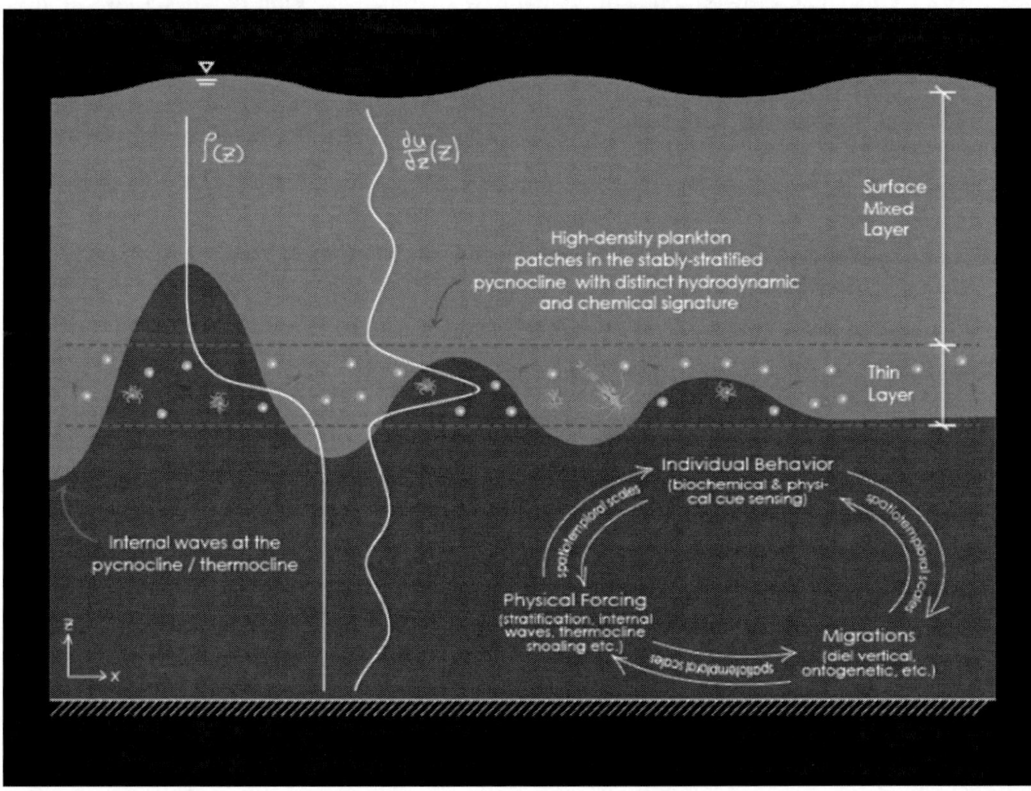

Figure 5. Thin layers are an example of biochemical-physical coupling in high-density plankton patches.

Foraging and Predation in a Turbulent Environment

Turbulence has a multi-faceted effect on copepod feeding and aggregation. At low levels, turbulence can act to increase prey encounter and ingestion rates for copepods simply by increasing the movement of prey items through the feeding current or into the sensory sphere of the copepod [29,147,163,164]. However, at higher levels, turbulence can disrupt feeding by increasing handling rates. The bell-shaped response of feeding by small plankters such as larval fish and copepods to turbulence during feeding also has implications for foraging and aggregation in regions of strong oceanographic gradients such as fronts and clines (Figures 3 and 5; [28]).

In the view of predator-prey interactions for copepods, turbulence acts roughly as a Brownian motion process by which particles are moved around at random and the degree to which this movement occurs is determined by the intensity level of turbulence, often reported as a dissipation rate [165]. At low levels, turbulence has no effect on feeding and ingestion rates among copepods. As turbulence increases due to changes in the flow environment, the contact rate of copepods with prey increases dramatically resulting in a concomitant increase in feeding and ingestion rates. However, as turbulence levels continue to increase, copepods themselves begin to be swept around, handling time for individual prey items increases, and consequently, feeding and ingestion rates begin to decline [28]. The bell-shaped nature of the effects of turbulence on feeding in copepods suggest an optimal feeding window that likely varies by species and may act as a modulator of species coexistence due to spatial variation in turbulence levels.

Moderate levels of turbulence enhance feeding rates for both filter feeding and direct prey capture copepods with stronger effects on the latter [33,166]. In fact, some species of copepods are known to switch dominant feeding mode based on intensity level of turbulence in the environment and can consequently affect the seasonal succession of phytoplankton communities [23,35,167]. This complex interaction between copepod behavior and turbulence provides yet another mechanism for maintaining diversity among copepod species.

Thin layers and fronts are often regions of increased turbulence within the range of enhanced copepod feeding (dissipation rates around 10^{-7} cm^2 s^{-3}) as opposed to wind-affected surface layers where dissipation rates can be detrimental to copepod feeding. Interestingly, in most coastal regions, winds are often minimal overnight when copepods ascend to the surface to feed. The hypothesis that copepods may be cued into diurnal variation in turbulence levels as a cue for reverse (nocturnal) DVM in copepods has not been extensively explored. Near fronts and clines, turbulence is also often elevated due to the shearing of the flow, but is quickly dissipated due to the stratification present [168]. This situation may provide not only an optimal environment for phytoplankton growth since nutrients are mixed, but also optimal conditions for copepod feeding where turbulence levels are typically maintained at intermediary levels [28].

For copepods, the effects of turbulence play a similar role for predators (e.g., [28]). Larval fish ingestion rates also increase with increasing turbulence up until a certain point and then decline [30]. However, the optimal levels of turbulence differ for varying size classes of larval fish. In addition, copepod thresholds for escape are species-specific (Table 1), thus suggesting differential susceptibility to predators based on turbulence levels. Similarly, different predator species are likely to deal with turbulence in varying capacities. For example, a small larval fish would be strongly affected by turbulence, while an anchovy might only be moderately affected, and a filter-feeding whale would not be affected at all. Thus, the spatiotemporal dynamics of turbulence in the environment also will regulate predation pressure. The role of turbulence in structuring pelagic communities is still a critical gap in our knowledge, and an active area of ongoing research.

Predation Pressure

It has long been assumed that predation pressure is extremely high and that predation therefore controls copepod population dynamics because of the extreme behaviors exhibited by many species of copepods associated with predator avoidance such as DVM and escape response. However, aggregations of copepods can lead to intense predation pressure as tertiary consumers are able to more readily exploit them similar to the case for copepod grazing on phytoplankton aggregations. In spite of this predation pressure, copepods continue to aggregate at fronts and clines, and thrive [1]. This suggests that, much like for phytoplankton, predation pressure does not play a major role in the population dynamics of copepods.

Individual fitness, however, is important from an evolutionary standpoint thus assuring that predation pressure is a dominant force in determining copepod behaviors at evolutionary time scales. Overall, the increased predation pressure of the population is outweighed by the increased production due to aggregating in or near regions of high phytoplankton densities. Consequently, copepod behaviors associated with predator avoidance may actually be detrimental to the success of the population as a whole where feeding alone could lead to intense secondary production and consequently fueling higher trophic level production. Regardless, copepod behaviors associated with predator avoidance have received considerable attention in the scientific literature due to the prevalence and easily observed effects of avoidance [19,31,169].

The apparent minimal effect of consumers on copepods at the population level mimics the effects of copepods on their prey where copepod predation can cause changes in the structure of the phytoplankton community, although it rarely severely impacts the overall population dynamics or abundance [29,167]. This suggests that in the ocean, very few systems are fully resource limited, and that the structure of the ocean (organizing the available resources into finite, dense patches) is the primary driver of productivity, at least for secondary consumers such as copepods [1,150,163]. It appears that predation has more effect on the population structure and not persistence or population density in plankton communities largely due to the immense productivity in the ocean.

Higher Trophic Level Production

Recently, studies are pointing towards the importance of prey size structure and distributions, and not prey abundance, as the determining factors in higher trophic level foraging success [170]. Spatial aggregations act to increase prey-predator covariance, a critical parameter for transfer of energy through the food web. At small scales, phytoplankton patchiness is largely governed by flow physics, which can lead to spatial aggregations across multiple temporal and spatial scales. The formation and persistence of phytoplankton and microzooplankton patches is controlled by several parameters, namely dispersion, turbulence, shear, and buoyancy, as reviewed in Durham & Stocker [22]. The size, population density, and longevity of prey patches then affects the ability of the next trophic level, copepods, to exploit the prey [2]. At the level of copepods, swimming and behavior are becoming more important in determining the spatiotemporal characteristics of aggregations. Beyond copepods, most organism aggregation is due to species-specific behaviors that involve foraging, schooling, and mating. Putting all of this together, however, continues to be a difficult task for oceanographers.

Foraging success of copepods can have dramatic effects on phytoplankton community structure [29,167,171]. Changes in the size structure of the phytoplankton community can drive similar changes in the copepod community presenting an interesting feedback loop within the plankton alone where grazing by copepods on particular phytoplankton species changes the prey size structure and thus changes the dominant predator [23,27,35]. The pattern of this feedback loop presents interesting questions regarding the effects of copepods in multi-level, complex food webs found in natural environments and especially on top predator production.

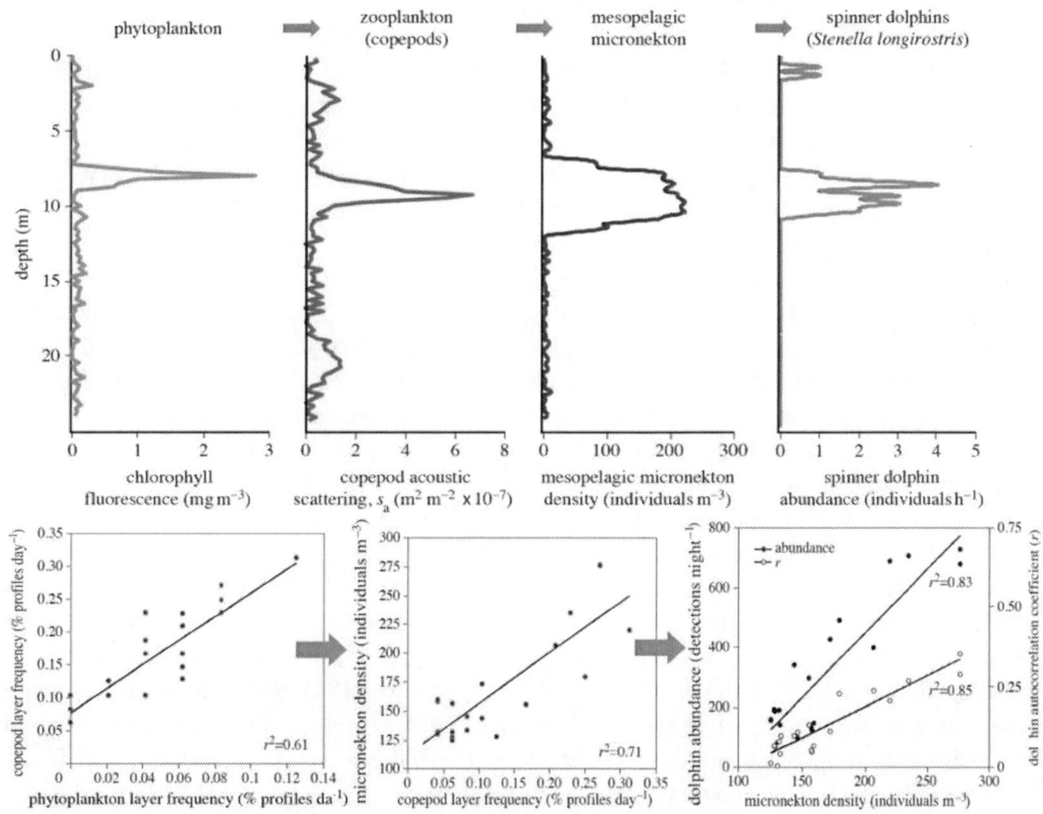

Figure 6. An example of covariation of predators and prey in Hawaii's nearshore pelagic ecosystem measured at 2300 on 4 May 2009 using instruments moored on the 25 m isobaths along with relationships between aggregation characteristics and predator-prey covariation. Adapted from [8].

Following the cookie allegory above, each trophic level requires a certain level of aggregation of its prey resource in order to survive in the ocean (Figure 6). As one moves up the trophic pyramid, the drivers of aggregation shift from physical to behavioral mechanisms [2]. Since copepods are uniquely positioned where their swimming capabilities are just beginning to match the scales of environmental cues in the ocean, they provide not only a critical link between phytoplankton and higher trophic levels, but they also provide the link between physical and biological drivers of production in marine food webs [1]. Regions that have the physical and flow characteristics to promote intense aggregations of phytoplankton at scales that can be exploited by copepods are often regions that are also highly productive across higher trophic levels [7]. This effect has been observed for a wide range of taxa from predatory zooplankters to whales [8,16,172]. Recent observations of copepod-fish-dolphin interactions in the Hawaiian Islands suggest that spatial aggregations lead to strong bottom-up control of the pelagic community consistent with the idea that copepods play a key role in trophic dynamics of these systems [8].

A Case Study: Harmful Algal Blooms, Copepod Ecology, and Ecological Implications

A case study is presented of particular ecological and economic relevance: copepod ecology as it relates to harmful algal bloom (HAB) dynamics. HABs (also often called '*red tides*') are dense blooms of phytoplankton in fresh and marine waters commonly due to dinoflagellates and diatoms that contain toxic compounds. The frequency, intensity, and geographic distribution of HAB events worldwide have been increasing in the past two decades, with associated adverse effects on ecological, human, and economic health [173-176]. In their critical role as trophic mediators, copepods link primary producers and higher order trophic levels; thus they can significantly influence HAB dynamics and modulate large-scale ecological effects through their physiological and behavioral interactions with toxic blooms. Understanding the relevant, coupled biological, chemical, and physical processes at the correct spatiotemporal scales is critical for building accurate early detection and predictive management tools (coupled biophysical Individual-Based Models, e.g., [177,178]).

Many chemical compounds (e.g., domoic acid, paralytic shellfish poison, dimethyl sulfide; see e.g., [179]) associated with HAB species are known to have deleterious effects on copepod fitness [180] not only because of toxicity-related responses but also from typically low nutritional values. Among studies of copepod grazers fed harmful algae, decreased grazing and fecundity are the most common results, although the causes of decreased grazing (physiological incapacitation, behavioral avoidance or lack of stimulation) and/or decreased fecundity (toxic versus nutritional effect) vary among studies [181]. Prince et al. [182] looked at the effects of toxic *Karenia brevis* on fitness for the copepod *A. tonsa*, and found that on diets rich in *K. brevis*, copepods experienced decreased survivorship and decreased egg production per female, but the percentage of eggs that hatched was unaffected. Similarly, egg production rates of *A. tonsa* fed toxic *K. brevis* strains were similar to those of starved copepods, whereas those of copepods fed non-toxic strains along with beneficial *Rhodomonas salina* were significantly higher [183]. Jiang et al. [180] found an interesting physiological response with evolutionary implications in the copepod *A. tonsa* responding to the harmful dinoflagellate *Cochlodinium polykrikoides:* a rapid gain and loss of evolutionary toxin resistance as a population. After four generations of chronic exposure to toxic phytoplankton, copepods had evolved toxin resistances three times greater than pre-exposure generations; within two generations lacking exposure, all increased toxin resistance was lost. This suggests that copepods can rapidly adapt to dissolved toxic compounds and likely play a critical role in implementing some form of top-down grazer control on toxic phytoplankton blooms. These findings also highlight how algal toxins may influence copepod feeding behavior and alter top-down control exerted by copepod grazers.

The idea of top-down grazer control (or the modification of it) is consistent across the currently available literature. Hong et al. [184] used digital holographic cinematography to compare feeding behavior of free-swimming copepods, *A. tonsa*, on nutritional prey (*Storeatula major*) to that occurring during exposure to toxic and non-toxic strains of *K. brevis* and *Karlodinium veneficum.* They found two different beating modes of the copepod's feeding appendages–a ''sampling beating'' of short durations and a longer duration ''grazing beating'' that generates feeding currents as well as a variety of behavioral regimes corresponding to the presence, absence, or combination of beneficial and harmful algal cells. This modification of top-down grazer control was also seen by Waggett et al. [183] whose data indicate that *K. brevis* influences copepod grazer populations via multiple synergistic mechanisms: (i) decreased ingestion rates, (ii) decreased egg production, and (iii) increased mortality of copepods through a combination of toxicity and nutritional inadequacy. In grazing and mortality experiments, Cohen et al. [185] found that *A. tonsa* exhibited minimal sublethal behavioral responses. However, there were significant effects on the swimming and photobehavior of *Temora turbinata* and *C. typicus* at the lowest sublethal concentrations tested (10^5 *K. brevis* cells l^{-1}), suggesting very low behavioral response thresholds. Their data suggest that sublethal effects of *K. brevis* and its brevetoxins on copepod behavior occur on a species-specific basis. Similarly, Schultz & Kiørboe [186] found that *A. tonsa* was capable of incredibly selective feeding behaviors by remotely identifying and rejecting toxic cells based on associated hydrodynamic or chemical signatures. In contrast, Leandro et al. [187] found that a lack of remote prey selection lead to the copepod *C. finmarchicus* becoming a preferential trophic vector for toxin accumulation, likely due to a lack of ability to remotely characterize toxic cells. Species-specific selective feeding behavior (or lack thereof) through active, remote prey selection can enhance or dampen bloom growth dynamics thus modulating ecological impacts on other trophic levels [186,188].

During toxic bloom conditions, anti-grazing deterrents act to lower predation rates by copepod grazers and sustain the bloom [189-192]. Teegarden et al. [193] showed that even though sloppy copepod grazing was effective at dispersing toxins to low concentration levels by aiding environmental advection and turbulent diffusion, toxins were still seen to bio-accumulate in copepods, acting as a preferential toxin accumulation vector. Extrapolating to higher trophic levels, Gannon et al. [194] examined the effect of HABs on nearshore fish communities through reduced fish abundance and health as well as changes in community structure. The effects are likely due to neurotoxic influences that cause zooplankton to act erratically, and thus become more conspicuous prey items and preferential vectors for trophic transfer of dissolved toxic compounds. These factors, on a species-specific basis, can act in concert to produce preferential vectors by which toxins accumulate at higher trophic levels.

Changes in copepod feeding and swimming behavior caused by toxic compounds can significantly influence predator, prey, and mate encounter rates by altering the fractality ("diffuseness" or "volume-fillingness", e.g., [195]) of a copepod's trajectory. Erratic, hydrodynamically conspicuous, diffuse swimming behavior (i.e., paths with high fractal dimensions) and inconspicuous, ballistic, steady swimming behavior (i.e., low path fractal dimensions) each have their ecological place for copepods; however, too much of one or the other can have deleterious effects on copepod life success (e.g., by decreasing mate or prey encounters or increasing conspicuity to predators). Changes in individual copepod behaviors are quickly extrapolated to population scale phenomena with important implications for HAB dynamics (intensity/toxicity, duration, trophic transfers, etc.) and overall ecological effects.

Finally, HABs are intricately linked to fronts and clines and thus to copepod ecology; often dissolved toxic compounds should be considered as another biochemical factor in the trophic and biogeochemical dynamics of fronts [196,197] and clines [198-201]. There are numerous biological (phyto-, zoo-, and icthyo-plankton species, algal growth dynamics, zooplankton grazing), chemical (toxic compound, toxicity levels), and physical (hydrography, advection, turbulent and molecular diffusion) factors coupled over wide spatiotemporal scales that affect HAB dynamics and determine overall ecological impacts. These factors interact in a complicated manner, but it is clear that ecological effects and trophic implications are highly species-specific and dependent on toxicity level, mode of action (pathway of toxicity), morphology, and monospecificity of a particular toxic algal species and its associated toxic compounds. In addition, species-specific copepod physiological and behavioral responses (e.g., fitness, foraging, grazing, sampling, mating) as well as local physics (turbulence, stratification, surface waves, internal waves, tides, etc.) play significant roles influencing the duration, density, toxicity, and overall ecological impacts of a particular HAB event.

CONCLUSION

As the largest collective of multicellular organisms on earth and direct consumers of primary producers, copepods are an integral component of the marine ecosystem and are a critical aspect of the spatiotemporal distribution of productivity in the world's oceans. The extent that gradients in oceanographic properties affect populations and communities is still not fully understood, but recent findings suggest that from the perspective of a copepod, the ocean is a structured and patterned habitat at scales much smaller than historically accepted [59,151,202].

Future Directions

The influence of individual behavior on pattern in pelagic habitats warrants further investigation at the fine-to-intermediate scales (mm to m). Emphasis on better resolution in field measurements combined with continued laboratory investigations is imperative to developing our understanding of copepod behavior and ecology at appropriate scales. A continued drive toward understanding and answering some of these large-scale issues will benefit from collaborations between field studies and experimental investigations where specific aspects can be accurately isolated and controlled. As we become more aware of the importance of fine-scale structure in the ocean, it is necessary to begin to test the impact of these features on larger scale ecological processes. This oceanographic structure leads to distinct behavioral responses by copepods that can have immense effects on aggregations and large-scale distributions. Such aggregations provide a critical link to higher trophic level production and interactions.

Copepod behavior remains an important topic in oceanographic research due to its importance in determining predator-prey interactions at the base of an immensely productive food web that is the primary source of sustenance for a large portion of the world's human population. Recent evidence suggesting that the immense role of structure and pattern in the marine environment (as a driver of production that determines trophic pathways) critically hinges on the ability of copepods to find and locate prey. We have identified through this chapter four areas that are important to furthering our understanding of the direct and indirect effects of copepod behavior on pelagic ecosystem dynamics. First, developing an understanding of how copepods may employ horizontal gradients in turbidity (light intensity) for position maintenance in regions where horizontal gradients are of appropriate scale (estuaries, coastal ocean fronts). Second, elucidation of the proximate cues for DVM as it relates to fine-scale foraging behavior. Third, the role of turbulence in structuring pelagic communities. Finally, the natural variability of predator-prey covariance affects the role of copepods in complex food webs. Many of these issues will require novel combinations of modeling and *in situ* observations. New emergent models of phytoplankton communities provide an important step in this direction [27,35,203]. Expanding these models to higher trophic level interactions should provide unique insights into zooplankton community dynamics. Although direct observation of copepod behavior in the field is still a difficult problem, some recent advances in imaging systems provide a ray of hope for this issue (Figure 7; [36,38,126]) that should allow testing and parameterization of large-scale ecosystem models in the near future.

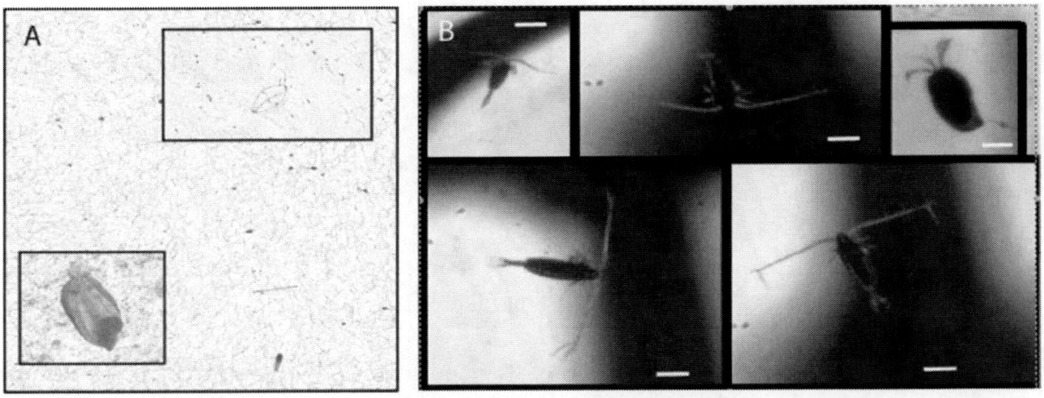

Figure 7. Images from (A) ISIIS (In Situ Ichthyoplankton Imaging System; courtesy of R.K. Cowen and A.T. Greer; RSMAS), and (B) ZOOPS (ZOOPlankton Sonar; courtesy of J. Jaffe and C. Briseno; Scripps), new technologies that are allowing quantitative estimation of predator-prey dynamics in copepod ecology. Scales in (B) correspond to 1 mm.

ACKNOWLEDGMENTS

The authors would like to thank R.K. Cowen (Oregon State Hatfield Marine Station), A.T. Greer (RSMAS, University of Miami), J. Jaffe (UCSD, Scripps), and C. Briseno (UCSD, Scripps) for providing ISIIS and ZOOPS images in Figure 7. C. B. Woodson was partially supported by NSF award # 1212124 during this work. D. R. Webster and A. C. True were partially supported by NSF award OCE-0728238.

References

[1] Kiørboe, T. (2011). What makes pelagic copepods so successful? *Journal of Plankton Research*, *33*, 677-685.

[2] McManus, M. & Woodson, C. B. (2012). Plankton distribution and ocean dispersal. *Journal of Experimental Biology*, *215*, 1008-1016.

[3] Mann, K. H. & Lazier, J. R. N. (1996). *Dynamics of Marine Ecosystems*. Blackwell Science Cambridge, Massachusetts.

[4] Mauchline, J. (1998). *The Biology of Calanoid Copepods*. Academic Press, San Diego.

[5] Bollens, S. M. & Frost, B. W. (1989). Zooplanktivorous fish and variable diel vertical migration in the marine planktonic copepod *Calanus pacificus*. *Limnology and Oceanography*, *34*, 1072-1083.

[6] Kann, L. M. & Wishner, K. (1995). Spatial and temporal patterns of zooplankton on baleen whale feeding grounds in the southern Gulf of Maine. *Journal of Plankton Research*, *17*, 235-262.

[7] Genin, A. (2004). Bio-physical coupling in the formation of zooplankton and fish aggregations over abrupt topographies. *Journal of Marine Systems*, *50*, 3-20.

[8] Benoit-Bird, K. J. & McManus, M. A. (2012). Bottom-up regulation of a pelagic community through spatial aggregations. *Biology Letters*, *8*, 813-816.

[9] Clarke, G. L. (1933). Diurnal migration of plankton in the Gulf of Maine and its correlation with changes in submarine irradiation. *Biological Bulletin*, *65*, 402-436.

[10] Clarke, G. L. (1934). Further observations on the diurnal migration of copepods in the Gulf of Maine. *Biological Bulletin*, *67*, 432–455.

[11] Johnson, W. H. & Raymont, J. E. G. (1939). The reactions of the planktonic copepod, Centropages typicus, to light and gravity. *Biological Bulletin*, *77*, 200-215.

[12] Cushing, D. (1951). The vertical migration of planktonic Crustacea. *Biological Reviews*, *26*, 158-192.

[13] Viitasalo, M., Kiørboe, T., Flinkman, J. et al. (1998). Predation vulnerability of planktonic copepods: consequences of predator foraging strategies and prey sensory abilities. *Marine Ecology Progress Series*, *175*, 129-142.

[14] Kiørboe, T. & Visser, A. (1999). Predator and prey perception in copepods due to hydromechanical signals. *Marine Ecology Progress Series*, *179*, 81-95.

[15] Woodson, C. B., Webster, D. R., Weissburg, M. J. & Yen, J. (2007). The prevalence and implications of copepod behavioral responses to oceanographic gradients and biological patchiness. *Integrative and Comparative Biology*, *47*, 831-846.

[16] Wishner, K., Durbin, E., Durbin, A. et al. (1988). Copepod patches and right whales in the Great South Channel off New England. *Bulletin of Marine Science*, *43*, 825-844.

[17] Cowles, T. J., Desiderio, R. A. & Neuer, S. (1993). *In situ* characterization of phytoplankton from vertical profiles of fluorescence emission spectra. *Marine Biology*, *115*, 217-222.

[18] Holliday, D. V., Greenlaw, C. F. & Donaghay, P. L. (2010). Acoustic scattering in the coastal ocean at Monterey Bay, CA, USA: Fine-scale vertical structures. *Continental Shelf Research*, *30*, 81-103.

[19] Fields, D. M. & Yen, J. (1997). The escape behavior of marine copepods in response to a quantifiable fluid mechanical disturbance. *Journal of Plankton Research*, *19*, 1289-1304.

[20] Buskey, E. J. (1998). Components of mating behavior in planktonic copepods. *Journal of Marine Systems*, *15*, 13-21.

[21] Clay, T. W., Bollens, S. M., Bochdansky, A. B. & Ignoffo, T. R. (2004). The effects of thin layers on the vertical distribution of larval Pacific herring, *Clupea pallasi*. *Journal of Experimental Marine Biology and Ecology*, *305*, 171-189.

[22] Durham, W. M. & Stocker, R. (2012). Thin phytoplankton layers: characteristics, mechanisms, and consequences. *Annual Review of Marine Science*, *4*, 177-207.

[23] Mariani, P., Andersen, K. H., Visser, A. W. et al. (2013). Control of plankton seasonal succession by adaptive grazing. *Limnology and Oceanography*, *58*, 173-184.

[24] Seuront, L., Schmitt, F. & Lagadeuc, Y. (2001). Turbulence intermittency, small-scale phytoplankton patchiness and encounter rates in plankton: where do we go from here? *Deep Sea Research I*, *48*, 1199-1215.

[25] Schmitt, F. G., Seuront, L., Hwang, J. S. et al. (2006). Scaling of swimming sequences in copepod behavior: data analysis and simulation. *Physica A*, *364*, 287-296.

[26] Murphy, D. W., Webster, D. R. & Yen, J. (2012). A high-speed tomographic PIV system for measuring zooplanktonic flow. *Limnology and Oceanography: Methods*, *10*, 1096-1112.

[27] Banas, N. S. (2011). Adding complex trophic interactions to a size-spectral plankton model: Emergent diversity patterns and limits on predictability. *Ecological Modeling*, *222*, 2663-2675.

[28] Rothschild, B. J. & Osborn, T. R. (1988). Small-scale turbulence and plankton contact rates. *Journal of Plankton Research*, *10*, 465-474.

[29] Kiørboe, T. (1993). Turbulence, phytoplankton cell size, and the structure of pelagic food webs. *Advances in Marine Biology*, *29*, 1-72.

[30] Mahjoub, M. S., Kumar, R., Souissi, S. et al. (2012). Turbulence effects on the feeding dynamics in European sea bass (*Dicentrarchus labrax*) larvae. *Journal of Experimental Marine Biology and Ecology*, *416*, 61-67.

[31] Titelman, J. (2001). Swimming and escape behavior of copepod nauplii: implications for predator-prey interactions among copepods. *Marine Ecology Progress Series*, *213*, 203-213.

[32] Fields, D. M. & Yen, J. (2002). Fluid mechanosensory stimulation of behaviour from a planktonic marine copepod, *Euchaeta rimana* Bradford. *Journal of Plankton Research*, *24*, 747-755.

[33] Kiørboe, T. & Jiang, H. (2013). To eat and not be eaten: optimal foraging behaviour in suspension feeding copepods. *Journal of the Royal Society Interface*, *10*, 20120693.

[34] Schmitt, F. G. & Seuront, L. (2001). Multifractal random walk in copepod behavior. Physica A, *301*, 375-396.

[35] Mariani, P. & Visser, A. W. (2010). Optimization and emergence in marine ecosystem models. Progress in Oceanography, *84*, 89-92.

[36] Davis, C. S., Gallager, S. M. & Solow, A. R. (1992). Microaggregations of oceanic plankton observed by towed video microscopy. *Science*, *257*, 230-232.

[37] Genin, A., Jaffe, J. S., Reef, R. et al. (2005). Swimming against the flow: a mechanism of zooplankton aggregation. *Science*, *308*, 860-862.

[38] Cowen, R. K., Greer, A. T., Guigand, C. M. et al. (2013). Evaluation of the In Situ Ichthyoplankton Imaging System (ISIIS): comparison with the traditional (bongo net) sampler. Fishery Bulletin, *111*, 1-12.

[39] True, A. C. (2011). *A Mechanism of Patchiness: Zooplankton Behavior in Finescale Vertical Shear Layers*. *MSc Thesis*, Georgia Institute of Technology.

[40] Castel, J. & Veiga, J. (1990). Distribution and retention of the copepod *Eurytemora affinis* hirundoides in a turbid estuary. *Marine Biology*, *107*, 119-128.

[41] Cohen, J. H. & Forward, R. B. (2009). Zooplankton diel vertical migration - A review of proximate control. *In*: Gibson RN, Atkinson RJA & Gordon JDM (eds.), *Oceanography and Marine Biology: An Annual Review*. *Taylor and Francis*, Vol. *47*, 77-110.

[42] Fields, D. M., Shema, S. D., Browman, H. I. et al. (2012). Light primes the escape response of the calanoid copepod, *Calanus finmarchicus*. *PloS One*, *7*, e39594.

[43] Stearns, D. E. & Forward, R. B. (1984). Copepod photobehavior in a simulated natural light environment and its relation to nocturnal vertical migration. *Marine Biology*, *82*, 91-100.

[44] Marcus, N. H. (1986). Population dynamics of marine copepods: the importance of photoperiodism. *American Zoologist*, *26*, 469-477.

[45] Forward, R. B. (1988). Diel vertical migration: Zooplankton photobiology and behavior. *Oceanography and Marine Biology: An Annual Review*, *26*, 361-393.

[46] Lampert, W. (1989). The adaptive significance of diel vertical migration of zooplankton. *Functional Ecology*, *3*, 21-27.

[47] Lampert, W. (1993). Ultimate causes of diel vertical migration of zooplankton: new evidence for the predator-avoidance hypothesis. *Arch Für Hydrobiol Ergeb Limnologie*, *339*, 79-88.

[48] Buskey, E. J., Baker, K. S., Smith, R. C. & Swift, E. (1989). Photosensitivity of the oceanic copepods *Pleuromamma gracilis* and *Pleuromamma xiphias* and its relationship to light penetration and daytime depth distribution. *Marine Ecology Progress Series*, *55*, 207-216.

[49] Cohen, J. H. & Forward, R. B. (2005). Diel vertical migration of the marine copepod *Calanopia americana*. I. Twilight DVM and its relationship to the diel light cycle. *Marine Biology*, *147*, 387-398.

[50] Cohen, J. H. & Forward, R. B. (2005). Diel vertical migration of the marine copepod *Calanopia americana*. II. Proximate role of exogenous light cues and endogenous rhythms. *Marine Biology*, *147*, 399-410.

[51] Stearns, D. E. & Forward, R. B. (1984). Photosensitivity of the calanoid copepod *Acartia tonsa*. *Marine Biology*, *82*, 85-89.

[52] Williamson, C. E., Fischer, J. M., Bollens, S. M. et al. (2011). Toward a more comprehensive theory of zooplankton diel vertical migration: integrating ultraviolet radiation and water transparency into the biotic paradigm. *Limnology and Oceanography*, *56*, 1603-1623.

[53] Cohen, J. H. & Forward, R. B. (2002). Spectral sensitivity of vertically migrating marine copepods. *Biological Bulletin*, *203*, 307-314.

[54] Bollens, S. M. & Frost, B. W. (1990). UV light and vertical distribution of the marine planktonic copepod *Acartia hudsonica* Pinhey. *Journal of Experimental Marine Biology and Ecology*, *137*, 89-93.

[55] Buskey, E. J., Peterson, J. O. & Ambler, J. W. (1995). The role of photoreception in the swarming behavior of the copepod *Dioithona oculata*. *Marine and Freshwater Behaviour and Physiology*, *26*, 273-285.

[56] Buskey, E. J. (1998). Energetic costs of swarming behavior for the copepod *Dioithona oculata*. *Marine Biology*, *130*, 425-431.

[57] Roman, M. R., Holliday, D. V. & Sanford, L. P. (2001). Temporal and spatial patterns of zooplankton in the Chesapeake Bay turbidity maximum. *Marine Ecology Progress Series*, *213*, 215-227.

[58] Lloyd, S. S., Elliott, D. T. & Roman, M. R. (2013). Egg production by the copepod, *Eurytemora affinis*, in Chesapeake Bay turbidity maximum regions. *Journal of Plankton Research*, *35*, 299-308.

[59] Cowles, T. J. (2004). Planktonic layers: Physical and biological interactions on the small scale. *In*: Seuront L & Strutton PG (eds.), *Handbook of Scaling Methods in Aquatic Ecology*. CRC Press, Boca Raton, 31-50.

[60] Woodson, C. B., Webster, D. R., Weissburg, M. J. & Yen, J. (2007). Cue hierarchy and foraging in calanoid copepods: ecological implications of oceanographic structure. *Marine Ecology Progress Series*, *330*, 163-177.

[61] Woodson, C. B., Webster, D. R., Weissburg, M. J. & Yen, J. (2005). Response of copepods to physical gradients associated with structure in the ocean. Limnology and Oceanography, *50*, 1552-1564.

[62] Lance, J. (1962). Effects of water of reduced salinity on the vertical migration of zooplankton. *Journal of the Marine Biological Association of the United Kingdom*, *42*, 131-154.

[63] Tiselius, P., Nielsen, G. & Nielsen, T. G. (1994). Microscale patchiness of plankton within a sharp pycnocline. *Journal of Plankton Research*, *16*, 543-554.

[64] Harder, W. (1968). Reactions of plankton organisms to water stratification. *Limnology and Oceanography*, *13*, 156-168.

[65] Lougee, L. A., Bollens, S. M. & Avent, S. R. (2002). The effects of haloclines on the vertical distribution and migration of zooplankton. *Journal of Experimental Marine Biology and Ecology*, *278*, 111-134.

[66] Hough, A. R. & Naylor, E. (1991). Field studies on retention of the planktonic copepod *Eurytemora affinis* in a mixed estuary. *Marine Ecology Progress Series*, *76*, 115-122.

[67] Kimmerer, W. J., Burau, J. R. & Bennett, W. A. (1998). Tidally oriented vertical migration and position maintenance of zooplankton in a temperate estuary. *Limnology and Oceanography*, *43*, 1697-1709.

[68] Seuront, L. (2006). Effect of salinity on the swimming behaviour of the estuarine calanoid copepod *Eurytemora affinis*. *Journal of Plankton Research*, *28*, 805-813.

[69] Michalec, F.-G., Holzner, M., Hwang, J. S. & Souissi, S. (2012). Three dimensional observation of salinity-induced changes in the swimming behavior of the estuarine calanoid copepod *Pseudodiaptomus annandalei*. *Journal of Experimental Marine Biology and Ecology*, *438*, 24-31.

[70] Moison, M., Schmitt, F. G. & Souissi, S. (2012). Effect of temperature on *Temora longicornis* swimming behaviour: illustration of seasonal effects in a temperate ecosystem. *Aquatic Biology*, *16*, 149-162.

[71] Bradley, B. P. (1975). The anomalous influence of salinity on temperature tolerances of summer and winter populations of the copepod *Eurytemora affinis*. *Biological Bulletin*, *148*, 26-34.

[72] Heuch, P. A. (1995). Experimental evidence for aggregation of salmon louse copepodids (*Lepeophtheirus salmonis*) in step salinity gradients. *Journal of the Marine Biological Association of the United Kingdom*, *75*, 927-940.

[73] Bochdansky, A. B. & Bollens, S. M. (2004). Relevant scales in zooplankton ecology: distribution, feeding, and reproduction of the copepod *Acartia hudsonica* in response to thin layers of the diatom *Skeletonema costatum*. *Limnology and Oceanography*, *49*, 625-636.

[74] Kundu, P. K. & Cohen, I. M. (2004). *Fluid Mechanics*. Elsevier, New York.

[75] Holzman, R. & Wainwright, P. C. (2009). How to surprise a copepod: Strike kinematics reduce hydrodynamic disturbance and increase stealth of suction-feeding fish. *Limnology and Oceanography*, *54*, 2201-2121.

[76] Strickler, J. R. & Balázsi, G. (2007). Planktonic copepods reacting selectively to hydrodynamic disturbances. *Philosophical Transactions of the Royal Society B: Biological Sciences*, *362*, 1947-1958.

[77] Yen, J. & Strickler, J. R. (1996). Advertisement and concealment in the plankton: what makes a copepod hydrodynamically conspicuous? *Invertebrate Biology*, *115*, 191-205.

[78] Moore, P. A., Fields, D. M. & Yen, J. (1999). Physical constraints of chemoreception in foraging copepods. *Limnology and Oceanography*, *44*, 166-177.

[79] Lenz, P. H. & Yen, J. (1993). Distal setal mechanoreceptors of the first antennae of marine copepods. *Bulletin of Marine Science*, *53*, 170-179.

[80] Fields, D. M. & Weissburg, M. J. (2004). Rapid firing rates from mechanosensory neurons in copepod antennules. *Journal of Comparative Physiology A*, *190*, 877-882.

[81] Yen, J., Lenz, P. H., Gassie, D. V. & Hartline, D. K. (1992). Mechanoreception in marine copepods: electrophysiological studies on the first antennae. *Journal of Plankton Research*, *14*, 495-512.

[82] Fields, D. M., Shaeffer, D. S. & Weissburg, M. J. (2002). Mechanical and neural responses from the mechanosensory hairs on the antennule of *Gaussia princeps*. *Marine Ecology Progress Series*, *227*, 173–186.

[83] Fields, D. M. & Weissburg, M. J. (2005). Evolutionary and ecological significance of mechanosenor morphology: copepods as a model system. *Marine Ecology Progress Series*, *287*, 263-307.

[84] Uttieri, M., Brown, E. R., Boxshall, G. A., Mazzocchi, M. G. (2008). Morphology of antennular sensors in *Clausocalanus furcatus* (Copepoda: Calanoida). *Journal of the Marine Biological Association of the United Kingdom*, *88*, 535-541.

[85] Yen, J. & Fields, D. M. (1992). Escape responses of *Acartia hudsonica* (Copepoda) nauplii from the flow field of *Temora longicornis* (Copepoda). *Archiv für Hydrobiologie Beih Ergebn Limnologie*, *36*, 123-134.

[86] Visser, A. (2001). Hydromechanical signals in the plankton. *Marine Ecology Progress Series*, *222*, 1-24.

[87] Kiørboe, T., Saiz, E. & Visser, A. (1999). Hydrodynamic signal perception in the copepod *Acartia tonsa*. *Marine Ecology Progress Series*, *179*, 97-111.

[88] Fields, D. M. & Yen, J. (1996). The escape behavior of *Pleuromamma xiphias* in response to a quantifiable fluid mechanical disturbance. *In*: Lenz PH, Hartline DK,

Purcell JE & Macmillan DL (eds.), *Zooplankton: Sensory Ecology and Physiology*. Gordon and Breach, Amsterdam, 323-340.

[89] Wong, C. K. (1996). Response of copepods to hydromechanical stimuli. *Crustaceana*, *69*, 853-859.

[90] Burdick, D. S., Hartline, D. K. & Lenz, P. H. (2007). Escape strategies in co-occurring calanoid copepods. *Limnology and Oceanography*, *52*, 2373-2385.

[91] Fields, D. M. (2010). Orientation affects the sensitivity of *Acartia tonsa* to fluid mechanical signals. *Marine Biology*, *157*, 505-514.

[92] Paffenhöfer, G. A. (1998). On the relation of structure, perception and activity in marine planktonic copepods. *Journal of Marine Systems*, *15*, 457-473.

[93] Jiang, H. & Paffenhöfer, G. A. (2008). Hydrodynamic signal perception by the copepod *Oithona plumifera*. *Marine Ecology Progress Series*, *373*, 37-52.

[94] Kjellerup, S. & Kiørboe, T. (2012). Prey detection in a cruising copepod. *Biology Letters*, *8*, 438-441.

[95] Jiang, H. & Osborn, T. R. (2004). Hydrodynamics of copepods: a review. *Surveys in Geophysics*, *25*, 339-370.

[96] Bagøien, E. & Kiørboe, T. (2005). Blind dating—mate finding in planktonic copepods. I. Tracking the pheromone trail of *Centropages typicus*. *Marine Ecology Progress Series*, *300*, 105-115.

[97] Mackas, D. L., Sefton, H., Miller, C. B. & Raich, A. (1993). Vertical habitat partitioning by large calanoid copepods in the oceanic subarctic Pacific during spring. *Progress in Oceanography*, *32*, 259-294.

[98] Franks, P. J. (2001). Turbulence avoidance: An alternate explanation of turbulence-enhanced ingestion rates in the field. *Limnology and Oceanography*, *46*, 959-963.

[99] Incze, L. S., Hebert, D., Wolff, N. et al. (2001). Changes in copepod distributions associated with increased turbulence from wind stress. *Marine Ecology Progress Series*, *213*, 229-240.

[100] Webster, D. R., Brathwaite, A. & Yen, J. (2004). A novel laboratory apparatus for simulating isotropic oceanic turbulence at low Reynolds number. *Limnology and Oceanography: Methods*, *2*, 1-12.

[101] Yen, J., Rasberry, K. D. & Webster, D. R. (2008). Quantifying copepod kinematics in a laboratory turbulence apparatus. *Journal of Marine Systems*, *69*, 283-294.

[102] Atema, J., Fay, R. R., Popper, A. N. & Tavolga, W. N. (1988). *Sensory Biology of Aquatic Animals*. Springer-Verlag, New York.

[103] Dusenberry, D. B. (1992). *Sensory Ecology: How Organisms Acquire and Respond to Information*. Freeman, New York.

[104] Yen, J., Weissburg, M. J. & Doall, M. H. (1998). The fluid physics of signal perception by mate-tracking copepods. *Philosophical Transactions of the Royal Society of London B: Biological Sciences*, *353*, 787-804.

[105] Zimmer, R. K. & Butman, C. A. (2000). Chemical signaling processes in the marine environment. *Biological Bulletin*, *198*, 168-187.

[106] Webster, D. R. & Weissburg, M. J. (2009). The hydrodynamics of chemical cues among aquatic organisms. *Annual Review of Fluid Mechanics*, *41*, 73-90.

[107] Poulet, S. A. a Marsot, P. (1978). Chemosensory grazing by marine calanoid copepods (Arthropoda: Crustacea). *Science*, *200*, 1403-1405.

[108] Jackson, G. A. & Kiørboe, T. (2004). Zooplankton use of chemodetection to find and eat particles. *Marine Ecology Progress Series*, *269*, 153-162.

[109] Lombard, F., Koski, M. & Kiørboe, T. (2013). Copepods use chemical trails to find sinking marine snow aggregates. *Limnology and Oceanography*, *58*, 185-192.

[110] Poulet, S. A. & Ouellet, G. (1982). The role of amino acids in the chemosensory swarming and feeding of marine copepods. *Journal of Plankton Research*, *4*, 341-361.

[111] Tiselius, P. (1992). Behavior of *Acartia tonsa* in patchy food environments. *Limnology and Oceanography*, *37*, 1640-1651.

[112] Kiørboe, T. & Bagøien, E. (2005). Motility patterns and mate encounter rates in planktonic copepods. *Limnology and Oceanography*, *50*, 1999-2007.

[113] Heuschele, J. & Kiørboe, T. (2012). The smell of virgins: mating status of females affects male swimming behaviour in *Oithona davisae*. *Journal of Plankton Research*, *34*, 929-935.

[114] Yen, J. & Lasley, R. (2010). Chemical communication between copepods: Finding the mate in a fluid environment. *In*: Breithaupt T & Thiel M (eds.), *Chemical Communication in Crustaceans*. Springer-Verlag, New York, 177-197.

[115] Doall, M. H., Colin, S. P., Strickler, J. R. & Yen, J. (1998). Locating a mate in 3D: the case of *Temora longicornis*. *Philosophical Transactions of the Royal Society of London B: Biological Sciences*, *353*, 681-689.

[116] Weissburg, M. J., Doall, M. H. & Yen, J. (1998). Following the invisible trail: kinematic analysis of mate-tracking in the copepod *Temora longicornis*. *Philosophical Transactions of the Royal Society of London B: Biological Sciences*, *353*, 701-712.

[117] Yen, J., Sehn, J. K., Catton, K. et al. (2011). Pheromone trail following in three dimensions by the freshwater copepod *Hesperodiaptomus shoshone*. *Journal of Plankton Research*, *33*, 907–916.

[118] Kiørboe, T., Bagøien, E. & Thygesen, U. H. (2005). Blind dating—mate finding in planktonic copepods. II. The pheromone cloud of *Pseudocalanus elongatus*. *Marine Ecology Progress Series*, *300*, 117-128.

[119] Kiørboe, T. (2007). Mate finding, mating, and population dynamics in a planktonic copepod *Oithona davisae*: There are too few males. *Limnology and Oceanography*, *52*, 1511-1522.

[120] Steinke, M., Stefels, J. & Stamhuis, E. (2006). Dimethyl sulfide triggers search behavior in copepods. *Limnology and Oceanography*, *51*, 1925-1930.

[121] Seuront, L. (2010). Zooplankton avoidance behaviour as a response to point sources of hydrocarbon-contaminated water. *Marine and Freshwater Research*, *61*, 263-270.

[122] Vet, L. E. (1999). From chemical to population ecology: infochemical use in an evolutionary context. *Journal of Chemical Ecology*, *25*, 31-49.

[123] Mullin, M. M. & Brooks, E. R. (1976). Some consequences of distributional heterogeneity of phytoplankton and zooplankton. *Limnology and Oceanography*, *21*, 784-796.

[124] Daro, M. H. (1988). Migratory and grazing behavior of copepods and vertical distribution of phytoplankton. *Bulletin of Marine Science*, *43*, 710-729.

[125] Derenbach, J. B., Astheimer, H., Hansen, H. P. & Leach, H. (1979). Vertical microscale distribution of phytoplankton in relation to the thermocline. *Marine Ecology Progress Series*, *1*, 187-193.

[126] Jaffe, J. S., Franks, P. J. & Leising, A. W. (1998). Simultaneous imaging of phytoplankton and zooplankton distributions. *Oceanography*, *11*, 24-29.

[127] Leising, A. W. & Franks, P. J. (2002). Does *Acartia clausi* (Copepoda: Calanoida) use an area-restricted search foraging strategy to find food? *Hydrobiologia*, *480*, 193-207.

[128] Buskey, E. J. (1984). Swimming pattern as an indicator of the roles of copepod sensory systems in the recognition of food. *Marine Biology*, *79*, 165-175.

[129] Tiselius, P & Jonsson, PR (1990). Foraging behaviour of six calanoid copepods: observations and hydrodynamic analysis. *Marine Ecology Progress Series*, *66*, 23-33.

[130] Leising, A. W., Pierson, J. J., Cary, S. & Frost, B. W. (2005). Copepod foraging and predation risk within the surface layer during night-time feeding forays. *Journal of Plankton Research*, *27*, 987-1001.

[131] Leising, A. W. & Franks, P. J. (2000). Copepod vertical distribution within a spatially variable food source: a simple foraging-strategy model. *Journal of Plankton Research*, *22*, 999-1024.

[132] McManus, M. A., Cheriton, O. M., Drake, P. J. et al. (2005). Effects of physical processes on structure and transport of thin zooplankton layers in the coastal ocean. *Marine Ecology Progress Series*, *301*, 199-215.

[133] Alldredge, A. L., Cowles, T. J., MacIntyre, S. et al. (2002). Occurrence and mechanisms of formation of a dramatic thin layer of marine snow in a shallow Pacific fjord. *Marine Ecology Progress Series*, *233*, 1-12.

[134] Sutor, M. M., Cowles, T. J., Peterson, W. T. & Pierce, S. D. (2005). Acoustic observations of finescale zooplankton distributions in the Oregon upwelling region. *Deep Sea Research II*, *52*, 109-121.

[135] Stacey, M. T., McManus, M. A. & Steinbuck, J. V. (2007). Convergences and divergences and thin layer formation and maintenance. *Limnology and Oceanography*, *52*, 1523-1532.

[136] Steinbuck, J. V., Genin, A., Monismith, S. G. et al. (2010). Turbulent mixing in fine-scale phytoplankton layers: Observations and inferences of layer dynamics. *Continental Shelf Research*, *30*, 442-455.

[137] Enright, J. T. & Hamner, W. M. (1967). Vertical diurnal migration and endogenous rhythmicity. *Science*, *157*, 937-941.

[138] Bollens, S. M., Frost, B. W. & Cordell, J. R. (1994). Chemical, mechanical and visual cues in the vertical migration behavior of the marine planktonic copepod *Acartia hudsonica*. *Journal of Plankton Research*, *16*, 555-564.

[139] Pierson, J. J., Frost, B. W. & Leising, A. W. (2013). Foray foraging behavior: seasonally variable, food-driven migratory behavior in two calanoid copepod species. *Marine Ecology Progress Series*, *475*, 49-64.

[140] Benoit-Bird, K. J. (2009). Dynamic 3-dimensional structure of thin zooplankton layers is impacted by foraging fish. *Marine Ecology Progress Series*, *396*, 61-76.

[141] Banas, N. S., Wang, D. P. & Yen, J. (2004). Experimental Validation of an-Individual-Based Model for Zooplankton Swarming. *In*: Seuront L & Strutton PG (eds.), *Handbook of Scaling Methods in Aquatic Ecology*. CRC Press, Boca Raton, 161-182.

[142] Okubo, A. & Anderson, J. J. (1984). Mathematical models for zooplankton swarms: Their formation and maintenance. *Eos, Transactions of the American Geophysical Union*, *65*, 731-732.

[143] Okubo, A. (1986). Dynamical aspects of animal grouping: swarms, schools, flocks, and herds. *Advances in Biophysics*, *22*, 1-94.

[144] Ambler, J. W. (2002). Zooplankton swarms: characteristics, proximal cues and proposed advantages. *Hydrobiologia*, *480*, 155-164.

[145] Folt, C. L. & Burns, C. W. (1999). Biological drivers of zooplankton patchiness. *Trends in Ecology and Evolution*, *14*, 300-305.

[146] Haury, L. R. & Yamazaki, H. (1995). The dichotomy of scales in the perception and aggregation behavior of zooplankton. *Journal of Plankton Research*, *17*, 191-197.

[147] Kils, U. (1993). Formation of micropatches by zooplankton-driven microturbulences. *Bulletin of Marine Science*, *53*, 160-169.

[148] Abraham, E. R. (1998). The generation of plankton patchiness by turbulent stirring. *Nature*, *391*, 577-580.

[149] Mackas, D. L., Denman, K. L. & Abbott, M. R. (1985). Plankton patchiness: biology in the physical vernacular. *Bulletin of Marine Science*, *37*, 652-674.

[150] Yamazaki, H. & Osborn, T. R. (1988). Review of oceanic turbulence: implications for biodynamics. *In*: Rothschild BJ (ed.), *Toward a Theory on Biological-Physical Interactions in the World Ocean*. Springer, New York, 215-234.

[151] Hamner, W. M. (1988). Behavior of plankton and patch formation in pelagic ecosystems. *Bulletin of Marine Science*, *43*, 752-757.

[152] Gallager, S. M., Yamazaki, H. & Davis, C. S. (2004). Contribution of fine-scale vertical structure and swimming behavior to formation of plankton layers on Georges Bank. *Marine Ecology Progress Series*, *267*, 27-43.

[153] Wolk, F., Yamazaki, H., Seuront, L. & Lueck, R. G. (2002). A new free-fall profiler for measuring biophysical microstructure. *Journal of Atmospheric and Oceanic Technology*, *19*, 780-793.

[154] Yamazaki, H., Mitchell, J. G., Seuront, L. et al. (2006). Phytoplankton microstructure in fully developed oceanic turbulence. *Geophysical Research Letters*, *33*, L01603.

[155] Dekshenieks, M. M., Donaghay, P. L., Sullivan, J. M. et al. (2001). Temporal and spatial occurrence of thin phytoplankton layers in relation to physical processes. *Marine Ecology Progress Series*, *223*, 61-71.

[156] Cheriton, O. M., McManus, M. A., Steinbuck, J. V. et al. (2010). Towed vehicle observations of thin layer structure and a low-salinity intrusion in Northern Monterey Bay, CA. *Continental Shelf Research*, *30*, 39-49.

[157] McManus, M. A., Alldredge, A. L., Barnard, A. H. et al. (2003). Characteristics, distribution and persistence of thin layers over a 48 hour period. *Marine Ecology Progress Series*, *261*, 1-19.

[158] Cowles, T. J., Desiderio, R. A. & Carr, M. E. (1998). Small-scale planktonic structure: persistence and trophic consequences. *Oceanography*, *11*, 4-9.

[159] Palacios, D. M., Bograd, S. J., Foley, D. G. & Schwing, F. B. (2006). Oceanographic characteristics of biological hot spots in the North Pacific: a remote sensing perspective. *Deep Sea Res II*, *53*, 250-269.

[160] Herman, A. W., Sameoto, D. D. & Longhurst, A. R. (1981). Vertical and horizontal distribution patterns of copepods near the shelf break south of Nova Scotia. *Canadian Journal of Fisheries and Aquatic Sciences*, *38*, 1065-1076.

[161] Uye, S.-I., Yamaoka, T. & Fujisawa, T. (1992). Are tidal fronts good recruitment areas for herbivorous copepods? *Fisheries Oceanography*, *1*, 216-226.

[162] Dubischar, C. D., Lopes, R. M. & Bathmann, U. V. (2002). High summer abundances of small pelagic copepods at the Antarctic Polar Front—implications for ecosystem dynamics. *Deep Sea Research II*, *49*, 3871-3887.

[163] Kiørboe, T. (2008). *A Mechanistic Approach to Plankton Ecology*. Princeton University Press, Princeton.

[164] Schmitt, F. G. & Seuront, L. (2008). Intermittent turbulence and copepod dynamics: increase in encounter rates through preferential concentration. *Journal of Marine Systems*, *70*, 263-272.

[165] Tennekes, H. & Lumley, J. L. (1972). *First Course in Turbulence*. MIT Press, Cambridge.

[166] Kiørboe, T., Andersen, A., Langlois, V. J. et al. (2009). Mechanisms and feasibility of prey capture in ambush-feeding zooplankton. *Proceedings of the National Acadademy of Sciences of the United States of America*, *106*, 12394-12399.

[167] Reichwaldt, E. S., Wolf, I. D. & Stibor, H. (2004). The effect of different zooplankton grazing patterns resulting from diel vertical migration on phytoplankton growth and composition: a laboratory experiment. *Oecologia*, *141*, 411-419.

[168] D'Asaro, E., Lee, C., Rainville, L. et al. (2011). Enhanced turbulence and energy dissipation at ocean fronts. *Science*, *332*, 318-322.

[169] Suchman, C. L. (2000). Escape behavior of Acartia hudsonica copepods during interactions with scyphomedusae. *Journal of Plankton Research*, *22*, 2307-2323.

[170] Golet, W. J., Cooper, A. B., Campbell, R. & Lutcavage, M. (2007). Decline in condition of northern bluefin tuna (*Thunnus thynnus*) in the Gulf of Maine. *Fishery Bulletin*, *105*, 390-395.

[171] Dagg, M. (1993). Grazing by the copepod community does not control phytoplankton production in the subarctic Pacific Ocean. *Progress in Oceanography*, *32*, 163-183.

[172] Pitchford, J. W. & Brindley, J. (2001). Prey patchiness, predator survival and fish recruitment. *Bulletin of Mathematical Biology*, *63*, 527-546.

[173] Wang, D. Z. (2008). Neurotoxins from marine dinoflagellates: a brief review. *Marine Drugs*, *6*, 349-371.

[174] Anderson, D. M. (1989). Toxic algal blooms: a global perspective. *In*: Okaichi T, Anderson DM & Nemoto T (eds.), *Red Tides: Biology, Environmental Science, and Toxicology*. Elsevier, New York, 11-16.

[175] Smayda, T. J. (1990). Novel and nuisance phytoplankton blooms in the sea - Evidence for a global epidemic. *In*: Graneli E, Sundstrom B, Elder L & Anderson DMT (eds.), *Toxic Marine Phytoplankton*. Elsevier, New York, 29-40.

[176] Anderson, D. M., Cembella, A. D. & Hallegraeff, G. M. (2012). Progress in understanding harmful algal blooms: paradigm shifts and new technologies for research, monitoring, and management. *Annual Review of Marine Science*, *4*, 143-176.

[177] Milroy, S. P., Dieterle, D. A., He, R. et al. (2008). A three-dimensional biophysical model of *Karenia brevis* dynamics on the west Florida shelf: a look at physical transport and potential zooplankton grazing controls. *Continental Shelf Research*, *28*, 112-136.

[178] Janowitz, G. S., Kamykowski, D. & Liu, G. (2008). A three-dimensional wind and behaviorally driven population dynamics model for *Karenia brevis*. *Continental Shelf Research*, *28*, 177-188.

[179] Prince, E. K., Poulson, K. L., Myers, T. L. et al. (2010). Characterization of allelopathic compounds from the red tide dinoflagellate *Karenia brevis*. *Harmful Algae*, *10*, 39-48.

[180] Jiang, X., Tang, Y., Lonsdale, D. J. & Gobler, C. J. (2009). Deleterious consequences of a red tide dinoflagellate *Cochlodinium polykrikoides* for the calanoid copepod *Acartia tonsa*. *Marine Eoclogy Progress Series*, *390*, 105-116.

[181] Breier, C. F. & Buskey, E. J. (2007). Effects of the red tide dinoflagellate, Karenia brevis, on grazing and fecundity in the copepod *Acartia tonsa*. *Journal of Plankton Research*, *29*, 115-126.

[182] Prince, E. K., Lettieri, L., McCurdy, K. J. & Kubanek, J. (2006). Fitness consequences for copepods feeding on a red tide dinoflagellate: deciphering the effects of nutritional value, toxicity, and feeding behavior. *Oecologia*, *147*, 479-488.

[183] Waggett, R. J., Hardison, D. R. & Tester, P. A. (2012). Toxicity and nutritional inadequacy of *Karenia brevis*: synergistic mechanisms disrupt top-down grazer control. *Marine Ecology Progress Series*, *444*, 15-30.

[184] Hong, J., Talapatra, S., Katz, J. et al. (2012). Algal toxins alter copepod feeding behavior. *PLoS One*, *7*, e36845.

[185] Cohen, J. H., Tester, P. A. & Forward, R. B. (2007). Sublethal effects of the toxic dinoflagellate *Karenia brevis* on marine copepod behavior. *Journal of Plankton Research*, *29*, 301-315.

[186] Schultz, M. & Kiørboe, T. (2009). Active prey selection in two pelagic copepods feeding on potentially toxic and non-toxic dinoflagellates. *Journal of Plankton Research*, *31*, 553-561.

[187] Leandro, L. F., Teegarden, G. J., Roth, P. B. et al. (2010). The copepod *Calanus finmarchicus*: a potential vector for trophic transfer of the marine algal biotoxin, domoic acid. *Journal of Experimental Marine Biology and Ecology*, *382*, 88-95.

[188] Solé, J., Garcia-Ladona, E. & Estrada, M. (2006). The role of selective predation in harmful algal blooms. *Journal of Marine Systems*, *62*, 46-54.

[189] Buskey, E. J. & Hyatt, C. J. (1995). Effects of the Texas (USA)'brown tide' alga on planktonic grazers. *Marine Ecology Progress Series*, *126*, 285-292.

[190] Kubanek, J., Snell, T. W. and Pirkle, C. (2007). Chemical defense of the red tide dinoflagellate *Karenia brevis* against rotifer grazing. *Limnology and Oceanography*, *52*, 1026-1035.

[191] Waggett, R. J., Tester, P. A. & Place, A. R. (2008). Anti-grazing properties of the toxic dinoflagellate *Karlodinium veneficum* during predator-prey interactions with the copepod *Acartia tonsa*. *Marine Ecology Progress Series*, *366*, 31-42.

[192] Graham, S. L. & Strom, S. L. (2010). Growth and grazing of microzooplankton in response to the harmful alga *Heterosigma akashiwo* in prey mixtures. *Aquatic Microbial Ecology*, *59*, 111-124.

[193] Teegarden, G. J., Cembella, A. D., Capuano, C. L. et al. (2003). Phycotoxin accumulation in zooplankton feeding on *Alexandrium fundyense*—vector or sink? *Journal of Plankton Research*, *25*, 429-443.

[194] Gannon, D. P., McCabe, E. B., Camilleri, S. A. et al. (2009). Effects of *Karenia brevis* harmful algal blooms on nearshore fish communities in southwest Florida. *Marine Ecology Progress Series*, *378*, 171-186.

[195] Seuront, L. (2011). Hydrocarbon contamination decreases mating success in a marine planktonic copepod. *PLoS One*, *6*, e26283.
[196] Stumpf, R. P., Litaker, R. W., Lanerolle, L. & Tester, P. A. (2008). Hydrodynamic accumulation of *Karenia* off the west coast of Florida. *Continental Shelf Research*, *28*, 189-213.
[197] Ryan, J. P., Fischer, A. M., Kudela, R. M. et al. (2009). Influences of upwelling and downwelling winds on red tide bloom dynamics in Monterey Bay, California. *Continental Shelf Research*, *29*, 785-795.
[198] Donaghay, P. L. & Osborn, T. R. (1997). Toward a theory of biological-physical control of harmful algal bloom dynamics and impacts. *Limnology and Oceanography*, *42*, 1283-1296.
[199] Rines, J. E. B., Donaghay, P. L., Dekshenieks, M. M. et al. (2002). Thin layers and camouflage: hidden *Pseudo-nitzschia* spp.(Bacillariophyceae) populations in a fjord in the San Juan Islands, Washington, USA. *Marine Ecology Progress Series*, *225*, 123-137.
[200] McManus, M. A., Kudela, R. M., Silver, M. W. et al. (2008). Cryptic blooms: Are thin layers the missing connection? *Estuaries and Coasts*, *31*, 396-401.
[201] Sullivan, J. M., McManus, M. A., Cheritom, O. M., Kelly, J. et al. (2010). Layered organization in the coastal ocean: An introduction to planktonic thin layers and the LOCO project. *Continental Shelf Research*, *30*, 1-6.
[202] Hutchinson, G. E. (1961). The paradox of the plankton. The American Naturalist, *95*, 137-145.
[203] Follows, M. J., Dutkiewicz, S., Grant, S. & Chisholm, S. W. (2007). Emergent biogeography of microbial communities in a model ocean. *Science*, *315*, 1843-1846.
[204] Waggett, R. J., Buskey, E. J. (2007). Copepod escape behavior in non-turbulent and turbulent hydrodynamic regimes. *Marine Ecology Progress Series*, *334*, 193-198.
[205] Haury, L. R., Kenyon, D. E. & Brooks, J. R. (1980). Experimental evaluation of the avoidance reaction of *Calanus finmarchicus*. *Journal of Plankton Research*, *2*, 187-202.
[206] Titelman, J. & Kiørboe, T. (2003). Predator avoidance by nauplii. *Marine Ecology Progress Series*, *247*, 137-149.

In: Copepods: Diversity, Habitat and Behavior
Editor: Laurent Seuront
ISBN: 978-1-63117-846-7

Chapter 11

COPEPOD CARCASSES: OCCURRENCE, FATE AND ECOLOGICAL IMPORTANCE

***Kam W. Tang*[1] *and David T. Elliott*[2]**
[1]Virginia Institute of Marine Science, College of William & Mary, Gloucester Point, VA, US
[2]Horn Point Laboratory, University of Maryland Center for Environmental Science, Cambridge, MD, US

ABSTRACT

Quantifying *in situ* copepod abundances is one of the most fundamental needs in copepod ecological research. Conventional field sampling does not distinguish between live and dead copepods in samples. Corollary to this oversight is that predation is often assumed to be the dominant cause of copepod mortality to the extent that non-predatory mortality is ignored. There is, however, no *a priori* reason to believe copepods are exempt from non-predatory mortality. Many studies have indeed shown that copepods succumb to non-predatory mortality from chemical contaminants, algal toxins, diseases, parasites, environmental stresses and food-related factors. A growing number of studies using various methods to collect and identify copepod carcasses have shown that carcasses are prevalent in the natural environment, indicating the potential importance of non-predatory mortality. A meta-analysis of literature data has suggested that, globally, non-predatory factors account for up to one-third of total mortality among marine epipelagic copepods. Identifying copepod carcasses in field samples can be a daunting task, but recent improvements to simple vital and mortal staining methods allow researchers to easily quantify carcasses in marine and freshwater samples. Although carcasses may be less abundant than live copepods *in situ*, ignoring them would introduce errors into estimation of demographic parameters, including mortality rates. These errors can be considerable, especially when the abundance of carcasses relative to live animals varies throughout ontogeny (e.g., higher percent dead in younger naupliar stages). Data on live-dead copepod composition can be used to estimate both *in situ* predatory and non-predatory mortality rates, and both mortality terms are required to produce realistic population growth predictions. The fate of carcasses is also of broader ecological importance. Newly deceased copepods tend to sink because of higher density than the surrounding water, but sinking is moderated by microbial decomposition, water column stratification and turbulence resuspension. Microbial decomposition remineralizes carcass organic materials and supports high bacterial production, in the process reducing carcass density and allowing the possibility for the carcass to approach neutral buoyancy in the water column. In contrast, fast sinking carcasses export carbon to the deep waters and benthos, and sediment trap studies have shown that copepod carcasses can be the dominant component of vertical flux outside of the phytoplankton growth period. Research into the occurrence and fate of copepod carcasses will therefore improve our understanding of not only copepod mortality and population dynamics, but also the link between copepods, microbial food web and biogeochemical fluxes.

INTRODUCTION

Death is very likely the single best invention of Life. It is Life's change agent. It clears out the old to make way for the new.

- Steve Jobs

One of the most basic needs in copepod research is accurate assessment of population abundances, from which researchers derive a multitude of ecological parameters such as grazing (feeding) impact, population growth, and food availability for higher trophic levels. In the tradition of field sampling, copepods are trapped and concentrated by nets into collectors (cod ends), and indiscriminately preserved and enumerated without knowledge of the original live/dead status of the animals sampled. Subsequently, researchers often assume that all copepods counted were originally live, and *in situ* mortality is caused solely by predation, which can be deduced from changes in abundances. However, there is no *a priori* reason to think copepods would be exempt from non-predatory mortality, which may leave intact carcasses behind. Without distinguishing between live and dead copepods in field samples, researchers will have overlooked evidence of non-predatory mortality and erred in understanding many potential ecological processes. Nevertheless, enumerating copepod carcasses costs extra time and effort and is therefore rarely done. It is then only fair to address some common skepticism when discussing this topic:

1. Copepod carcasses and non-predatory copepod mortality do not exist in nature;
2. Carcasses exist in field samples, but they are results of poor sample handling (artifact mortality), not natural causes;
3. Carcasses and non-predatory mortality exist in nature, but they are so trivial relative to live copepods and predatory mortality that we can safely ignore them.

The purpose of this chapter is to address these issues and discuss the importance of considering not only the live, but also the dead, when studying copepods.

EARLY OBSERVATIONS OF COPEPOD CARCASSES

Our dead are never dead to us until we have forgotten them.

- George Eliot

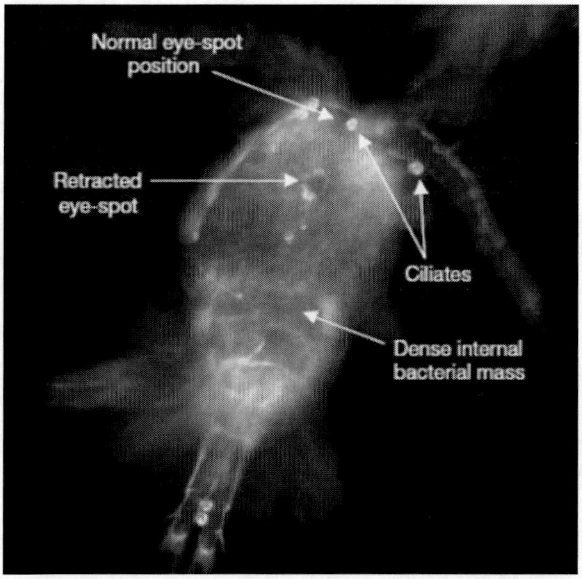

Figure 1. A decomposing carcass of *Mesocyclops lecukartis* collected from Lake Dagow, Germany. Under dark-field microscopy, it is obvious that the carcass has already lost most of its internal tissues, and internal decomposition has caused the eyespot to retract from its normal position. Large amount of internal bacteria turn the carcass opaque, and a few ciliates can be seen hovering around, presumably feeding on the bacteria. Reproduced from *Tang* et al. [84] with permission from Inter-Research.

Copepod carcasses can be distinguished from live copepods before preservation in various ways, such as (i) lack of movement even upon stimulation (probing), ii) lack of heart beats, (iii) lack of gut peristaltic movement, and (iv) signs of decomposition such as development of rigor mortis, missing body tissues, or colonization by microbes (Figure 1). Carcasses are also clearly different from exuviae because carcasses contain traces of tissues even at the advanced stage of decomposition [1]. These methods, however, require close and lengthy examination of individual copepods, and are therefore impractical in field surveys. Such examinations are more commonly used in investigations of the effects of specific non-predatory mortality factors, such as chemical contaminants [2], toxins from harmful algae [3,4], parasites and diseases [5,6], other environmental stressors [7,8] and food quality/ quantity [9-11].

While experimental studies have repeatedly demonstrated that many non-predatory factors can cause copepod mortality, other researchers have worked to painstakingly identify partially decomposed copepod carcasses in field samples by microscopy. For example, in the upwelling region south of Cap Blanc, N.W. Africa, copepod carcasses (*Temora* spp.) amounted to as much as 28% of the population [12]. Of the field studies enumerating copepod carcasses in the open ocean, many have reported substantial number of carcasses as well as conspicuous patterns in carcass occurrence, such as higher relative abundances of carcasses at greater depths [13-18], accumulation of carcasses near the transition between water masses such as the thermocline [12,13,18], inter-specific differences in carcass abundances and greater relative carcass abundances in smaller copepod species [15,16,19,20]. There also have been several reports of accumulation of copepod carcasses near bottom topographic structures, providing a potential food source for deep-sea detritivores [21,22]. In fact, recent work using stable isotopes suggests that copepod carcasses may be an important component in the diets of the omnivorous euphausiid *Euphausia pacifica* [23].

In sedimentation flux studies where researchers deploy traps to measure gravitational deposition of materials into the deep waters, a common issue is the presence of copepods and other zooplankton in the traps, especially when preservatives are used [24]. Do these individuals represent carcasses that have sunk into the traps? Or, are they 'swimmers' that have accidentally entered the traps, been exposed to preservative and subsequently died there? This is not a trivial question. Copepod bodies represent highly concentrated sources of organic matter relative to many of the other small particles collected in traps (e.g., fecal pellets, detrital aggregates), and the decision to include (carcasses) or exclude (swimmers) them can substantially alter vertical flux calculations.

Several sediment trap designs have been proposed to exclude 'swimmers' from passively sinking materials (e.g., [25-27]), although their use has not been very wide spread. In a recent review paper of sedimentation trap studies, the issue of 'swimmers' vs. carcasses was mentioned only very briefly [28]. Nevertheless, using specially designed traps, further aided by visual examination for partial decomposition, researchers have been able to recover copepod carcasses from sediment traps in the Mediterranean [29,30]. Other researchers were also able to identify copepod carcasses in sediment traps in Arctic waters, based on the post-mortem morphology and posture of copepods in trap samples [31,32].

These studies—laboratory experiments of non-predatory mortality factors, visual examination of carcasses in field samples, and sediment trap collection of carcasses—together provide answers to some of our earlier questions: (i) copepods can suffer non-predatory mortality, and (ii) carcasses resulting from natural (non-artifact) causes can be identified in field samples, either microscopically (albeit labor intensively) or by collection in specially designed sediment traps.

In practice, however, one may choose to ignore carcasses and non-predatory mortality if they prove to be a trivial portion of the total. Unfortunately, there is no easy way to determine their importance precisely because there have been so few studies quantifying carcasses and measuring non-predatory mortality in copepods. This is a dilemma in research: If we don't know enough of an issue, it is hard to be convinced of its importance, and therefore it becomes even harder to justify investing valuable resources to learn more about the issue! Nevertheless, a meta-analysis study comparing copepod mortality derived from laboratory-measured life history parameters and field-measured (predatory) mortality has concluded that predation accounts for ⅔ - ¾ of the total mortality of epi-pelagic copepods [33]. It follows that ¼ - ⅓ of the mortality is due to non-predatory causes. This is a non-trivial portion of the total mortality, and, as we will discuss later in the chapter, ignoring this mortality term (and by extension, the resulting carcasses) would lead to incorrect understanding of copepod population dynamics and other related pelagic processes.

Perhaps one of the most intriguing observations of copepod carcasses was recently made by Dr. Morten Hjorth of Denmark, who in 2010 found a large number of *Calanus hyperboreus*, a key species in Arctic fjords, laid dead along the beach near Young Sound in eastern Greenland (Figure 2). According to Dr. Hjorth, the copepods were brought to shore by incoming tides, trapped in rock pools when the tides receded, and subsequently died of salinity shock due to freshwater injection from melting ice. Upon the following tidal flushing some of the carcasses were dispersed along the sandy beach, where seabirds could be seen feasting on them. This is an example of physical processes causing non-predatory mortality and providing a trophic short-cut from copepods directly to birds! Imagine what we would have missed if we ignored copepod carcasses.

Figure 2. Carcasses of *Calanus hyperboreus* along the shore near Young Sound of eastern Greenland. The copepods were believed to have been brought in by tides, trapped within rock pools during receding tides, and subsequently killed by meltwater influx. The carcasses were then dispersed by following tidal flushing along the sandy beach, where seabirds could be seen feasting on them. Photo credit: Dr. Morten Hjorth, COWI, Denmark.

VITAL AND MORTAL STAINING METHODS

Life and death are balanced as it were on the edge of a razor.

- Homer

For researchers to more widely study copepod carcasses, a fast, easy and inexpensive method for sorting live and dead copepods is highly desirable. A variety of histological stains can be adapted for this purpose, among which two have been successfully used in copepod field studies: Neutral Red and Aniline Blue.

Neutral Red ($C_{15}H_{17}ClN_4$) is a vital stain that is actively taken up by lysosomes of live cells, but not by dead cells. It is widely used in testing cell viability and cytoxicity assays [34]. Although Neutral Red is generally considered a non-toxic vital stain, there have been reports of some negative effects on stained organisms [35]. For sorting live and dead copepods, samples only need to be stained prior to preservation and therefore any negative effects the stain may have on the live animals are not a concern. Nevertheless, if researchers want to use stained copepods for experiments (for example, as a way to distinguish between two copepod populations), a more careful investigation of potential negative effects of the stain will be needed.

A laboratory protocol for distinguishing between dead and live marine copepods based on their differential uptake of Neutral Red was first published in 1972 [36]. Later work proposed that Neutral Red could be applied to other types of estuarine zooplankton [37], and preservation techniques were refined so as to increase potential storage time of stained samples [38]. However, application of Neutral Red staining to copepods has since been largely limited to vital assessment in laboratory and small-scale field experiments. More recently, the protocol has been modified and refined for broader application in field studies on copepods across a range of working conditions such as different temperatures, salinities and methods of preservation [1,39]. Subsequent studies have compared Neutral Red with several alternative methods for making live and dead determination on copepods and other zooplankton. In field samples of mesozooplankton collected from the Black Sea, Neutral Red staining results were highly correlated with those obtained using an alternative stain, fluorescein diacetate (FDA) [40]. Using natural North Sea mesozooplankton, the accuracy of live and dead determinations was assessed for cell digestion assay, SYTOX® Green nucleic acid staining, and Neutral Red [41]. Neutral Red staining was found to be the most robust, was broadly applicable to many taxa, and performed well on the copepods tested. Live copepods that have taken up Neutral Red stain appear bright red under the microscope, whereas dead copepods are only slightly stained or unstained (see [39] for photos). The clear color contrast makes counting live and dead copepods in field samples quick and easy.

When making live and dead determinations in field collected samples, researchers must also take care to avoid artifact mortality; i.e., copepods killed by rough handling of the samples. To address this concern, we tested that short and gentle net tows, plus careful handling of the samples on board, is sufficient to prevent artifact mortality [1,39]. For biomass quantification, researchers often rigorously wash down the net into the cod end, but this should be avoided for samples used in live/dead determination. It is therefore recommended that a separate, gentle, net tow is done, without washing, for determining the relative proportions of live and dead copepods. Alternatively, the net can be washed down into a separate container. A detailed protocol of the Neutral Red field method can be found in [39].

Because Neutral Red does not work well in freshwater, an alternative stain, Aniline Blue, is needed for live and dead determinations of freshwater copepods. Aniline Blue ($C_{32}H_{27}N_3O_9S_3Na_2$) is a stain commonly used in histology for visualization of plant tissues [42,43] and as a fluorochrome in epifluorescence imaging [44]. When applied to freshwater zooplankton, the blue stain penetrates dead zooplankton but not live ones [45]. Based on this property, Aniline Blue can be used as a mortal stain to help distinguish between dead (stained blue) and live (unstained) freshwater zooplankton (e.g., [46,47]). As with Neutral Red, the method has been further tested and refined for field applications, and a detailed protocol can be found in [48].

Both Neutral Red and Aniline Blue staining methods are simple, easy and inexpensive, and when used properly they allow researchers to quantify live and dead copepods in field samples. These staining methods, however, do have their limitations. In laboratory testing, Neutral Red staining accuracy decreased with decreasing temperature (down to 5°C) [39]. This makes its application in cold regions questionable. The very lipid rich copepod species in polar and sub-polar waters also seem to yield more false positive Neutral Red staining results, likely due to the lipophilic nature of the stain (our own observations). When using Aniline Blue, a highly decomposed freshwater copepod carcass may fail to take up the stain, yielding a false negative [49].

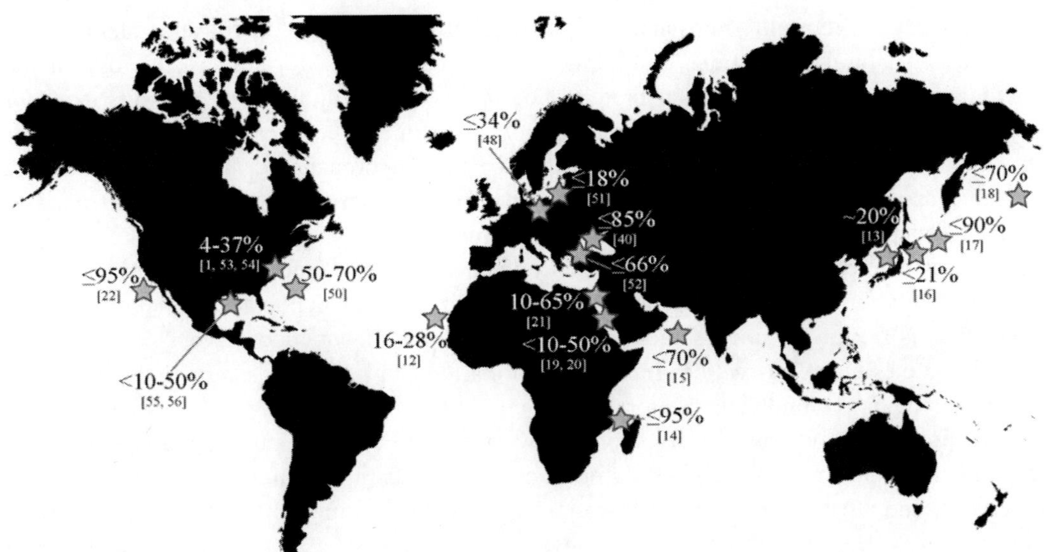

Figure 3. Occurrence of copepod carcasses as percent total in samples. Numbers in brackets indicate data sources; bibliographic information is given in the references. In all cases, carcasses were examined directly by microscopy with a few exceptions where Neutral Red staining [1,40,53-56], Aniline Blue staining [48, 51], or fluorescein diacetate staining [40] was used to aid carcass identification.

Based on staining and other methods (e.g., microscopic examination, sediment trap collection), data are emerging on the occurrence of copepod carcasses in different freshwater and marine systems (Figure 3). These data cover only a limited number of locations, and the sampling resolution was often not high enough to resolve any spatial and temporal variabilities of carcass abundances. Nevertheless, they clearly demonstrate that copepod carcasses are commonly present in the environments. In the tributaries of Chesapeake Bay, copepod carcasses are more abundant and make up a larger fraction of the total copepods collected during the late spring, summer, and early fall, coincident with and immediately following peaks in the abundance of live copepods [54], and indicative of death due to senescence or other density-dependent mortality in the populations. Senescence following completion of the reproductive cycle, or at the end of a bloom period, has also been invoked to explain peaks in the downward flux of copepod carcasses in both the Arctic Ocean and the Mediterranean during winter (January-March), as observed in sediment trap samples [30-32]. In areas with heavy boat traffic, propeller-generated turbulence may also add to the variability in carcass abundance on small spatial and temporal scales [53]. Understanding what factors determine the temporal and spatial variations in carcass abundance will be an important task in future research.

Mortality and Population Dynamics

To the living we owe respect, but to the dead we owe only the truth.

- Voltaire

Knowledge of mortality rates is one of the most crucial components for understanding the dynamics of natural populations, including those of copepods. Mortality is also one of the most challenging parameters to accurately measure *in situ*. A number of methods exist for estimating mortality rates in copepod populations, and some of the most widely used methods can be classified as either horizontal or vertical life table approaches [57]. The application of life table approaches is not limited to mortality rate estimation as they are also used to estimate copepod stage durations and total recruitment [57]. Detailed treatment of copepod mortality estimation methods is beyond the scope of this chapter, and for this we refer readers to other available resources (e.g., [57-61]). However, a brief explanation of the life table approaches is useful for understanding how the presence of copepod carcass can influence mortality rate estimates, as well as how that information can be used to estimate the predatory and non-predatory components of mortality.

The horizontal life table (HLT) approach works by following a single group or cohort of copepods through time, quantifying temporal changes in the abundance of individuals within the cohort, and attributing observed losses from the cohort to mortality. Horizontal methods require a high degree of temporal coverage, and it is necessary to follow the same cohort through its entire development [57]. As such, these methods are appropriate only when high temporal resolution sampling can be achieved over the lifetime of the cohort without disruption by processes such as advection.

The vertical life table (VLT) approach estimates mortality rate using the abundance ratio between pairs of copepod developmental stages within single time point field samples. The method represents an alternative to the horizontal approach when high temporal resolution sampling of a cohort is not feasible. When using the vertical approach it is advisable to collect a large number of replicate samples to account for inter-sample variability in relative abundances. The VLT approach assumes that different developmental stages of the sampled copepods are influenced equally by advection, and that recruitment into each stage is relatively constant over the duration of the stages pair being considered [57,62]; see also [61] for further explanation of important assumptions.

Because both HLT and VLT approaches rely on accurate abundance measurements, failure to separate intact carcasses from live animals will bias the abundance data and lead to errors in estimates of mortality rate as well as other relevant population parameters. Although the abundance of intact carcasses may be small relative to the abundance of live copepods in the samples, large errors in mortality rate estimation using the VLT approach will still result when the percentage of dead animals is disproportionate between consecutive developmental stages, as demonstrated by our data on *Acartia tonsa* in Chesapeake Bay [63]. When treating carcasses as alive in samples, VLT tended to slightly underestimate mortality rate of *A. tonsa* copepodites (Figure 4) due to a higher percentage of dead individuals in the adult stage compared to younger copepodites [63]. Conversely, the mortality rate of nauplii was consistently overestimated when carcasses were counted as live individuals (Figure 4) due to a disproportionally higher percentage of dead individuals among the younger naupliar stages [63]. Because copepod population models are sensitive to the mortality term (e.g., [63,64]), even small bias in mortality rate can have substantial consequences for simulated population dynamics. For example, in Chesapeake Bay, *A. tonsa* copepodite mortality rates were only slightly underestimated when carcasses were counted as live individuals (Figure 4), but this small underestimation, when propagated over the entire growth season, resulted in a large difference in population model predictions (Figure 5).

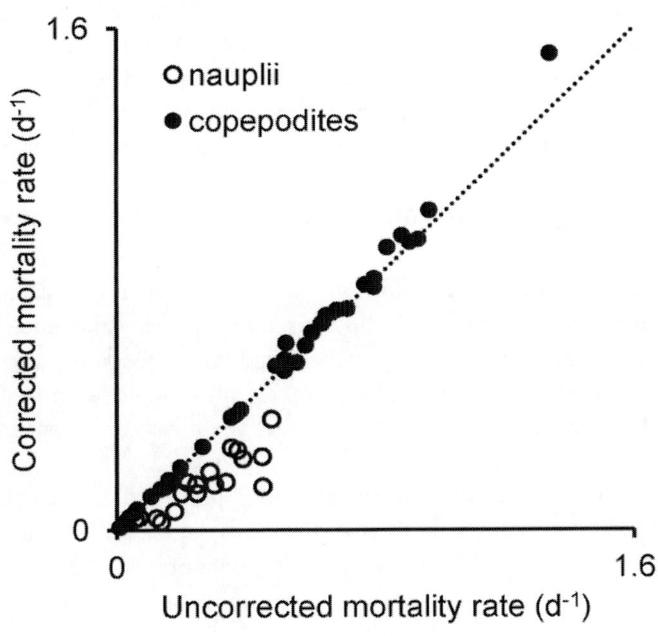

Figure 4. Vertical life table estimates of *Acartia tonsa* naupliar and copepodite mortality rates based on abundances prior to removing carcasses (uncorrected) and after removal of carcasses (corrected). The dotted line indicates the 1:1 line where the two estimates would be identical; see Elliott & Tang [63] for details on methods and data used.

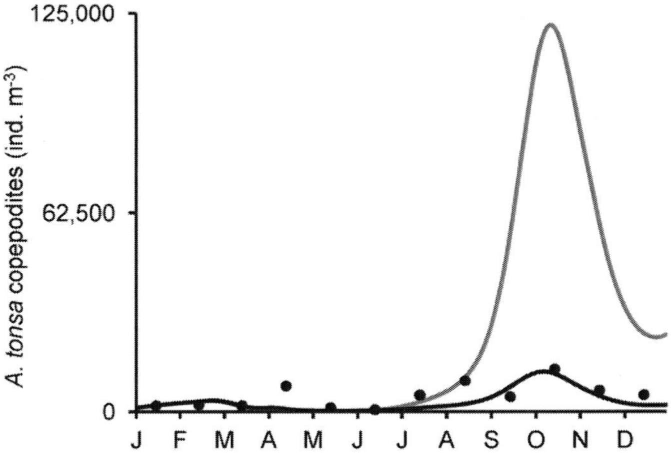

Figure 5. Predicted abundances of *Acartia tonsa* copepodites from a population dynamics model with and without correcting mortality rate estimates for carcasses (Figure 4). Solid black line: corrected mortality rates; solid gray line: uncorrected copepodite mortality rate (e.g., copepodite carcasses treated as alive); individual plot points: monthly mean abundances of *Acartia* spp. copepodites from Chesapeake Bay Program; see Elliott & Tang [63] for details on methods and data used.

Information on the abundance of live copepods and carcasses can also be used to estimate the relative contributions of predatory and non-predatory factors to mortality. Such an approach is reasonable when carcasses appear to represent instances of non-predatory mortality, that is, when carcasses are generally intact without signs of predation related injury. (On the other hand, a wound could be inflicted post mortem, but this is beyond our scope of discussion.) There are several options for calculating predatory and non-predatory mortality rates, each requires an estimate of the amount of time that a copepod carcass remains in the water that will be sampled by the collection method, referred to here as the carcass turnover time (τ). Estimation of τ rests on knowledge of the fate of copepod carcasses (*see next section*). Once τ is known, the rate of carcass production is simply the carcass abundance divided by τ. Assuming steady state conditions, an estimate of non-predatory mortality rate can be obtained by dividing carcass production rate by the abundance of live animals, yielding the rate at which live animals in the population become carcasses [1]. If total mortality rate is known from a life table or other appropriate approach, then predatory mortality rate can be calculated as the difference between total and non-predatory mortality rates.

An alternative method for estimating predatory and non-predatory mortality rates applies the VLT approach. In this method, both live copepods and intact carcasses are treated as survivors of predatory mortality, and the rate of predatory mortality is estimated using modified VLT equations [63]. Non-predatory mortality rate is then calculated as the difference between total and predatory mortality rates, both derived from VLT calculations.

Past studies of copepod population dynamics have often focused largely on predatory mortality, and some methods for estimating copepod mortality implicitly ignore non-predatory factors, such as those based on predation experiments or predator gut contents. While predation usually accounts for the larger portion of mortality experienced by copepods, a global meta-analysis study estimated that non-predatory factors account for between 25%

and 33% of total mortality in marine epi-pelagic copepods, with non-predatory rates ranging from 0.02 d^{-1} to 0.065 d^{-1} [33]. These estimates agree well with other studies: For example, based on sediment trap collections and temporal data on copepod biomass and growth rates, researchers have estimated biomass specific non-predatory mortality rates of < 0.01 d^{-1} to 0.15 d^{-1} in several freshwater, coastal and open ocean areas [30]. Using live and dead copepod abundances and a modified VLT approach, we estimated an average non-predatory mortality rate of 0.035 d^{-1} for naupliar stages and 0.064 d^{-1} for copepodites of *A. tonsa* in the lower Chesapeake Bay, which accounted for 25% and 12% of the total naupliar and copepodite mortality, respectively [63].

There appears to be a positive relationship between temperature and both the total mortality rate [33,65,66] and non-predatory mortality rate [33,63]. Because copepods are poikilotherms, metabolic processes are dependent on environmental temperature. As such, higher temperatures may increase the risk of starvation [67], decrease maximum lifespan [68], and cause a general increase in the animal's sensitivity to environmental stressors. Higher temperatures may also promote disease and parasitism in natural populations [69].

Although non-predatory mortality rate is generally lower than predatory mortality rate, its importance in controlling copepod population dynamics is not to be overlooked. Using a simple population dynamics model, we demonstrated that when non-predatory mortality was ignored, predation mortality alone failed to control population growth of *A. tonsa* in Chesapeake Bay. The modeled population increased to 10^9 copepodites m^{-3} by the end of the growing season, which was five orders of magnitude higher than the observed abundance [63]. Considering a conservative scenario where non-predatory factors account for an average of only 6% of total mortality (*cf.* a minimum of 12% observed in [63]), the error due to ignoring even such a small non-predatory mortality rate is still substantial—the model overestimates copepod abundance by more than an order of magnitude over the growing season (Figure 6).

Because copepods are the major trophic link between microplankton and fish, they occupy a prominent place in many ecosystem and food web models, and the traditional emphasis on predation in copepod demographic research may stem from a motivating interest of linking copepod production to energy flow and fishery yield. Nevertheless, such a narrow view of copepod mortality is unsubstantiated. Quoting R.G. Wetzel [70] on aquatic ecosystems:

> The common idea that the living net production in organisms, particularly algae and microbes but also metazoans, dies largely by ingestive predation is not only unsubstantiated but intuitively unreasonable.
>
> Many if not most organisms, particularly bacteria and fungi but also algae and higher organisms simply mature physiologically, senesce, and die, and then enter the combined particulate and dissolved detrital pool for microbial heterotrophic utilization.
>
> Just because we cannot measure non-predatory mortality well does not mean it does not exist or even dominate at most times of the year.

Here, we have described ways of using carcass abundance data to estimate non-predatory mortality of copepods. Further improvements to this approach are needed, particularly for estimating carcass turnover time (τ) in various aquatic environments. Nonetheless, this approach will allow researchers to begin to study non-predatory copepod mortality in greater details.

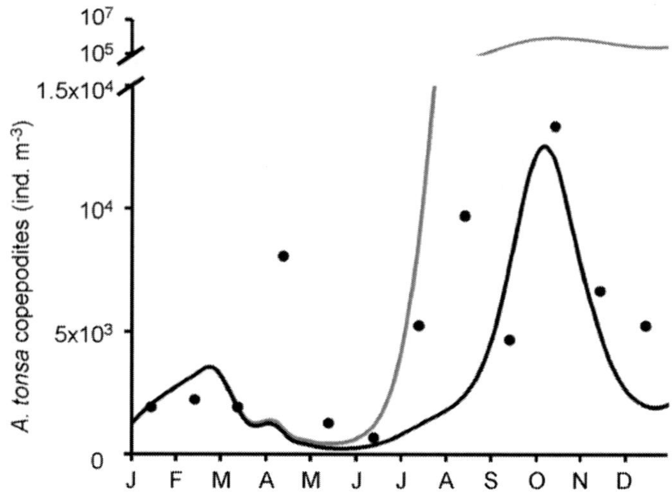

Figure 6. Predicted abundances of *Acartia tonsa* copepodites from a population dynamics model with and without inclusion of non-predatory mortality rate. Solid black line: total mortality (predatory plus non-predatory; same as black line in Figure 5); solid gray line: a hypothetical case where non-predatory mortality accounted for only 6% of total mortality, but was excluded from the model; individual plot points: monthly mean abundances of *Acartia* spp. copepodites from Chesapeake Bay Program; see Elliott & Tang [63] for details on methods and data used.

Fate of Carcasses

When you're dead, you're dead. That's it.

- Marlene Dietrich

Copepod carcasses are dense packages of organic matter and their fate therefore has important implications for material and energy fluxes. Here we will focus on two possible fates: microbial decomposition and sedimentation.

Detrital particles such as phytoplankton aggregates and fecal pellets are often densely colonized by bacteria [71] and support elevated bacterial exoenzyme activites [72] and growth [73-75; but also see [76]). The carbon-to-nitrogen ratio of these detrital particles ranges 5-20, with particulate combined amino acids accounting for 8-51% of their particulate organic content [71]. By comparison, a calanoid copepod body has an average carbon-to-nitrogen ratio of ~3, and 30-70% protein by dry weight [77]. Copepod carcasses are therefore higher quality organic matter than detrital particles for bacteria. In a study using ^{14}C-labeled copepod carcasses poisoned by $HgCl_2$, it was estimated that 35% of the carcass carbon was lost within the first 24 hours, which could be attributed to leaching through autolysis [78]. In the un-poisoned treatment, the leached carbon appeared to be rapidly consumed and the total carbon loss rate was also higher, indicating direct microbial decomposition [78]. In addition to organic carbon, trace elements were also being remineralized during microbial decomposition of copepod carcasses [79].

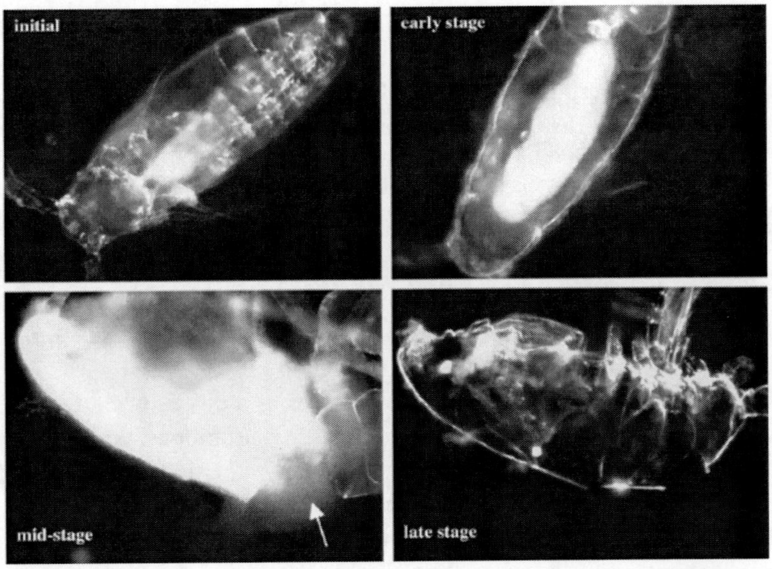

Figure 7. Stages of decomposition of an *Acartia tonsa* carcass as observed by dark-field microscopy. The body of a newly deceased copepod is transparent. As decomposition progresses, the interior of the carcass turns opaque due to accumulation of bacteria. At mid-stage the carapace begins to fracture and releases bacteria from the inside (arrow). At the end of the decomposition most of the internal tissues and bacteria have disappeared, leaving the carapace with a trace amount of tissues. Reproduced from Tang et al. [1] with permission from Elsevier.

Although live copepods naturally carry dense communities of bacteria on their bodies [80], experiments have shown that decomposition of copepod carcasses is driven by exogenous, not endogenous, bacteria [81-82]. In laboratory studies, when copepod carcasses were exposed to natural bacteria, the bacteria quickly colonized and multiplied inside the carcasses (Figure 7). At the peak of decomposition, bacterial production rate, protease and lipase activities were 5-7 times higher inside the carcasses relative to the ambient water [83,84]. The high protease activity led to 70% loss of carcass protein content within the first 8 hours, and a concurrent reduction of protein molecule size from ≥100 kD to ~60 kD. The even smaller protein fraction (45 kD) persisted for >24 hours [82], representing a more refractory pool of carcass protein [79]. In the later stage of decomposition the carcasses became increasingly fractured releasing bacteria from the inside; hence, the decomposing carcasses functioned as a source of new bacteria to the surrounding water. In addition to elevated bacterial activities, denaturing gradient gel electrophoresis (DGGE) of extracted bacterial DNA (16S rRNA gene fragments) showed a change in the bacterial community composition both on and around the carcasses, suggesting that a subset of the ambient bacteria was better able to exploit carcass organic matter, leading to a shift in the bacterial community composition [83,84].

The rate of carcass decomposition appears to be temperature dependent. Based on an empirical equation predicting change in copepod carcass dry weight as a function of temperature (Eq. 4 in [85]), it can be predicted that the majority of decomposition would occur in several days at 25°C, but could take a week or more at 15°C, and an indefinite amount of time at 5°C. This temperature function yields a Q_{10} of 2.15 for bacteria decomposing copepod carcasses, which is similar to the measured Q_{10} value of 2.4 for bacteria decomposing a daphnid carcass [84], and is within the common range of Q_{10} values (1.2-4.8) for aquatic bacterial growth and production [86,87]. In the Japan Sea, an accumulation of carcasses of the copepod *Calanus cristatus* was consistently found below the thermocline, where the low ambient temperature (2°C) might preserve the carcasses for over a year [13].

Toward the end of the decomposition, most of the tissues and the bacteria will have disappeared, leaving behind the chitinous exoskeleton and trace amounts of internal tissues (Figure 7). Although chitinolytic bacteria are present in lakes and estuaries [88-90], chitin degradation by bacteria in nature tends to be very slow. For example, *in situ* incubation of purified chitin particles showed a loss rate of <1% per day [89,91,92].

Other microbes may also exploit copepod carcasses as a direct or indirect food source. In an experiment when bacterial activities were inhibited by antibiotics, copepod carcasses were colonized and decomposed by epiphytic fungi [83]. Many aquatic fungi are able to hydrolyze chitin [93]; therefore, although fungi may not act as fast as bacteria in hydrolyzing carcass organic matter, they may be more able to remineralize the carcass exoskeleton toward the end of the decomposition process.

Research on detrital aggregates has shown that bacteria are usually the first to colonize and exploit the detrital organic matter, and the development of dense bacterial populations on the detritus subsequently turns the detrital particles into a favorable feeding ground for bacterivorous protozoans [74,94,95]. The succession of microbes (from bacteria to bacterivorous protozoans) on copepod carcasses is less obvious, perhaps due to the fact that copepod carcass organic matter tends to be more labile and is turned over faster than the time needed to establish a microbial food chain. Nevertheless, both field and lab observations have confirmed that some ciliates are able to exploit the dense bacterial mass on decomposing copepod carcasses (Figure 1). Another group of protozoan that is of particular interest is histophagous ciliates. 'Histophagous' translates to 'tissue eating,' meaning that the ciliates can directly feed on other organisms' tissues. Among copepods, histophagous ciliates are normally regarded as parasites feeding on the tissues of a live (usually wounded) host [96,97]. Nevertheless, when offered zooplankton carcasses in laboratory experiments, histophagous ciliates quickly colonize and feed on the dead tissues (our own observations). Their role in turning over copepod carcasses in nature, however, is still unclear.

Rapid decomposition and remineralization by microbes will channel carcass materials to the microbial food web within the pelagic zone. The opposite side of the coin is deposition of carcass materials to the benthos. Because a copepod body is denser than the surrounding water, it tends to sink. In quiescent water, copepod carcasses may sink as much as >100 m per day [85, 98]. In reality, however, carcass sinking rate is moderated by a number of factors, some of which are still not well understood. In a stratified water column, the ambient water density and viscosity increase across the pycnocline, which will slow sinking and increase the settling time of the carcasses. For a typical temperate water column with a thermocline gradient of ~15°C, stratification can nearly double the settling time of a carcass [98]. Turbulence fluctuations within the water column can further modify the carcass falling trajectory, but these effects can be complex and variable. A modeling study based on *in situ* vertical profiles of shear microstructure in thermally stratified lakes showed that turbulence diffusivity was highest at the surface, but decreased to a minimum in the metalimnion, then increased again in the hypolimnion [98]. Overall, the effect of turbulence on the average carcass sinking rate appeared to be small; nevertheless, the individual carcass sinking trajectories became highly variable within the upper mixed layer such that some carcasses could remain there for up to 5 times the average duration.

Increased carcass residence time due to stratification and turbulence mixing will prolong microbial decomposition of the carcasses within the water column, during which loss of internal carcass tissues will further reduce the carcass mass density. Based on the carcass' initial excess density and water column temperature profile, it is possible to predict the time it takes a carcass to reach neutral buoyancy (Figure 8). The reported body density ranges 1048-1109 kg m^{-3} among different copepod species [99]; it is therefore less likely for copepod carcasses to attain neutral buoyancy in shallow freshwater lakes where ambient water density is ~1000 kg m^{-3}. On the other hand, the lipid-rich species *Calanus finmarchicus* has an excess density of only ~20 kg m^{-3} in seawater [100], and could potentially attain neutral buoyancy within the water column. This could also explain the accumulation of carcasses of the lipid-rich *Calanus cristatus* within the mesopelagic and bathypelagic zones in the Japan Sea [13].

Sediment trap studies have shown repeatedly that copepod bodies often make their way to the deep water layers, although there is still much debate as to whether they are carcasses or 'swimmers' [28]. Identifying carcasses in sediment trap samples can be difficult, but is necessary for accurate carbon flux calculations. The importance of copepod carcass fluxes is illustrated in several recently published sediment trap studies.

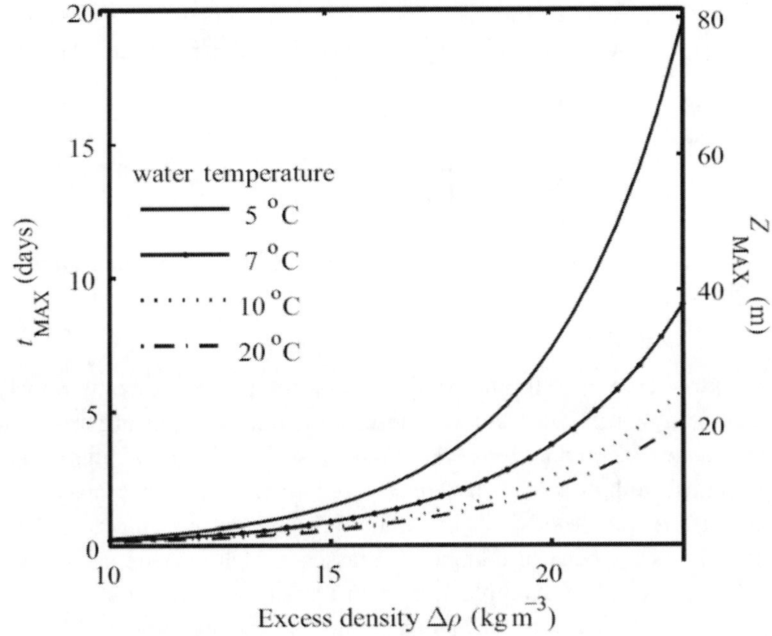

Figure 8. Predicted time required for copepod carcasses to achieve neutral buoyancy (t_{MAX}) and the corresponding terminal depth (Z_{MAX}) as functions of excess density ($\Delta\rho$) at different water temperatures. Reproduced from Kirillin et al. [98] with permission from Association for the Sciences of Limnology and Oceanography. Copyright (2012) by the Association for the Sciences of Limnology and Oceanography, Inc.

In the Aegean Sea, zooplankton carcasses accounted for 11-41% of the total particulate organic carbon (POC) flux to 16 m, and copepod carcasses alone accounted for up to 23% of the POC flux in the month of April [29]. In the Bay of Calvi, Western Mediterranean, carbon fluxes to the seafloor (~38 m) due to fecal pellets and copepod carcasses were measured at 0.05-2.69 mg m^{-2} d^{-1} and 0.42-4.37 mg m^{-2} d^{-1}, respectively [30]. Carcass carbon flux almost always exceeded fecal carbon flux except during the spring phytoplankton bloom, and both constituted an important nitrogen source to the benthos [30].

In the Beaufort Sea of the Canadian Arctic, carcasses accounted for an annual average of only 0.8-5.0% of the total copepods (including 'swimmers') collected in a sediment trap set at 70 m (station depth 505 m) [31]. Nevertheless, these carcasses represented a main component of the sedimentation flux on an annual basis, accounting for 12%, 36% and 34% of the dry weight, POC and particulate organic nitrogen (PON) fluxes, respectively. Contributions of copepod carcasses to sedimentation flux were particularly significant in late fall and winter, accounting for 66-86% of POC flux and 56-83% of PON flux [31]. Similarly, a study in the Amundsen Gulf showed that POC flux due to copepod carcasses was low (1-30%) during the productive seasons, but generally high (16-91%) in winter to early spring [32]. The highest copepod carcass flux (14.6 mg C m^{-2} d^{-1}) was recorded in February when other particulate flux was at the lowest (1.5 mg C m^{-2} d^{-1}), and was equivalent to nearly 18% of the traditionally measured particle flux over the year.

The number of sinking copepod carcasses may be small, and identifying them can be tedious. Nevertheless, as demonstrated in these studies, ignoring even a small number of carcasses in sediment trap samples out of convenience can greatly underestimate the sedimentation flux.

Conclusion

> However many ways there may be of being alive, it is certain that there are vastly more ways of being dead.
>
> - Richard Dawkins

Because plankton have very limited range of mobility, they are continuously influenced by local environmental conditions, and are therefore useful sentinel organisms for detecting environmental changes. Zooplankton sampling is a key part of many environmental monitoring programs, although the live/dead composition of zooplankton is often not considered. The different methods discussed in this chapter provide researchers with the means to monitor the occurrence of copepod carcasses, which may be used as a diagnosis of the state of the ecosystem. For example, drastic increase in water temperature, both natural and artificial, has been linked to an increase in carcass abundance [1,7]. Quantitative sampling of copepod carcasses, coupled with water temperature measurements, may therefore provide useful information about ocean warming effects on the planktonic food web. Likewise, localized increase in copepod carcasses may signal a deterioration of environmental quality due to harmful algal bloom [51] or other natural as well as anthropogenic causes [3,8, 53].

As we have shown, carcass abundance data may be used to estimate *in situ* non-predatory mortality rate, which may help reconcile the difference between total mortality and predatory mortality of pelagic copepods globally [33]. Although such an effort will improve our understanding of copepod life history and population dynamics, it does not tell us the actual cause(s) of mortality, which is needed for making predictions and linking copepod mortality to other ecological processes. Because non-predatory mortality is defined by what it is not—mortality not due to predation, there are virtually countless possible causes. While researchers have focused on some of the more likely candidates mentioned earlier, such as pollution, algal toxin, environmental stresses, food quality and quantity, there are surely more to be discovered. Studying the relative importance of various non-predatory mortality factors and their temporal and spatial variations will be a major challenge in copepod ecology, but will be much needed to advance this field of research.

Studying the fate of copepod carcasses can also be a daunting task. In this chapter we have focused on microbial decomposition and sedimentation. Carcass sedimentation rate is strongly influenced by physical gradients and turbulence, of which we have only limited understanding. Other carcass removal mechanisms, such as ingestion by other organisms (i.e., necrophagy) within the water column, have been rarely studied, but could significantly alter the fluxes of carcass materials and energy [85].

Our natural fascination with the living world is what motivates many of us to become biologists, but biology is the study of *life*, not just *the living*. Death is a natural part of life. To fully understand life, we ought to consider both the living and the dead. Copepods may well be the most numerous group of metazoans on earth [99], and it is not unreasonable to posit that copepod carcasses could therefore represent a form of detritus that is important on a global scale. In copepod ecology scientists still have much to learn, and much to gain, by studying copepod carcasses. This chapter, we hope, has helped bring this research topic—to life!

ACKNOWLEDGMENTS

Funding support was provided by the U.S. National Science Foundation (OCE-0352125, OCE-0814458, DGE-0840804), Jeffress Memorial Trust (J-697, J-895), German Leibniz Foundation, and Humboldt Fellowship for Experienced Researchers. Dr. Morten Hjorth of Denmark generously provided the photos used in Figure 2. Many students and collaborators have contributed to the research described in this chapter: Christopher Schweitzer, Curtis Freund, Joseph Hammond, Samantha Bickel, Maureen Lynch, Courtney Harris, Hans-Peter Grossart, Kristine Hutalle, Claudia Dziallas and Georgiy Kirillin.

REFERENCES

[1] Tang, K. W., Freund, C. S. & Schweitzer, C. L. (2006). Occurrence of copepod carcasses in the lower Chesapeake Bay and their decomposition by ambient microbes. *Estuarine, Coastal and Shelf Science*, *68*, 499-508.

[2] Hall, L. W. & Alden, R. W. (1997). A review of concurrent ambient water column and sediment toxicity testing in the Chesapeake Bay watershed: 1990-1994. *Environmental Toxicology and Chemistry*, *16*, 1606-1617.

[3] Delgado, M. & Alcaraz, M. (1999). Interactions between red tide microalgae and herbivorous zooplankton: the noxious effects of *Gyrodinium corsicum* (Dinophyceae) on *Acartia grani* (Copepdoa: Calanoida). *Journal of Plankton Research*, *21*, 2361-2371.

[4] Sopanen, S., Koski, M., Kuuppo, P. et al. (2007). Toxic haptophyte Prymnesium parvum affects grazing, survival, egestion and egg production of the calanoid copepods *Eurytemora affinis* and *Acartia bifilosa*. *Marine Ecology Progress Series*, *327*, 223-232.

[5] Burns, C. W. (1985). Fungal parasitism in a copepod population: The effects of Aphanomyces on the population dynamics of *Boeckella dilatata* Sars. *Journal of Plankton Research*, *7*, 201-205.

[6] Kimmerer, W. J. & McKinnon, A. D. (1990). High mortality in a copepod population caused by a parasitic dinoflagellate. *Marine Biology*, *107*, 449-452.

[7] Carpenter, E. J., Peck, B. B. & Anderson, S. J. (1974). Survival of copepods passing through a nuclear power station on northeastern Long Island Sound, USA. *Marine Biology*, *24*, 49-55.

[8] Roman, M. R., Gauzens, A. L., Rhinehart, W. K. & White, J. R. (1993). Effects of low oxygen waters on Chesapeake Bay zooplankton. *Limnology and Oceanography*, *38*, 1603-1614.

[9] Byron, E. R., Folt, C. L. & Goldman, C. R. (1984). Copepod and cladoceran success in an oligotrophic lake. *Journal of Plankton Research*, *6*, 45-65.

[10] Arnott, G. H., Brand, G. W. & Kos, L. C. (1986). Effects of food quality and quantity on the survival, development and egg production of *Gladioferens pectinatus* (Brady) (Copepoda: Calanoida). *Australian Journal of Marine and Freshwater Research*, *37*, 467-473.

[11] Koski, M. & Klein Breteler, W. C. M. (2003). Influence of diet on copepod survival in the laboratory. *Marine Ecology Progress Series*, *264*, 73-82.

[12] Weikert, H. (1977). Copepod carcasses in the upwelling region south of Cap Blanc, N.W. Africa. *Marine Biology*, *42*, 351-355.

[13] Terazaki, M. & Wada, M. (1988). Occurrence of large numbers of carcasses of the large, grazing copepod *Calanus cristatus* from the Japan Sea. *Marine Biology*, *97*, 177-183.

[14] Geptner, M. V., Zaikin, A. N. & Rudyakov, Y. A. (1990). Dead copepods in plankton: facts and hypotheses. *Oceanology*, *30*, 99-102.

[15] Böttger-Schnack, R. (1996). Vertical structure of small metazoan plankton, especially noncalanoid copepods. I. Deep Arabian Sea. *Journal of Plankton Research*, *18*, 1073-1101.

[16] Yamaguchi, A. & Ikeda, T. (2001). Abundance and population structure of three mesopelagic *Paraeuchaeta* species (Copepoda: Calanoida) in the Oyashio region, western subarctic Pacific Ocean with notes on their carcasses and epizoic ciliates. *Plankton Biology and Ecology*, *48*, 104-113.

[17] Yamaguchi, A., Watanabe, Y., Ishida, H. et al. (2002). Community and trophic structures of pelagic copepods down to greater depths in the western sub-arctic Pacific (WEST-COSMIC). *Deep-Sea Research I*, *49*, 1007-1025.

[18] Homma, T. & Yamaguchi, A. (2010). Vertical changes in abundance, biomass and community structure of copepods down to 3000 m in the southern Bering Sea. *Deep-Sea Research I*, *57*, 965-977.

[19] Böttger-Schnack, R. (1990). Community structure and vertical distribution of cyclopoid copepods in the Red Sea. I. Central Red Sea, autumn 1980. *Marine Biology*, *106*, 473-485.

[20] Böttger-Schnack, R. (1995). Summer distribution of micro- and small meso-zooplankton in the Red Sea and Gulf of Aden, with special reference to non-calanoid copepods. *Marine Ecology Progress Series*, *118*, 81-102.

[21] Genin, A., Gal, G. & Haury, L. (1995). Copepod carcasses in the ocean. II. Near coral reefs. *Marine Ecology Progress Series*, *123*, 65-71.

[22] Haury, L., Fey, C., Gal, G. et al. (1995). Copepod carcasses in the ocean. I. Over seamounts. *Marine Ecology Progress Series*, *123*, 57-63.

[23] Park, J. I., Kang, C. K. & Suh, H. L. (2011). Ontogenetic diet shift in the euphausiid *Euphausia pacifica* quantified using stable isotope analysis. *Marine Ecology Progress Series*, *429*, 103-109.

[24] Lee, C., Wakeham, S. G. & Hedges, J. I. (1988). The measurement of oceanic particle flux — Are "swimmers" a problem? *Oceanography*, *2*, 34-36.

[25] Coale, K. H. (1990). Labyrinth of Doom: a device to minimize the "swimmer component" in sediment trap collections. *Limnology and Oceanography*, *35*, 1376-1381.

[26] Peterson, M. L., Hernes, P. J., Thoreson, D. S. et al. (1993). Field evaluation of a valved sediment trap. *Limnology and Oceanography*, *38*, 1741-1761.

[27] Hansel, D. A. & Newton, J. A. (1994). Design and evaluation of a "swimmer"-segregating particle interceptor trap. *Limnology and Oceanography*, *39*, 1487-1495.

[28] Buesseler, K. O., Antia, A. N., Chen, M. et al. (2007). An assessment of the use of sediment traps for estimating upper ocean particle fluxes. *Journal of Marine Research*, *65*, 345-416.

[29] Frangoulis, C., Psarra, S., Zervakis, V. et al. (2010). Connecting export fluxes to plankton food-web efficiency in the Black Sea waters inflowing into the Mediterranean Sea. *Journal of Plankton Research*, *32*, 1203-1216.

[30] Frangoulis, C., Skliris, N., Lepoint, G. et al. (2011). Importance of copepod carcasses versus faecal pellets in the upper water column of an oligotrophic sea. *Estuarine, Coastal and Shelf Science*, *92*, 456-463.

[31] Sampei, M., Sasaki, H., Hattori, H. et al. (2009). Significant contribution of passively sinking copepods to the downward export flux in Arctic waters. *Limnology and Oceanography*, *54*, 1894-1900.

[32] Sampei, M., Sasaki, H., Forest, A. & Fortier, L. (2012). A substantial export flux of particulate organic carbon linked to sinking dead copepods during winter 2007-2008 in the Amundsen Gulf (southeastern Beaufort Sea, Arctic Ocean). *Limnology and Oceanography*, *57*, 90-96.

[33] Hirst, A. G. & Kiørboe, T. (2002). Mortality of marine planktonic copepods: global rates and patterns. *Marine Ecology Progress Series*, *230*, 195-209.

[34] Repetto, G., del Peso, A. & Zurita, J. L. (2008). Neutral red uptake assay for the estimation of cell viability/cytotoxicity. *Nature Protocols*, *3*, 1125-1131.

[35] Barbosa, P. & Peters, T. M. (1971). The effects of vital dyes on living organisms with special reference to neutral red and methylene blue. *Histochemistry Journal*, *3*, 71-93.

[36] Dressel, D. M., Heinle, D. R. & Grote, M. C. (1972). Vital staining to sort dead and live copepods. *Chesapeake Science*, *13*, 156-159.

[37] Crippen, R. W. & Perrier, J. L. (1974). The use of neutral red and Evans blue for live-dead determinations of marine plankton (with comments on the use of rotenone for inhibition of grazing). *Biotechnic & Histochemistry*, *49*, 97-104.

[38] Fleming, J. M. & Coughlan, J. (1978). Preservation of vitally stained zooplankton for live/dead sorting. *Estuaries*, *1*, 135-137.

[39] Elliott, D. T. & Tang, K. W. (2009). Simple staining method for differentiating live and dead marine zooplankton in field samples. *Limnology and Oceanography: Methods*, *7*, 585-594.

[40] Litvinyuk, D. A., Altukhov, D. A., Mukhanov V. S. & Popova E. V. (2011). Dynamics of live Copepoda in plankton of Sevastopol Bay and open coastal waters (the Black Sea) in 2010-2011. *Marine Ecological Journal*, *10*, 56-65.

[41] Zetsche, E. M. & Meysman, F. J. R. (2012). Dead of alive? Viability assessment of micro- and mesoplankton. *Journal of Plankton Research*, *34*, 493-509.

[42] Currier, H. B. (1957). Callose substance in plant cells. *American Journal of Botany*, *44*, 478-488.

[43] Martin, F. W. (1959). Staining and observing pollen tubes in the style by means of fluorescence. *Biotechnic & Histochemistry*, *34*, 125-128.

[44] Smith, M. M. & McCully, M. E. (1978). A critical evaluation of the specificity of aniline blue induced fluorescence. *Protoplasma*, *95*, 229-254.

[45] Seepersad, B. & Crippen, R. W. (1978). Use of aniline blue for distinguishing between live and dead freshwater zooplankton. *Journal of Fisheries Research Board of Canada*, *35*, 1363-1366.

[46] Dubovskaya, O., Gladyshev, M., Gubanov, V. G. & Makhutova, O. N. (2003). Study of non-consumptive mortality of crustacean zooplankton in a Siberian reservoir using staining for live/dead sorting and sediment traps. *Hydrobiologia*, *504*, 223-227.

[47] Gladyshev, M. I., Dubovskaya, O. P., Gubanov, V. G. & Makhutova, O. N. (2003). Evaluation of non-predatory mortality of two *Daphnia* species in a Siberian reservoir. *Journal of Plankton Research*, *25*, 999-1003.

[48] Bickel, S. L., Tang, K. W. & Grossart, H. P. (2009). Use of aniline blue to distinguish live and dead crustacean zooplankton composition in freshwaters. *Freshwater Biology*, *54*, 971-981.

[49] Dubovskaya, O. P. (2008). Evaluation of possible causes of non-predatory mortality of crustacean zooplankton in a small Siberian reservoir. *Journal of Siberian Federal University-Biology*, *1*, 3-18.

[50] Wheeler, E. H. (1967). Copepod detritus in the deep sea. *Limnology and Oceanography*, *12*, 697-701.

[51] Semenova, A. S. (2011). Proportion of dead individuals in the zooplankton of the Curonian Lagoon of the Baltic Sea. *Inland Water Biology*, *4*, 332-340.

[52] Hubareva, E., Svetlichny, L., Kideys, A. & Isinibilir, M. (2008). Fate of the Black Sea *Acartia clausi* and *Acartia tonsa* (Copepoda) penetrating into the Marmara Sea through the Bosphorus. *Estuarine, Coastal and Shelf Science*, *76*, 131-140.

[53] Bickel, S. L., Hammond, J. D. M. & Tang, K. W. (2011). Boat-generated turbulence as a potential source of mortality among copepods. *Journal of Experimental Marine Biology and Ecology*, *410*, 105-109.

[54] Elliott, D. T. & Tang, K. W. (2011). Spatial and temporal distributions of live and dead copepods in the lower Chesapeake Bay (Virginia, USA). *Estuaries and Coasts*, *34*, 1039-1048.

[55] Kimmel, D. G., Boicourt, W. C., Pierson, J. J. et al. (2009). A comparison of the mesozooplankton response to hypoxia in Chesapeake Bay and the northern Gulf of Mexico using the biomass size spectrum. *Journal of Experimental Marine Biology and Ecology*, *381*, S65-S73.

[56] Kimmel, D. G., Boicourt, W. C., Pierson, J. J. et al. (2010). The vertical distribution and diel variability of mesozooplankton biomass, abundance and size in response to hypoxia in the northern Gulf of Mexico USA. *Journal of Plankton Research*, *32*, 1185-1202.

[57] Aksnes, D. L., Miller, C. B., Ohman, M. D. & Wood, S. N. (1997). Estimation techniques used in studies of copepod population dynamics – a review of underlying assumptions. *Sarsia*, *82*, 279-296.

[58] Nisbet, R. M. & Wood, S. N. (1996). Estimation of copepod mortality rates. *Ophelia*, *44*, 157-169.

[59] Carlotti F., Giske, J. & Werner, F. (2000). Modeling zooplankton dynamics. *In*: Harris R, Wiebe P, Lenz J et al. (eds.), *ICES Zooplankton Methodology Manual*. Academic Press, London, 571-667.

[60] Ohman, M. D. (2012). Estimation of mortality for stage-structured zooplankton populations: what is to be done? *Journal of Marine Systems*, *93*, 4-10.

[61] Gentleman, W. C., Pepin, P. & Doucette, S. (2012). Estimating mortality: Clarifying assumptions and sources of uncertainty in vertical methods. *Journal of Marine Systems*, *105-108*, 1-19.

[62] Aksnes, D. L. & Ohman, M. D. (1996). A vertical life table approach to zooplankton mortality estimation. *Limnology and Oceanography*, *41*, 1461-1469.

[63] Elliott, D. T. & Tang, K. W. (2011). Influence of carcass abundance on estimates of mortality and assessment of population dynamics in *Acartia tonsa*. *Marine Ecology Progress Series*, *427*, 1-12.

[64] Plourde, S., Maps, F. & Joly, P. (2009a). Mortality and survival in early stages control recruitment in *Calanus finmarchicus*. *Journal of Plankton Research*, *31*, 371-388.

[65] Hirst, A. G., Bonnet, D. & Harris, R. P. (2007). Seasonal dynamics and mortality rates of *Calanus helgolandicus* over two years at a station in the English Channel. *Marine Ecology Progress Series*, *340*, 189-205.

[66] Plourde, S., Pepin, P. & Head, E. J. H. (2009b). Long-term seasonal and spatial patterns in mortality and survival of *Calanus finmarchicus* across the Atlantic Zone Monitoring Programme region, northwest Atlantic. *ICES Journal of Marine Science*, *66*, 1942-1958.

[67] Tsuda, A. (1994). Starvation tolerance of a planktonic marine copepod *Pseudocalanus newmani* Frost. *Journal of Experimental Marine Biology and Ecology*, *181*, 81-89.

[68] Gophen, M. (1976). Temperature effect on lifespan, metabolism, and development time of *Mesocyclops leuckarti* (Claus). *Oecologia*, *25*, 271-277.

[69] Harvell, C. D., Mitchell, C. E., Ward, J. R., et al. (2002). Climate warming and disease risks for terrestrial and marine biota. *Science*, *296*, 2158-2162.

[70] Wetzel, R. G. (1995). Death, detritus, and energy flow in aquatic ecosystems. *Freshwater Biology*, *33*, 83-89.

[71] Simon, M., Grossart, H. P., Schweitzer, B. & Ploug, H. (2002). Microbial ecology of organic aggregates in aquatic ecosystems. *Aquatic Microbial Ecology*, *28*, 175-211.

[72] Smith, D. C., Simon, M., Alldredge, A. L. & Azam, F. (1992). Intense hydrolytic enzyme activity on marine aggregates and implications for rapid particle dissolution. *Nature*, *359*, 139-142.

[73] Grossart, H. P. & Ploug, H. (2000). Bacterial production and growth efficiencies: Direct measurements on riverine aggregates. *Limnology and Oceanography*, *45*, 436-445.

[74] Kiørboe, T., Tang, K., Grossart, H. P. & Ploug, H. (2003). Dynamics of microbial communities on marine snow aggregates: colonization, growth, detachment and grazing mortality of attached bacteria. *Applied and Environmental Microbiology*, *69*, 3036-3047.

[75] Grossart, H. P., Tang, K. W., Kiørboe, T. & Ploug, H. (2007). Comparison of cell-specific activity between free-living and attached bacteria using isolates and natural assemblages. *FEMS Microbiology Letters*, *266*, 194-200.

[76] Alldredge, A. L., Cole, J. J. & Caron, D. A. (1986). Production of heterotrophic bacteria inhabiting macroscopic organic aggregates (marine snow) from surface waters. *Limnology and Oceanography*, *31*, 68-78.

[77] Båmstedt, U. (1986). Chemical composition and energy content. *In*: Corner EDS & O'Hara SCM (eds.), *The Biological Chemistry of Marine Copepods*. Clarendon Press, Oxford, 1-58.

[78] Lee, B. G. & Fisher, N. S. (1992). Decomposition and release of elements from zooplankton debris. *Marine Ecology Progress Series*, *88*, 117-128.

[79] Reinfelder, J. R., Fisher, N. S., Fowler, S. W. & Teyssié, J. L. (1993). Release rates of trace elements and protein from decomposing planktonic debris. 2. Copepod carcasses and sediment trap particulate matter. *Journal of Marine Research*, *51*, 423-442.

[80] Tang, K. W., Turk, V. & Grossart, H. P. (2010). Linkage between crustacean zooplankton and aquatic bacteria. *Aquatic Microbial Ecology*, *61*, 261-277.

[81] Harding, G. C. H. (1973). Decomposition of marine copepods. *Limnology and Oceanography*, *18*, 670-673.

[82] Bickel, S. L. & Tang, K. W. (2010). Microbial decomposition of proteins and lipids in copepod versus rotifer carcasses. *Marine Biology*, *157*, 1613-1624.

[83] Tang, K. W., Hutalle, K. M. L. & Grossart, H. P. (2006). Microbial abundance, composition and enzymatic activity during decomposition of copepod carcasses. *Aquatic Microbial Ecology*, *45*, 219-227.

[84] Tang, K. W., Bickel, S. L., Dziallas, C. & Grossart, H. P. (2009). Microbial activities accompanying decomposition of cladoceran and copepod carcasses under different environmental conditions. *Aquatic Microbial Ecology*, *57*, 89-100.

[85] Elliott, D. T., Harris, C. K. & Tang, K. W. (2010). Dead in the water: the fate of copepod carcasses in the York River estuary, Virginia. *Limnology and Oceanography*, *55*, 1821-1834.

[86] Lovell, C. R. & Konopka, A. (1985). The effects of temperature on bacterial production in a dimictic eutrophic lake. *FEMS Microbiology Ecology*, *31*, 135-140.

[87] Apple, J. K., del Giorgio, P. A. & Kemp, W. M. (2006). Temperature regulation of bacterial production, respiration, and growth efficiency in a temperate salt-marsh estuary. *Aquatic Microbial Ecology*, *43*, 243-254.

[88] Cottrell, M. T., Moore, J. A. & Kirchman, D. L. (1999). Chitinases from uncultured marine microorganisms. *Applied and Environmental Microbiology*, *65*, 2553-2557.

[89] Kirchman, D. L. & White, J. (1999). Hydrolysis and mineralization of chitin in the Delaware Estuary. *Aquatic Microbial Ecology*, *18*, 187-196.

[90] Brzezinska, M. S. & Donderski, W. (2006). Chitinolytic bacteria in two lakes of different trophic status. *Polish Journal of Ecology*, *54*, 295-301.

[91] Hood, M. A. & Meyers, S. P. (1977). Rates of chitin degradation in an estuarine environment. *Journal of Oceanography*, *33*, 328-334.

[92] Kirchner, M. (1995). Microbial colonization of copepod body surfaces and chitin degradation in the sea. *Helgoländer Meeresuntersuchungen*, *49*, 201-212.

[93] Wurzbacher, C. M., Bärlocher, F. & Grossart, H. P. (2010). Fungi in lake ecosystems. *Aquatic Microbial Ecology, 59*, 125-149.

[94] Ploug, H. & Grossart, H. P. (2000). Bacterial growth and grazing on diatom aggregates: respiratory carbon turnover as a function of aggregate size and sinking velocity. *Limnology and Oceanography*, *45*, 1467-1475.

[95] Tang, K. W., Grossart, H. P., Yam, E. M. et al. (2006). Mesocosm study of particle dynamics and control of particle-associated bacteria by flagellate grazing. *Marine Ecology Progress Series*, *325*, 15-27.

[96] Grimes, B. H. & Bradbury, P. C. (1992). The biology of *Vampyrophrya pelagica* (Chatton & Lwoff, 1930), a histophagous apostome ciliate associated with marine calanoid copepods. *Journal of Eurkaryotic Microbiology*, *39*, 65-79.

[97] Ohtsuka, S., Hora, M., Suzaki, T. et al. (2004). Morphology and host-specificity of the apostome ciliate *Vampyrophrya pelagica* infecting pelagic copepods in the Seto Inland Sea, Japan. *Marine Ecology Progress Series*, *282*, 129-142.

[98] Kirillin, G., Grossart, H. P. & Tang, K. W. (2012). Modeling sinking rate of zooplankton carcasses: effects of stratification and mixing. *Limnology and Oceanography*, *57*, 881-894.

[99] Mauchline, J. (1998). *The Biology of Calanoid Copepods*. Academic Press, San Diego.

[100] Visser, A. W. & Jónasdóttir, S. H. (1999). Lipids, buoyancy and the seasonal vertical migration of *Calanus finmarchicus*. *Fisheries Oceanography*, *8*, 100-106.

INDEX

#

A

B

C

D

E

F

G

N

O

P

Q

R

S

T

U

V

W

Y

Z